MW00477708

STATISTICAL METHODS
FOR HANDLING
INCOMPLETE DATA

STATISTICAL METHODS FOR HANDLING INCOMPLETE DATA

Jae Kwang Kim

Department of Statistics
Iowa State University, USA

Jun Shao

Department of Statistics
University of Wisconsin – Madison, USA

CRC Press is an imprint of the
Taylor & Francis Group, an **informa** business
A CHAPMAN & HALL BOOK

CRC Press
Taylor & Francis Group
6000 Broken Sound Parkway NW, Suite 300
Boca Raton, FL 33487-2742

Printed on acid-free paper
Version Date: 20130531

International Standard Book Number-13: 978-1-4398-4963-7 (Hardback)

Library of Congress Cataloging-in-Publication Data

Kim, Jae Kwang, 1968-
Statistical methods for handling incomplete data / Jae Kwang Kim and Jun Shao.
pages cm
Includes bibliographical references and index.
ISBN 978-1-4398-4963-7 (hardback : acid-free paper)
1. Missing observations (Statistics) 2. Multiple imputation (Statistics) I. Shao, Jun (Statistician) II. Title.

QA276.8.K55 2014
519.5'4--dc23 2013010114

Visit the Taylor & Francis Web site at
http://www.taylorandfrancis.com

and the CRC Press Web site at
http://www.crcpress.com

To Timothy, Jenny, and Jungwoo

and

To Jason, Annie, and Guang

Contents

Preface

Missing data is frequently encountered in statistics. Statistical analysis with missing data is an area of extensive research for the last two decades. Many statistical problems assuming a latent variable can also be viewed as missing data problems. Furthermore, with the advances in statistical computing, there has been a rapid development of techniques and applications in missing data analysis inspired by theoretical findings in this area. This book aims to cover the most up-to-date statistical theories and computational methods for analyzing incomplete data.

The main features of the book can be summarized as follows:

1. Rigorous treatment of statistical theories on likelihood-based inference with missing data.
2. Comprehensive treatment of computational techniques and theories on imputation.
3. Most up-to-date treatment of methodologies involving propensity score weighting, nonignorable missing, longitudinal missing, survey sampling application, and statistical matching.

This book is developed under the frequentist framework and puts less emphasis on Bayesian methods and nonparametric methods. Apart from some real data examples, many artificial examples are presented to help with the understanding of the methodologies introduced. The book is suitable for use as a textbook for a graduate course in statistics departments. Materials in Chapter 2 - Chapter 6 can be covered systematically in the course. Materials in Chapter 7 - Chapter 9 are more advanced and can perhaps serve for future reference. To be comfortable with the materials the reader should have completed courses in statistical theory and in linear models. This book can also be used as a reference book for those interested in this area. Some of the research ideas introduced in the book can be developed further for specific applications.

We would like to thank those who made enormous contribution to this book. Jae Kwang Kim would like to thank professor Wayne Fuller for his comments and suggestions, former students Ji-young Kim, Ming Zhou, Sixia Chen and Minsun Riddles and collaborators Cindy Yu, J.N.K. Rao, Mingue Park, David Haziza, and Dong Wan Shin for their contribution in the development of the researches on this area, Jongho Im and Shu Yang for their computational supports, and Jie Li for editorial supports. His research at Iowa State University was partially supported by a Cooperative Agreement between the US Department of Agriculture Natural Resources Conservation Service and Iowa State University. Last but not least, we would like to thank Rob Calver, Rachel Holt, and Marsha Hecht at CRC Press for their support during the production of this book. We take full responsibilities for all errors and omissions in the book.

Jae Kwang Kim
Jun Shao

List of Tables

Chapter 1

Introduction

1.1 Introduction

Missing data, or incomplete data, is frequently encountered in many disciplines. Statistical analysis with missing data has been an area of considerable interest in the statistical community. Many tools, generic or tailor-made, have already been developed, and many more will be forthcoming to handle missing data problems. Missing data is particularly useful because many statistical issues can be treated as special cases of the missing data problem. For example, data with measurement error can be viewed as a special case of missing data where an imperfect measurement is available instead of true measurement. Two-phase sampling can also be viewed as a missing data problem where the key items are observed only in the second-phase sample. Combining two independent surveys with some common items is a special case of two-phase sampling. Many statistical problems assuming a latent variable can also be viewed as missing data problem. Furthermore, the advances in statistical computing have made the computational aspects of the missing data analysis techniques more feasible. This book aims to cover the most up-to-date statistical theories and computational methods of the missing data analyses.

Generally speaking, let z be the study variable with distribution function $f(z;\theta)$. We are interested in estimating the parameter θ. If z were observed throughout the sample, then θ would be able to be obtained by the maximum likelihood method. Instead of observing z, however, we only observe $y = T(z,\delta)$ and δ, where $y = T(z,\delta)$ is an incomplete version of z satisfying $T(z,\delta = 1) = z$ and δ is an indictor function that takes either one or zero. Parameter estimation of θ from the observation of (y,δ) is the core of the problem in missing data analyses.

To handle this problem, the marginal density function of (y,δ) needs to be expressed as a function of the original distribution $f(z;\theta)$. Maximum likelihood estimation can be obtained under some identifying assumptions and statistical theories can be developed for the maximum likelihood estimator obtained from the observed sample. Computational tools for producing the maximum likelihood estimator need to be introduced. How to assess the uncertainty of the resulting maximum likelihood estimator is also important in the area of missing data analyses.

When z is a vector, there will be more complications. Because several random variables are subject to missingness, the missing data pattern can figure in to simply modeling and estimation. The monotone missing pattern refers to the situation where the set of respondents in one variable is always a subset of the set of respondents for another variable, which may host further subsetting. See Table 1.1 for an illustration of the monotone missing pattern.

1.2 Outline

Maximum likelihood estimation with missing data serves as the starting point of this book. Chapter 2 is about defining the observed likelihood function from the marginal density of the observed part of the data, finding the maximum of the observed likelihood by solving the mean score equation, and obtaining the observed information matrix from the observed likelihood. Chapter 3 deals with computational tools to arrive at the maximum likelihood estimator, especially the EM algorithm.

Imputation, covered in Chapter 4, is also a popular tool for handling missing data. Imputation

Table 1.1 Monotone missing pattern

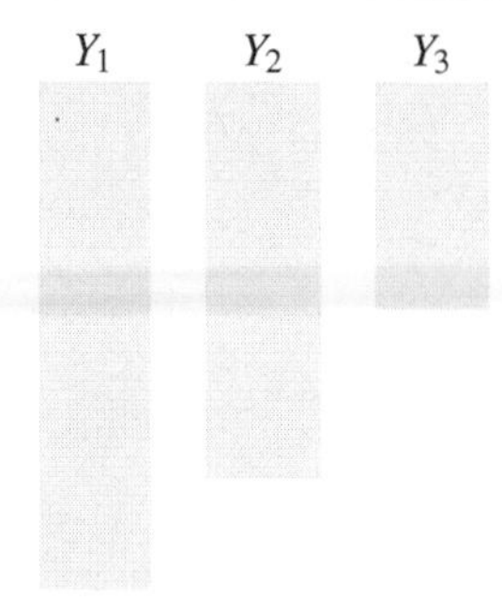

can be viewed as a computational technique for the Monte Carlo approximation of the conditional expectation of the original complete-sample estimator given the observed data. As for variance estimation of the imputation estimator, an important subject in missing data analyses, the Taylor linearization or replication method can be used. Multiple imputation has been proposed as a general tool for imputation and simplified variance estimation but it requires some special conditions, called *congeniality* and *self-efficiency*. Fractional imputation is an alternative general-purpose estimation tool for imputation.

Propensity score weighting, covered in Chapter 5, is another tool for handling missing data. Basically the responding units are assigned with propensity score weights so that the weighted analysis can lead to valid inference. The propensity score weighting method is often based on an assumption about the response mechanism and the resulting estimator can be made more efficient by properly taking into account of the auxiliary information available from the whole sample. The propensity score weighted estimator can further incorporate the outcome model for imputation and in turn make the resulting estimator doubly protected against the failure of the assumed models.

Nonignorable missing data occurs in the challenging situation where the response mechanism depends on the study variable that is subject to missingness. Nonignorable missing data is also an important area of research that is yet to be thoroughly investigated. Chapter 6 covers some important topics in the analysis of nonignorable missing data. Nonresponse instrumental variables can be used to identify the parameters in some nonignorable missing data and obtain consistent parameter estimates. The generalized method of moments (GMM) and the pseudo likelihood method are useful tools for parameter estimation with nonresponse instrumental variables. Also covered is the exponential tilting model whereby the conditional distribution of the nonrespondents can be expressed via the conditional distribution of the respondents, which eventually enables the derivation of nonparametric maximum likelihood estimators. Instead of using nonresponse instrumental variables, callback samples can also be used to provide consistent parameter estimates for nonignorable missing data.

The rest of the book discusses specific applications of the covered methodology for analyzing missing data. The confluence between missing data and longitudinal analysis, survey sampling, and statistical matching are the subject of Chapter 7, Chapter 8 and Chapter 9, respectively.

1.3 How to use this book

This book assumes a basic understanding of graduate-level mathematical statistics and linear models. Being slightly more technical than Little and Rubin (2002), it is written at the mathematical level of a second-year Ph.D. course in statistics.

Table 1.2 provides a sample outline for a 15-week semester-long course in the graduate course offering of the Department of Statistics at Iowa State University. Core materials are from Chapter 2 to Chapter 5 and can be covered in about 11-12 weeks. Chapter 6 (excluding Section 6.7 and

Section 6.8) can be covered in 2 weeks. In the remaining weeks, topics can be chosen from Chapters 7 through 9 at the discretion of the instructor. Exercises are provided at the end of each chapter, with the exclusion of Chapter 7 and Chapter 9.

Table 1.2 Outline of a 15-week lecture

Weeks	Chapter	Topic
1-3	1-2	Introduction. Likelihood-based approach
4-6	3	Computation
7-9	4	Imputation
10-11	5.1-5.4	Propensity scoring approach
12-13	6.1-6.6	Nonignorable missing data
14 -15	7-9	Special topics (Choose from Chapter 7 - Chapter 9)

For readers using this book as reference material, the chapters have been written to be as self-contained as possible. The core theory of maximum likelihood estimation with missing data is covered in Chapter 2 in a concise manner. Chapter 3 may overlap with Little and Rubin (2002) to some extent, but Section 3.2 contains newer materials. Chapter 4 is technically oriented, but Section 4.2 still contains the main theoretical results on the imputation estimator, which was originally covered by Wang and Robins (1998). Furthermore, this chapter presents fractional imputation as an alternative tool after providing a critical review of multiple imputation.

The main novelty of the book comes from the materials in Chapter 5 and after. Topics covered in Chapter 5 through Chapter 9 are either not fully addressed or totally uncharted previously. Researchers in this area may find the materials interesting and useful to their own research.

Chapter 2

Likelihood-based approach

2.1 Introduction

In this chapter, we discuss the likelihood-based approach in the analysis of missing data. To do this, we first review the likelihood-based methods in the case of complete response. Let $\mathbf{y} = (y_1, y_2, \cdots y_n)$ be a realization of the random sample from an infinite population with density $f(y)$ so that $P(Y \in B) = \int_B f(y) d\mu(y)$ for any measurable set B where $\mu(y)$ is a σ-finite dominating measure. Assume that the true density $f(y)$ belongs to a parametric family of densities $\mathcal{P} = \{f(y;\theta) : \theta \in \Omega\}$ indexed by $\theta \in \Omega$. That is, there exists a $\theta_0 \in \Omega$ such that $f(y;\theta_0) = f(y)$ for all y. Once the parametric density is specified, the likelihood function and the maximum likelihood estimator can be defined formally as follows.

Definition 2.1. *The likelihood function of θ, denoted by $L(\theta)$, is defined as the probability density (mass) function of the observed data $\mathbf{y}$ considered as a function of θ. That is,*

$$L(\theta) = f(\mathbf{y};\theta)$$

where $f(\mathbf{y};\theta)$ is the joint density function of $\mathbf{y}$.

Definition 2.2. *Let $\hat{\theta}$ be the maximum likelihood estimator (MLE) of θ_0 if it satisfies*

$$L(\hat{\theta}) = \max_{\theta \in \Omega} L(\theta).$$

If $y_1, y_2, \ldots, y_n$ are independently and identically distributed (IID),

$$L(\theta) = \prod_{i=1}^{n} f(y_i;\theta).$$

Also, if $\hat{\theta}$ is the MLE of θ_0, then $g(\hat{\theta})$ is the MLE of $g(\theta_0)$. The MLE is not necessarily unique. To guarantee uniqueness of the MLE, we require that the family of densities is identified. The definition of an identifiable distribution is given as follows:

Definition 2.3. *A parametric family of densities, given by $\mathcal{P} = \{f(y;\theta);\theta \in \Omega\}$, is called* identifiable (or identified) *if*

$$f(y;\theta_1) \neq f(y;\theta_2) \quad \textit{for every } \theta_1 \neq \theta_2$$

for all $\mathbf{y}$ in the support of $\mathcal{P}$.

Under the identifiability condition, the uniqueness of the MLE follows from the following lemma.

Lemma 2.1. *If $\mathcal{P} = \{f(y;\theta);\theta \in \Omega\}$ is identifiable and $E\{|\ln f(Y;\theta)|\} < \infty$ for all θ, then*

$$M(\theta) = -E_{\theta_0} \ln\left\{\frac{f(y;\theta)}{f(y;\theta_0)}\right\} \geq 0 \tag{2.1}$$

with equality at $\theta = \theta_0$.

Proof. Let $Z = f(y;\theta)/f(y;\theta_0)$. Using the strict version of Jensen's inequality

$$-\ln\{E_{\theta_0}(Z)\} < E_{\theta_0}\{-\ln(Z)\}.$$

Because $E_{\theta_0}(Z) = \int f(y;\theta)\,d\mu(y) = 1$, we have $\ln\{E_{\theta_0}(Z)\} = 0$. □

In Lemma 2.1, $M(\theta)$ is called the *Kullback–Leibler divergence measure* of $f(y;\theta)$ from $f(y;\theta_0)$. It is often considered a measure of distance between two densities. If $\mathcal{P} = \{f(y;\theta);\theta \in \Omega\}$ is not identifiable, then $M(\theta)$ may not have a unique minimizer and $\hat{\theta}$ may not converge (in probability) to a single point.

The following two theorems present some asymptotic properties of the maximum likelihood estimator (MLE): (weak) consistency and asymptotic normality. To discuss the asymptotic properties, let $\hat{\theta}$ be any solution of

$$Q_n(\hat{\theta}) = \min_{\theta\in\Omega} Q_n(\theta)$$

where

$$Q_n(\theta) = -\frac{1}{n}\sum_{i=1}^{n}\log f(y_i,\theta).$$

Also, define $Q(\theta) = -E_{\theta_0}\{\log f(Y;\theta)\}$ to be the probability limit of $Q_n(\theta)$ evaluated at θ_0. By Lemma 2.1, $Q(\theta)$ is minimized at $\theta = \theta_0$. Thus, under some conditions, we may expect that $\hat{\theta} = \arg\min Q_n(\theta)$ converges to $\theta_0 = \arg\min Q(\theta)$. The following theorem presents a formal result.

Theorem 2.1. *Assume the following two conditions:*

1. *Identifiability: $Q(\theta)$ is uniquely minimized at θ_0. That is, for any $\varepsilon > 0$, there exists a $\delta > 0$ such that $\theta \notin B_\varepsilon(\theta_0)$ implies $Q(\theta) - Q(\theta_0) \geq \delta$, where $B_\varepsilon(\theta_0) = \{\theta \in \Omega; |\theta - \theta_0| < \varepsilon\}$.*
2. *Uniform weak convergence:*

$$\sup_{\theta\in\Omega}|Q_n(\theta) - Q(\theta)| \xrightarrow{p} 0$$

for some nonstochastic function $Q(\theta)$

Then, $\hat{\theta} \xrightarrow{p} \theta_0$.

Proof. For any $\varepsilon > 0$, we can find $\delta > 0$ such that

$$\begin{aligned} 0 \leq P\left[\hat{\theta} \notin B_\varepsilon(\theta_0)\right] &\leq P\left[Q(\hat{\theta}) - Q_n(\hat{\theta}) + Q_n(\hat{\theta}) - Q(\theta_0) \geq \delta\right] \\ &\leq P\left[Q(\hat{\theta}) - Q_n(\hat{\theta}) + Q_n(\theta_0) - Q(\theta_0) \geq \delta\right] \\ &\leq P[2\sup|Q_n(\theta) - Q(\theta)| \geq \delta] \to 0. \end{aligned}$$

□

In Theorem 2.1, $Q_n(\theta)$ is the negative log-likelihood of θ obtained from $f(y;\theta)$. The function $Q(\theta)$ is the uniform probability limit of $Q_n(\theta)$ and θ_0 is the unique minimizer of $Q(\theta)$. In Theorem 2.1, it is assumed that $\hat{\theta}$ is not necessarily uniquely determined, but θ_0 is. Uniform weak convergence of $Q_n(\theta)$ is stronger than pointwise convergence of $Q_n(\theta)$. One simple set of sufficient conditions for it is that $Q_n(\theta)$ converges pointwise to $Q(\theta)$ and that $Q_n(\theta)$ is continuous and has a unique minimizer at $\theta = \hat{\theta}$.

Theorem 2.2. *Assume the following regularity conditions:*

1. *θ_0 is in the interior of Ω.*
2. *$Q_n(\theta)$ is twice continuously differentiable on some neighborhood $\Omega_0(\subset \Omega)$ of θ_0 almost everywhere.*

3. *The first-order partial derivative of Q_n satisfies*

$$\sqrt{n}\frac{\partial}{\partial\theta}Q_n(\theta_0) \xrightarrow{d} N(0, A_0)$$

for some positive definite A_0.

4. *The second-order partial derivative of Q_n satisfies*

$$\sup_{\theta\in\Omega_o} \left\| \frac{\partial^2}{\partial\theta\partial\theta'}Q_n(\theta) - B(\theta) \right\| \xrightarrow{p} 0$$

for some $B(\theta)$ continuous at θ_0 and $B_0 = B(\theta_0)$ is nonsingular.

Furthermore, assume that $\hat{\theta}$ satisfies

5. $\hat{\theta} \xrightarrow{p} \theta_0$.
6. $\sqrt{n}\partial Q_n(\hat{\theta})/\partial\theta = o_p(1)$.

Then

$$\sqrt{n}\left(\hat{\theta} - \theta_0\right) \xrightarrow{d} N\left(0, B_0^{-1}A_0B_0^{-1'}\right).$$

Proof. By assumption 6 and the mean value theorem,

$$\begin{aligned} \partial Q_n(\hat{\theta})/\partial\theta &= \partial Q_n(\theta_0)/\partial\theta + \left[\partial^2 Q_n(\theta^*)/\partial\theta\partial\theta'\right]\left(\hat{\theta} - \theta_0\right) \\ &= o_p\left(n^{-1/2}\right) \end{aligned}$$

for some θ^* between $\hat{\theta}$ and θ_0. By assumption 5, $\theta^* \xrightarrow{p} \theta_0$. Hence, by assumptions 2 and 4,

$$\frac{\partial^2 Q_n}{\partial\theta\partial\theta'}(\theta^*) = B(\theta_0) + o_p(1).$$

Thus, by the invertibility of B_0,

$$o_p(1) = B_0^{-1}\sqrt{n}\frac{\partial Q_n(\theta_0)}{\partial\theta} + [1 + o_p(1)]\sqrt{n}\left(\hat{\theta} - \theta_0\right)$$

By Slutsky's theorem,

$$\sqrt{n}\left(\hat{\theta} - \theta_0\right) = -B_0^{-1}\sqrt{n}\frac{\partial Q_n(\theta_0)}{\partial\theta} + o_p(1)$$

and the asymptotic normality follows from assumption 3. □

In assumption 3, the partial derivative of the log-likelihood is called the *score function*, denoted by

$$S(\theta) = \frac{\partial}{\partial\theta}\ln L(\theta).$$

Under the model $y_1, \cdots, y_n \overset{i.i.d.}{\sim} f(y; \theta_0)$, condition 3 in Theorem 2.2 can be expressed as

$$n^{-1/2}S(\theta_0) \xrightarrow{d} N(0, A_0)$$

where $A_0 = n^{-1}E_{\theta_0}\{S(\theta)S(\theta)'\}$. In assumption 4, $B(\theta)$ is essentially the probability limit of the second-order partial derivatives of the log-likelihood. This is the expected Fisher information matrix based on a single observation, defined by

$$\mathcal{I}(\theta_0) = -E_{\theta_0}\left\{\frac{\partial^2}{\partial\theta\partial\theta'}\ln f(y;\theta) \mid \theta = \theta_0\right\}.$$

That is, $B_0 = \mathcal{I}(\theta_0)$. We now summarize the definitions associated with the score function.

Definition 2.4. *1. Score function:*

$$S(\theta) = \frac{\partial}{\partial \theta} \ln L(\theta)$$

2. Fisher information (representing curvature of the log-likelihood function)

$$I(\theta) = -\frac{\partial^2}{\partial \theta \partial \theta'} \ln L(\theta) = -\frac{\partial}{\partial \theta'} S(\theta)$$

3. Observed (Fisher) information: $I(\hat{\theta})$, where $\hat{\theta}$ is the MLE.

4. Expected (Fisher) information: $\mathcal{I}(\theta) = E_\theta \{I(\theta)\}$.

Because of the definition of the MLE, the observed Fisher information is always positive. The expected information is meaningful as a function of θ across the admissible values of θ, but $I(\theta)$ is only meaningful in the neighborhood of $\hat{\theta}$. The observed information applies to a single dataset. In contrast, the expected information is an average quantity over all possible datasets generated at the true value of the parameter. For exponential families of distributions, we have $\mathcal{I}(\hat{\theta}) = I(\hat{\theta})$. In general, $I(\hat{\theta})$ is preferred for variance estimation of $\hat{\theta}$. The use of the observed information in assessing the accuracy of the MLE is advocated by Efron and Hinkley (1978).

Example 2.1. *1. Let $x_1, \cdots, x_n$ be an IID sample from $N(\theta, \sigma^2)$ with σ^2 known. We have*

$$\begin{aligned} V_\theta \{S(\theta)\} &= V_\theta \left\{ \sum_{i=1}^n (x_i - \theta)/\sigma^2 \right\} = n/\sigma^2 \\ I(\theta) &= -\partial S(\theta)/\partial \theta = n/\sigma^2. \end{aligned}$$

In this case, $\mathcal{I}(\theta) = I(\theta)$, a happy coincidence in any exponential family model with canonical parameter θ.

2. Let $x_1, \cdots, x_n$ be an IID sample from Poisson(θ). In this case,

$$\begin{aligned} V_\theta \{S(\theta)\} &= V_\theta \left\{ \sum_{i=1}^n (x_i - \theta)/\theta \right\} = n/\theta \\ I(\theta) &= -\partial S(\theta)/\partial \theta = n\bar{x}/(\theta^2). \end{aligned}$$

Thus, $\mathcal{I}(\theta) \neq I(\theta)$, but $\mathcal{I}(\hat{\theta}) = I(\hat{\theta})$ at $\hat{\theta} = \bar{x}$. This is true for the exponential family. It means that we can estimate the variance of the score function by either $\mathcal{I}(\hat{\theta})$ or $I(\hat{\theta})$.

3. If $x_1, \cdots, x_n$ are an IID sample from Cauchy(θ), then

$$\begin{aligned} \mathcal{I}(\theta) &= n/2 \\ I(\theta) &= -\sum_{i=1}^n \frac{2\{(x_i - \theta)^2 - 1\}}{\{(x_i - \theta)^2 + 1\}^2} \end{aligned}$$

and so $\mathcal{I}(\hat{\theta}) \neq I(\hat{\theta})$.

We now expand what we have found out through the above examples and present two important equalities of the score function. The equality in (2.3) is often called the (second-order) *Bartlett identity*.

Theorem 2.3. *Under regularity conditions allowing the exchange of the order of integration and differentiation,*

$$E_\theta \{S(\theta)\} = 0 \tag{2.2}$$

and

$$V_\theta \{S(\theta)\} = \mathcal{I}(\theta). \tag{2.3}$$

Proof.

$$\begin{aligned} E_\theta\{S(\theta)\} &= E_\theta\left\{\frac{\partial}{\partial\theta}\ln L(\theta)\right\} \\ &= E_\theta\left\{\frac{\partial f(\mathbf{y};\theta)/\partial\theta}{f(\mathbf{y};\theta)}\right\} \\ &= \int\frac{\partial}{\partial\theta}f(\mathbf{y};\theta)\,d\mu(\mathbf{y}). \end{aligned}$$

By assumption,

$$\int\frac{\partial}{\partial\theta}f(\mathbf{y};\theta)\,d\mu(\mathbf{y}) = \frac{\partial}{\partial\theta}\int f(\mathbf{y};\theta)\,d\mu(\mathbf{y}).$$

Since $\int f(\mathbf{y};\theta)\,d\mu(\mathbf{y}) = 1$,

$$\frac{\partial}{\partial\theta}\int f(\mathbf{y};\theta)\,d\mu(\mathbf{y}) = 0$$

and (2.2) is proven.

To prove (2.3), note that since $E_\theta\{S(\theta)\} = 0$, equality (2.3) is equivalent to

$$E_\theta\{S(\theta)S(\theta)'\} = -E_\theta\left\{\frac{\partial}{\partial\theta'}S(\theta)\right\}. \tag{2.4}$$

To show (2.4), taking the partial derivative of (2.2) with respect to θ, we get

$$\begin{aligned} 0 &= \frac{\partial}{\partial\theta'}\int S(\theta;\mathbf{y})f(\mathbf{y};\theta)\,d\mu(\mathbf{y}) \\ &= \int\left\{\frac{\partial}{\partial\theta'}S(\theta;\mathbf{y})\right\}f(\mathbf{y};\theta)\,d\mu(\mathbf{y}) + \int S(\theta;\mathbf{y})\left\{\frac{\partial}{\partial\theta'}f(\mathbf{y};\theta)\right\}d\mu(\mathbf{y}) \\ &= E_\theta\left\{\frac{\partial}{\partial\theta'}S(\theta)\right\} + E_\theta\{S(\theta)S(\theta)'\}, \end{aligned}$$

and we have shown (2.3). □

Equality (2.3) is a special case of the general equality

$$Cov\{g(\mathbf{y};\theta),S(\theta)\} = -E\{\partial g(\mathbf{y};\theta)/\partial\theta'\} \tag{2.5}$$

for any $g(\mathbf{y};\theta)$ such that $E\{g(\mathbf{y};\theta)\} = 0$. Under the model $y_1,\cdots,y_n \overset{i.i.d.}{\sim} f(y;\theta_0)$, Theorem 2.2 states that the limiting distribution of the MLE is

$$\sqrt{n}\left(\hat{\theta}-\theta_0\right) \xrightarrow{d} N\left(0, n\mathcal{I}(\theta_0)^{-1}A_0\mathcal{I}(\theta_0)^{-1}\right)$$

where $A_0 = E_{\theta_0}\{S(\theta_0)S(\theta_0)'\}$. By Theorem 2.3, $A_0 = \mathcal{I}(\theta_0)$, then the limiting distribution of the MLE is

$$\sqrt{n}\left(\hat{\theta}-\theta_0\right) \xrightarrow{d} N\left(0, n\mathcal{I}^{-1}(\theta_0)\right).$$

The MLE also satisfies

$$-2\ln\left[\frac{L(\theta_0)}{L(\hat{\theta})}\right] \xrightarrow{d} \chi^2_p$$

which can be used to develop likelihood-ratio (LR) confidence intervals for θ_0. The level α LR confidence intervals (CI) are constructed by

$$\left\{\theta; L(\theta) > k_\alpha \times L(\hat{\theta})\right\}$$

for some k_α which is the upper α quantile of the chi-square distribution with p degrees of freedom. The LR confidence interval is more attractive than the Wald confidence interval in two aspects: (i) A Wald CI can often produce interval estimates beyond the parameter space. (ii) A LR interval is invariant with respect to parameter transformation. For example, if (θ_L,θ_U) is the 95% CI for θ, then $(g(\theta_L), g(\theta_U))$ is the 95% CI for a monotone increasing function $g(\theta)$.

2.2 Observed likelihood

We now derive the likelihood function under the existence of missing data. Roughly speaking, the marginal density for the observed data is called the *observed likelihood*. To formally define the observed likelihood, let $\mathbf{y} = (y_1, \cdots, y_p)$ be a p-dimensional random vector with probability distribution function $f(\mathbf{y}; \theta)$. We are interested in estimating the parameter θ. If n independent realizations of $\mathbf{y}$ were observed throughout the sample, then estimation of θ would be obtained by the maximum likelihood method.

Now suppose that, for unit i in the sample, we observe only a part of $\mathbf{y}_i = (y_{i1}, \cdots, y_{ip})$. Let δ_{ij} be the response indicator function of y_{ij}, defined by

$$\delta_{ij} = \begin{cases} 1 & \text{if } y_{ij} \text{ is observed} \\ 0 & \text{otherwise.} \end{cases}$$

To estimate the parameter θ, we assume response probability model $\Pr(\delta \mid \mathbf{y})$, where $\delta = (\delta_1, \cdots, \delta_p)$.

In general, we have the following two models:

1. Original sample distribution: $f(\mathbf{y}; \theta)$
2. Response mechanism: $\Pr(\delta \mid \mathbf{y})$. Often, it is assumed that $\Pr(\delta \mid \mathbf{y}) = P(\delta \mid \mathbf{y}; \phi)$ with ϕ being the parameter of this response probability model.

Let $(\mathbf{y}_{i,\text{obs}}, \mathbf{y}_{i,\text{mis}})$ be the observed part and missing part of $\mathbf{y}_i$, respectively. For each unit i, we observe $(\mathbf{y}_{i,\text{obs}}, \delta_i)$ instead of observing $\mathbf{y}_i$. Given the above models, we can derive the marginal density of $(\mathbf{y}_{i,\text{obs}}, \delta_i)$ as

$$\tilde{f}(\mathbf{y}_{i,\text{obs}}, \delta_i; \theta, \phi) = \int f(\mathbf{y}_i; \theta) P(\delta_i \mid \mathbf{y}_i; \phi) \, d\mu(\mathbf{y}_{i,\text{mis}}). \tag{2.6}$$

Under the IID assumption, we can write the joint density as

$$\tilde{f}(\mathbf{y}_{\text{obs}}, \delta; \theta, \phi) = \prod_{i=1}^{n} \tilde{f}(\mathbf{y}_{i,\text{obs}}, \delta_i; \theta, \phi) \tag{2.7}$$

where $\mathbf{y}_{\text{obs}} = (\mathbf{y}_{1,\text{obs}}, \cdots, \mathbf{y}_{n,\text{obs}})$ and $\tilde{f}(\mathbf{y}_{i,\text{obs}}, \delta_i; \theta, \phi)$ is defined in (2.6). The joint density in (2.7) as a function of parameter (θ, ϕ) can be called the *observed likelihood*. The observed likelihood is the marginal density of the observation, expressed as a function of the parameters. To give a formal definition of the observed likelihood function, let

$$\mathcal{R}(\mathbf{y}_{\text{obs}}, \delta) = \{\mathbf{y} : \mathbf{y}_{\text{obs}}(\mathbf{y}_i, \delta_i) = \mathbf{y}_{i,\text{obs}}, \ i = 1, \cdots, n\} \tag{2.8}$$

be the set of all possible values of $\mathbf{y}$ with the same realized value of $\mathbf{y}_{\text{obs}}$, for a given δ, where $\mathbf{y}_{\text{obs}}(\mathbf{y}_i, \delta_i)$ is a function that gives the value of y_{ij} for $\delta_{ij} = 1$. That is, for given $\mathbf{y}_i = (y_{i1}, \cdots, y_{ip})$, the j-th component of $\mathbf{y}_{\text{obs}}(\mathbf{y}_i, \delta_i)$ is equal to the j-th component of $\mathbf{y}_{i,\text{obs}}$.

Definition 2.5. *Let $f(\mathbf{y}; \theta)$ be the joint density of the original observation $\mathbf{y} = (\mathbf{y}_1, \cdots, \mathbf{y}_n)$ and let $P(\delta \mid \mathbf{y}; \phi)$ be the conditional probability of $\delta = (\delta_1, \cdots, \delta_n)$ given $\mathbf{y}$. The observed likelihood of (θ, ϕ) based on the realized observation $(\mathbf{y}_{\text{obs}}, \delta)$ is given by*

$$L_{\text{obs}}(\theta, \phi) = \int_{\mathcal{R}(\mathbf{y}_{\text{obs}}, \delta)} f(\mathbf{y}; \theta) P(\delta \mid \mathbf{y}; \phi) \, d\mu(\mathbf{y}). \tag{2.9}$$

Under the IID setup, the observed likelihood is given by

$$L_{\text{obs}}(\theta, \phi) = \prod_{i=1}^{n} \left[\int f(\mathbf{y}_i; \theta) P(\delta_i \mid \mathbf{y}_i; \phi) \, d\mu(\mathbf{y}_{i,\text{mis}}) \right]$$

where it is understood that, if $\mathbf{y}_i = \mathbf{y}_{i,\text{obs}}$ and $\mathbf{y}_{i,\text{mis}}$ is empty then there is nothing to integrate out.

Parameter ϕ can be viewed as a nuisance parameter in the sense that we are not directly interested in estimating ϕ, but an estimate of ϕ is needed to estimate θ, which is the parameter of interest in $f(y;\theta)$.

In the special case of scalar y, the observed likelihood can be written as

$$L_{\text{obs}}(\theta,\phi) = \prod_{\delta_i=1} [f(y_i;\theta)\pi(y_i;\phi)] \times \prod_{\delta_i=0} \left[\int f(y;\theta)\{1-\pi(y;\phi)\}d\mu(y)\right]. \tag{2.10}$$

where $\pi(y;\phi) = P(\delta = 1 \mid y;\phi)$.

Example 2.2. *Consider the following regression model*

$$y_i = x_i'\beta + \varepsilon_i, \quad \varepsilon_i \sim N\left(0,\sigma^2\right)$$

and, instead of y_i, we observe

$$y_i^* = \begin{cases} y_i & \text{if } y_i > 0 \\ 0 & \text{if } y_i \le 0. \end{cases}$$

The observed log-likelihood is

$$l_{\text{obs}}\left(\beta,\sigma^2\right) = -\frac{1}{2}\sum_{y_i^*>0}\left[\ln 2\pi + \ln\sigma^2 + \frac{(y_i^* - x_i'\beta)^2}{\sigma^2}\right] + \sum_{y_i^*=0}\ln\left[1-\Phi\left(\frac{x_i'\beta}{\sigma}\right)\right]$$

where $\Phi(x)$ is the cumulative distribution function of the standard normal distribution. This model is sometimes called the Tobit model in the econometric literature (Amemiya, 1985), termed after Tobin (1958).

Example 2.3. *Let $t_1,t_2,\cdots,t_n$ be an IID sample from a distribution with density $f_\theta(t) = \theta e^{-\theta t}I(t>0)$. Instead of observing t_i, we observe (y_i,δ_i) where*

$$y_i = \begin{cases} t_i & \text{if } \delta_i = 1 \\ c & \text{if } \delta_i = 0 \end{cases}$$

and

$$\delta_i = \begin{cases} 1 & \text{if } t_i \le c \\ 0 & \text{if } t_i > c, \end{cases}$$

where c is a known censoring time. The observed likelihood for θ can be derived as

$$\begin{aligned} L_{\text{obs}}(\theta) &= \prod_{i=1}^{n}\left[\{f_\theta(t_i)\}^{\delta_i}\{P(t_i>c)\}^{1-\delta_i}\right] \\ &= \theta^{\sum_{i=1}^{n}\delta_i}\exp(-\theta\sum_{i=1}^{n}y_i). \end{aligned}$$

In Example 2.2 and Example 2.3, the missing mechanism is known in the sense that the censoring time is known. In general, the missing mechanism is unknown and it often depends on some unknown parameter ϕ. In this case, the observed likelihood can be expressed as in (2.10) or (2.9). Suppose the observed likelihood can be expressed as a product of two terms:

$$L_{\text{obs}}(\theta,\phi) = L_1(\theta)L_2(\phi). \tag{2.11}$$

In this case, the value of θ that maximizes the observed likelihood does not depend on ϕ. Thus, to compute the MLE of θ, we have only to find $\hat{\theta}$ that maximizes $L_1(\theta)$ and the modeling for the missing mechanism can be avoided in this simple situation. To find a sufficient condition for (2.11), we need to define the concept of missing at random (MAR), which was introduced by Rubin (1976).

Definition 2.6. *Let the joint density of* δ *given* $\mathbf{y}$ *be* $P(\delta \mid \mathbf{y})$. *Let* $\mathbf{y}_{\mathrm{obs}}$ *be a random vector that is defined through* $\mathbf{y}_{\mathrm{obs}} = \mathbf{y}_{\mathrm{obs}}(\mathbf{y}, \delta)$, *such that*

$$y_{i,\mathrm{obs}} = \begin{cases} y_i & \text{if } \delta_i = 1 \\ * & \text{if } \delta_i = 0. \end{cases}$$

The response mechanism is called missing at random (MAR) *if*

$$P(\delta \mid \mathbf{y}_1) = P(\delta \mid \mathbf{y}_2), \tag{2.12}$$

for all $\mathbf{y}_1$ *and* $\mathbf{y}_2$ *satisfying* $\mathbf{y}_{\mathrm{obs}}(\mathbf{y}_1, \delta) = \mathbf{y}_{\mathrm{obs}}(\mathbf{y}_2, \delta)$.

Condition (2.12) can be expressed as

$$P(\delta \mid \mathbf{y}) = P(\delta \mid \mathbf{y}_{\mathrm{obs}}). \tag{2.13}$$

Thus, the MAR condition means that the response mechanism $P(\delta \mid \mathbf{y})$ depends on $\mathbf{y}$ only through $\mathbf{y}_{\mathrm{obs}}$, the observed part of $\mathbf{y}$. By the construction of $\mathbf{y}_{\mathrm{obs}}$, we have $\sigma(\mathbf{y}_{\mathrm{obs}}) \subset \sigma(\mathbf{y})$ and

$$P(\delta \mid \mathbf{y}, \mathbf{y}_{\mathrm{obs}}) = P(\delta \mid \mathbf{y}).$$

Here, $\sigma(\mathbf{y})$ is the sigma-algebra generated by $\mathbf{y}$ (Billingsley, 1986). By Bayes' theorem,

$$P(\mathbf{y} \mid \mathbf{y}_{\mathrm{obs}}, \delta) = \frac{P(\delta \mid \mathbf{y}, \mathbf{y}_{\mathrm{obs}})}{P(\delta \mid \mathbf{y}_{\mathrm{obs}})} P(\mathbf{y} \mid \mathbf{y}_{\mathrm{obs}}).$$

Therefore, the MAR condition (2.13) is equivalent to

$$P(\mathbf{y} \mid \mathbf{y}_{\mathrm{obs}}, \delta) = P(\mathbf{y} \mid \mathbf{y}_{\mathrm{obs}}).$$

The MAR condition is essentially the conditional independence of δ and $\mathbf{y}$ given $\mathbf{y}_{\mathrm{obs}}$. Using the notation of Dawid (1979),

$$\delta \perp \mathbf{y} \mid \mathbf{y}_{\mathrm{obs}}.$$

The following theorem, originally proved by Rubin (1976), shows likelihood can be factorized under MAR.

Theorem 2.4. *Let the joint density of* δ *given* $\mathbf{y}$ *be* $P(\delta \mid \mathbf{y}; \phi)$ *and the joint density of* $\mathbf{y}$ *be* $f(\mathbf{y}; \theta)$. *Under the following two conditions*

1. *the parameters* θ *and* ϕ *are distinct*
2. *the MAR condition holds,*

the observed likelihood can be written as in (2.11) and the MLE of θ *can be obtained by maximizing* $L_1(\theta)$.

Proof. Under MAR, we have (2.13) and the observed likelihood (2.9) can be written as (2.11) where

$$L_1(\theta) = \int_{\mathcal{R}(\mathbf{y}_{\mathrm{obs}}, \delta)} f(\mathbf{y}; \theta)\, d\mu(\mathbf{y})$$

and $L_2(\phi) = P(\delta \mid \mathbf{y}_{\mathrm{obs}}; \phi)$. □

Example 2.4. *Consider bivariate data* (x_i, y_i) *with pdf* $f(x, y) = f_1(y \mid x) f_2(x)$, *where* x_i *is always observed and* y_i *is subject to missingness. Assume that the response status variable* δ_i *of* y_i *satisfies*

$$P(\delta_i = 1 \mid x_i, y_i) = \Lambda_1(\phi_0 + \phi_1 x_i + \phi_2 y_i)$$

for $\Lambda_1(x) = 1 - \{1 + \exp(x)\}^{-1}$. *Let* θ *be the parameter of interest in the regression model*

$f_1(y \mid x;\theta)$. Let α be the parameter in the marginal distribution of x, denoted by $f_2(x_i;\alpha)$. Define $\Lambda_0(x) = 1 - \Lambda_1(x)$. Then, the observed likelihood can be written as

$$\begin{aligned} L_{\text{obs}}(\theta,\alpha,\phi) &= \left[\prod_{\delta_i=1} f_1(y_i \mid x_i;\theta) f_2(x_i;\alpha) \Lambda_1(\phi_0+\phi_1 x_i+\phi_2 y_i)\right] \\ &\times \left[\prod_{\delta_i=0} \int f_1(y_i \mid x_i;\theta) f_2(x_i;\alpha) \Lambda_0(\phi_0+\phi_1 x_i+\phi_2 y_i)\, dy_i\right] \\ &= L_1(\theta,\phi) \times L_2(\alpha) \end{aligned}$$

where

$$\begin{aligned} L_1(\theta,\phi) &= \prod_{\delta_i=1} f_1(y_i \mid x_i;\theta) \Lambda_1(\phi_0+\phi_1 x_i+\phi_2 y_i) \\ &\quad \times \prod_{\delta_i=0} \int f_1(y_i \mid x_i;\theta) \Lambda_0(\phi_0+\phi_1 x_i+\phi_2 y_i)\, dy_i \end{aligned}$$

and

$$L_2(\alpha) = \prod_{i=1}^{n} f_2(x_i;\alpha).$$

If $\phi_2 = 0$, then MAR holds and

$$L_1(\theta,\phi) = L_{1a}(\theta) \times L_{1b}(\phi) \tag{2.14}$$

where

$$L_{1a}(\theta) = \prod_{\delta_i=1} f_1(y_i \mid x_i;\theta)$$

and

$$L_{1b}(\phi) = \prod_{\delta_i=1} \Lambda_1(\phi_0+\phi_1 x_i) \times \prod_{\delta_i=0} \Lambda_0(\phi_0+\phi_1 x_i).$$

Thus, under MAR, the MLE of θ can be obtained by maximizing $L_{1a}(\theta)$, which is obtained by ignoring the missing part of the data.

In Example 2.4, instead of y_i subject to missingness, if x_i is subject to missingness, then the observed likelihood becomes

$$\begin{aligned} L_{\text{obs}}(\theta,\phi,\alpha) &= \left[\prod_{\delta_i=1} f_1(y_i \mid x_i;\theta) f_2(x_i;\alpha) \Lambda_1(\phi_0+\phi_1 x_i+\phi_2 y_i)\right] \\ &\times \left[\prod_{\delta_i=0} \int f_1(y_i \mid x_i;\theta) f_2(x_i;\alpha) \Lambda_0(\phi_0+\phi_1 x_i+\phi_2 y_i)\, dx_i\right] \\ &\neq L_1(\theta,\phi) \times L_2(\alpha). \end{aligned}$$

If $\phi_1 = 0$ then

$$L_{\text{obs}}(\theta,\alpha,\phi) = L_1(\theta,\alpha) \times L_2(\phi)$$

and MAR holds. Although we are not interested in the marginal distribution of x, we have to specify the model for the marginal distribution of x. If $\phi_1 = \phi_2 = 0$ holds then

$$L_{\text{obs}}(\theta,\alpha,\phi) = L_{1a}(\theta) \times L_{1b}(\phi) \times L_2(\alpha)$$

and this is called *missing at completely random (MCAR)*.

2.3 Mean score approach

The observed likelihood is simply the marginal likelihood of the observation $(\mathbf{y}_{\text{obs}}, \delta)$. Express $\eta = (\theta, \phi)$ and the observed likelihood can be written as follows:

$$\begin{aligned} L_{\text{obs}}(\eta) &= \int_{\mathcal{R}(\mathbf{y}_{\text{obs}}, \delta)} f(\mathbf{y};\theta) P(\delta \mid \mathbf{y};\phi)\, d\mu(\mathbf{y}) \\ &= \int f(\mathbf{y};\theta) P(\delta \mid \mathbf{y};\phi)\, d\mu(\mathbf{y}_{\text{mis}}) \\ &= \int f(\mathbf{y},\delta;\eta)\, d\mu(\mathbf{y}_{\text{mis}}) \end{aligned}$$

where $\mathcal{R}(\mathbf{y}_{\text{obs}}, \delta)$ is defined in (2.8) and $\mathbf{y}_{\text{mis}}$ is the missing part of $\mathbf{y}$. To find the MLE that maximizes the observed likelihood, we often need to solve

$$S_{\text{obs}}(\eta) \equiv \partial \ln L_{\text{obs}}(\eta) / \partial \eta = 0. \tag{2.15}$$

The score equation in (2.15) is called the *observed score equation* because it is based on the observed likelihood. The function, $S_{\text{obs}}(\eta)$, can be called the *observed score function*. Working with the observed score function can be computationally challenging because the observed likelihood is in integral form.

To overcome this difficulty, we can establish the following theorem, which was originally proposed by Fisher (1922) and also discussed by Louis (1982).

Theorem 2.5. *Under some regularity conditions,*

$$S_{\text{obs}}(\eta) = \bar{S}(\eta) \tag{2.16}$$

where

$$\begin{aligned} \bar{S}(\eta) &= E\{S_{\text{com}}(\eta) \mid \mathbf{y}_{\text{obs}}, \delta\}, \\ S_{\text{com}}(\eta) &= \partial \ln f(\mathbf{y},\delta;\eta)/\partial \eta, \end{aligned}$$

and

$$f(\mathbf{y},\delta;\eta) = f(\mathbf{y};\theta) P(\delta \mid \mathbf{y};\phi). \tag{2.17}$$

Proof.

$$\frac{\partial}{\partial \eta} \ln\{L_{\text{obs}}(\eta)\} = \frac{\partial L_{\text{obs}}(\eta)/\partial \eta}{L_{\text{obs}}(\eta)} = \frac{\int [\partial f(\mathbf{y},\delta;\eta)/\partial \eta]\, d\mu(\mathbf{y}_{\text{mis}})}{L_{\text{obs}}(\eta)}.$$

Next, consider the numerator only.

$$\begin{aligned} \int [\partial f(\mathbf{y},\delta;\eta)/\partial\eta]\, d\mu(\mathbf{y}_{\text{mis}}) &= \int \frac{[\partial f(\mathbf{y},\delta;\eta)/\partial\eta]}{f(\mathbf{y},\delta;\eta)} \frac{f(\mathbf{y},\delta;\eta)}{L_{\text{obs}}(\eta)} d\mu(\mathbf{y}_{\text{mis}}) \times L_{\text{obs}}(\eta) \\ &= E\{\partial \ln f(\mathbf{y},\delta;\eta)/\partial\eta \mid \mathbf{y}_{\text{obs}}, \delta\} \times L_{\text{obs}}(\eta) \\ &= E\{S_{\text{com}}(\eta) \mid \mathbf{y}_{\text{obs}}, \delta\} \times L_{\text{obs}}(\eta), \end{aligned}$$

we then have the result. □

The function $\bar{S}(\eta)$ is called the *mean score function*. It is computed by taking the conditional expectation of the complete-sample score function given the observation. The mean score function is easier to compute than the observed score function.

Remark 2.1. *An alternative proof for Theorem 2.5 can be made as follows. In connection with the aforementioned proof, it also plays with the form of the conditional density,*

$$\frac{f(\mathbf{y},\delta;\eta)}{L_{\text{obs}}(\eta)} = f(\mathbf{y}_{\text{mis}} \mid \mathbf{y}_{\text{obs}}, \delta).$$

Here, we can express

$$f(\mathbf{y}_{\text{mis}} \mid \mathbf{y}_{\text{obs}}, \delta) = f(\mathbf{y}, \delta \mid \mathbf{y}_{\text{obs}}, \delta),$$

as we can decompose $\mathbf{y} = (\mathbf{y}_{\text{obs}}, \mathbf{y}_{\text{mis}})$. *Since* $L_{\text{obs}}(\eta) = f(\mathbf{y}, \delta; \eta) / f(\mathbf{y}, \delta \mid \mathbf{y}_{\text{obs}}, \delta; \eta)$, *we have*

$$\frac{\partial}{\partial \eta} \ln L_{\text{obs}}(\eta) = \frac{\partial}{\partial \eta} \ln f(\mathbf{y}, \delta; \eta) - \frac{\partial}{\partial \eta} \ln f(\mathbf{y}, \delta \mid \mathbf{y}_{\text{obs}}, \delta; \eta), \tag{2.18}$$

taking conditional expectation of the above equation over the conditional distribution of $(\mathbf{y}, \delta)$ *given* $(\mathbf{y}_{\text{obs}}, \delta)$, *we have*

$$\begin{aligned} \frac{\partial}{\partial \eta} \ln L_{\text{obs}}(\eta) &= E\left\{ \frac{\partial}{\partial \eta} \ln L_{\text{obs}}(\eta) \mid \mathbf{y}_{\text{obs}}, \delta \right\} \\ &= E\{S(\eta) \mid \mathbf{y}_{\text{obs}}, \delta\} - E\left\{ \frac{\partial}{\partial \eta} \ln f(\mathbf{y}, \delta \mid \mathbf{y}_{\text{obs}}, \delta; \eta) \mid \mathbf{y}_{\text{obs}}, \delta \right\}. \end{aligned}$$

Here, the first equality holds because $L_{\text{obs}}(\eta)$ *is a function of* $(\mathbf{y}_{\text{obs}}, \delta)$ *only. The second equality follows from (2.18). The last term is equal to zero by Theorem 2.3 applied to the conditional distribution, which states that the expected value of the score function is zero and the reference distribution in this case is the conditional distribution of* $(\mathbf{y}, \delta)$ *given* $(\mathbf{y}_{\text{obs}}, \delta)$.

By the form of the joint density in (2.17), we can express

$$\bar{S}(\eta) = [E\{S_1(\theta) \mid \mathbf{y}_{\text{obs}}, \delta\}, E\{S_2(\phi) \mid \mathbf{y}_{\text{obs}}, \delta\}],$$

where

$$\begin{aligned} S_1(\theta) &= \partial \ln f(\mathbf{y}; \theta) / \partial \theta \\ S_2(\phi) &= \partial \ln P(\delta \mid \mathbf{y}; \phi) / \partial \phi. \end{aligned}$$

Example 2.5. *Suppose that the study variable* y *is randomly distributed as Bernoulli* (p_i), *where*

$$p_i = p_i(\beta) = \frac{\exp(\mathbf{x}_i'\beta)}{1 + \exp(\mathbf{x}_i'\beta)}$$

for some unknown parameter β *and* $\mathbf{x}_i$ *is a vector of covariates in the logistic regression model for* y_i. *Let* δ_i *be the response indicator function for* y_i *with distribution Bernoulli*(π_i) *where*

$$\pi_i = \frac{\exp(\mathbf{x}_i'\phi_0 + y_i\phi_1)}{1 + \exp(\mathbf{x}_i'\phi_0 + y_i\phi_1)}.$$

We assume that $\mathbf{x}_i$ *is always observed, but* y_i *is missing if* $\delta_i = 0$.

Under complete response, the score function for β *is*

$$S_1(\beta) = \sum_{i=1}^{n} (y_i - p_i(\beta))\mathbf{x}_i$$

and the score function for ϕ *is*

$$S_2(\phi) = \sum_{i=1}^{n} (\delta_i - \pi_i(\phi))(\mathbf{x}_i', y_i)'.$$

With missing data, the mean score function for β *becomes*

$$\bar{S}_1(\beta, \phi) = \sum_{\delta_i = 1} \{y_i - p_i(\beta)\}\mathbf{x}_i + \sum_{\delta_i = 0} \sum_{y=0}^{1} w_i(y; \beta, \phi)\{y - p_i(\beta)\}\mathbf{x}_i, \tag{2.19}$$

where

$$w_i(y;\beta,\phi) = \frac{P(y_i = y \mid \mathbf{x}_i;\beta) P(\delta_i = 0 \mid y_i = y, \mathbf{x}_i;\phi)}{\sum_{z=0}^{1} P(y_i = z \mid \mathbf{x}_i;\beta) P(\delta_i = 0 \mid y_i = z, \mathbf{x}_i;\phi)}$$

Thus, $\bar{S}_1(\beta,\phi)$ is also a function of ϕ. If the response mechanism is MAR so that $\phi_1 = 0$, then

$$w_i(y;\beta,\phi) = \frac{P(y_i = y \mid \mathbf{x}_i;\beta)}{\sum_{z=0}^{1} P(y_i = z \mid \mathbf{x}_i;\beta)} = P(y_i = y \mid \mathbf{x}_i;\beta)$$

and so

$$\bar{S}_1(\beta,\phi) = \sum_{\delta_i=1} \{y_i - p_i(\beta)\}\mathbf{x}_i = \bar{S}_1(\beta).$$

Similarly, under missing data, the mean score function for ϕ is

$$\begin{aligned}\bar{S}_2(\beta,\phi) &= \sum_{\delta_i=1} \{\delta_i - \pi(\phi; x_i, y_i)\}(\mathbf{x}_i, y_i) \\ &\quad + \sum_{\delta_i=0}\sum_{y=0}^{1} w_i(y;\beta,\phi)\{\delta_i - \pi_i(\phi;\mathbf{x}_i, y)\}(\mathbf{x}_i, y).\end{aligned}$$

Thus, $\bar{S}_2(\beta,\phi)$ is also a function of β. In general, finding the solution to $[\bar{S}_1(\beta,\phi), \bar{S}_2(\beta,\phi)] = (0,0)$ is not easy because the weight $w_i(y;\beta,\phi)$ is a function of the unknown parameters.

In the general missing data problem, we have

$$\bar{S}(\theta,\phi) = [E\{S_1(\theta) \mid \mathbf{y},\delta\}, E\{S_2(\phi) \mid \mathbf{y},\delta\}] = [\bar{S}_1(\theta,\phi), \bar{S}_2(\theta,\phi)]$$

and, under MAR, we can write

$$S_{\text{obs}}(\theta,\phi) = [\bar{S}_1(\theta), \bar{S}_2(\theta,\phi)].$$

Thus, under MAR, the MLE of θ can be obtained by solving $\bar{S}_1(\theta) = 0$.

Example 2.6. *Suppose that the study variable y follows a normal distribution with mean $\mathbf{x}'\beta$ and variance σ^2. The score equations for β and σ^2 under complete response are*

$$S_1(\beta,\sigma^2) = \sum_{i=1}^{n} (y_i - \mathbf{x}_i'\beta)\mathbf{x}_i/\sigma^2 = \mathbf{0}$$

and

$$S_2(\beta,\sigma^2) = -n/(2\sigma^2) + \sum_{i=1}^{n} (y_i - \mathbf{x}_i'\beta)^2/(2\sigma^4) = 0.$$

Assume that y_i are observed only for the first r elements and the MAR assumption holds. In this case, the mean score function reduces to

$$\bar{S}_1(\beta,\sigma^2) = \sum_{i=1}^{r} (y_i - \mathbf{x}_i'\beta)\mathbf{x}_i/\sigma^2$$

and

$$\bar{S}_2(\beta,\sigma^2) = -n/(2\sigma^2) + \sum_{i=1}^{r} (y_i - \mathbf{x}_i'\beta)^2/(2\sigma^4) + (n-r)/(2\sigma^2).$$

The maximum likelihood estimator obtained by solving the mean score equations is

$$\hat{\beta} = \left(\sum_{i=1}^{r} \mathbf{x}_i\mathbf{x}_i'\right)^{-1} \sum_{i=1}^{r} \mathbf{x}_i y_i$$

and

$$\hat{\sigma}^2 = \frac{1}{r}\sum_{i=1}^{r}\left(y_i - \mathbf{x}_i'\hat{\beta}\right)^2.$$

The resulting estimators can also be obtained by simply ignoring the missing part of the sample.

2.4 Observed information

We discuss some statistical properties of the observed score function in the missing data setup. Before we derive the main theory, it is necessary to give a definition of the information matrix associated with the observed score function. The following definition is an extension to Definition 2.4.

Definition 2.7. *1. Observed score function:*

$$S_{\text{obs}}(\eta) = \frac{\partial}{\partial \eta} \ln L_{\text{obs}}(\eta)$$

2. Fisher information (representing curvature of the log-likelihood) from the observed likelihood

$$I_{\text{obs}}(\eta) = -\frac{\partial^2}{\partial \eta \partial \eta'} \ln L_{\text{obs}}(\eta) = -\frac{\partial}{\partial \eta'} S_{\text{obs}}(\eta)$$

3. Expected (Fisher) information from the observed likelihood: $\mathcal{I}_{\text{obs}}(\eta) = E_\eta \{I_{\text{obs}}(\eta)\}$.

The following theorem presents the basic properties of the observed score function.

Theorem 2.6. *Under the conditions of Theorem 2.3, we have*

$$E\{S_{\text{obs}}(\eta)\} = 0 \tag{2.20}$$

and

$$V\{S_{\text{obs}}(\eta)\} = \mathcal{I}_{\text{obs}}(\eta) \tag{2.21}$$

where $\mathcal{I}_{\text{obs}}(\eta) = E_\eta \{I_{\text{obs}}(\eta)\}$ *is the expected information from the observed likelihood.*

Proof. Equality (2.20) follows because, by (2.16),

$$\begin{aligned} E\{S_{\text{obs}}(\eta)\} &= E[E\{S_{\text{com}}(\eta) \mid \mathbf{y}_{\text{obs}}, \delta\}] \\ &= E\{S_{\text{com}}(\eta)\} \end{aligned}$$

which is equal to zero by (2.2). Equality (2.21) can be derived using the same argument for proving (2.4), with the observed likelihood at concern this time. □

By Theorem 2.2, under some regularity conditions, the solution to $\bar{S}(\eta) = 0$ is consistent to η_0 and has the asymptotic variance $\mathcal{I}_{\text{obs}}(\eta_0)^{-1}$ where

$$\mathcal{I}_{\text{obs}}(\eta) = E\left\{-\frac{\partial S_{\text{obs}}(\eta)}{\partial \eta'}\right\} = E\left\{S_{\text{obs}}^{\otimes 2}(\eta)\right\} = E\{\bar{S}^{\otimes 2}(\eta)\}$$

with $B^{\otimes 2}$ denoting BB'.

For variance estimation of the estimate obtained by solving $\bar{S}(\eta) = 0$, one can use $\mathcal{I}_{\text{obs}}(\hat{\eta})^{-1}$, the inverse of the expected Fisher information applied to the observed data. Under the IID setup, Redner and Walker (1984) proposed using

$$\left[\hat{H}(\hat{\eta})\right]^{-1} = \left\{\sum_{i=1}^{n} \bar{S}_i^{\otimes 2}(\hat{\eta})\right\}^{-1}$$

to estimate the variance of $\hat{\eta}$, where $\bar{S}_i(\eta) = E\{S_i(\eta) \mid \mathbf{y}_{i,\text{obs}}, \delta_i\}$. Meilijson (1989) termed $\hat{H}(\hat{\eta})$ the empirical Fisher information and touted its computational convenience.

The following theorem, first proved by Louis (1982), presents a way of computing the observed information obtained from the observed likelihood. The formula in (2.22) is often called *Louis' formula.*

Theorem 2.7. *Let $L_{\text{com}}(\eta) = f(\mathbf{y}, \delta; \eta)$ be the complete sample likelihood and let*

$$I_{\text{com}}(\eta) = -\frac{\partial}{\partial \eta'} S_{\text{com}}(\eta) = -\frac{\partial^2}{\partial \eta \partial \eta'} \ln L_{\text{com}}(\eta)$$

be the Fisher information of $L_{\text{com}}(\eta)$.

Under the conditions of Theorem 2.3,

$$I_{\text{obs}}(\eta) = E\{I_{\text{com}}(\eta) \mid \mathbf{y}_{\text{obs}}, \delta\} + \bar{S}(\eta)^{\otimes 2} - E\{S_{\text{com}}^{\otimes 2}(\eta) \mid \mathbf{y}_{\text{obs}}, \delta\} \tag{2.22}$$

or

$$I_{\text{obs}}(\eta) = E\{I_{\text{com}}(\eta) \mid \mathbf{y}_{\text{obs}}, \delta\} - V\{S_{\text{com}}(\eta) \mid \mathbf{y}_{\text{obs}}, \delta\}, \tag{2.23}$$

where $I_{\text{obs}}(\eta)$ is the Fisher information from the observed likelihood and $\bar{S}(\eta) = E\{S_{\text{com}}(\eta) \mid \mathbf{y}_{\text{obs}}, \delta\}$ is the mean score function defined in (2.16).

Proof. Since

$$L_{\text{obs}}(\eta) = L_{\text{com}}(\eta) / f(\mathbf{y}, \delta \mid \mathbf{y}_{\text{obs}}, \delta; \eta),$$

we have

$$-\frac{\partial^2}{\partial \eta \partial \eta'} \ln L_{\text{obs}}(\eta) = -\frac{\partial^2}{\partial \eta \partial \eta'} \ln L_{\text{com}}(\eta) + \frac{\partial^2}{\partial \eta \partial \eta'} \ln f(\mathbf{y}, \delta \mid \mathbf{y}_{\text{obs}}, \delta; \eta). \tag{2.24}$$

The first term on the right side of the equality in (2.24) is equal to $I_{\text{com}}(\eta)$, the observed information for the full likelihood. For the second term, define

$$S_{\text{mis}}(\eta) = \frac{\partial}{\partial \eta} \ln f(\mathbf{y}, \delta \mid \mathbf{y}_{\text{obs}}, \delta; \eta). \tag{2.25}$$

Applying the same argument for proving (2.4) to $S_{\text{mis}}(\eta)$, we can show that

$$E\left\{-\frac{\partial}{\partial \eta'} S_{\text{mis}}(\eta) \mid \mathbf{y}_{\text{obs}}, \delta\right\} = E\left\{S_{\text{mis}}(\eta)^{\otimes 2} \mid \mathbf{y}_{\text{obs}}, \delta\right\}. \tag{2.26}$$

Thus, taking the conditional expectation of (2.24) given $(\mathbf{y}_{\text{obs}}, \delta)$, we have

$$I_{\text{obs}}(\eta) = E\{I_{\text{com}}(\eta) \mid \mathbf{y}_{\text{obs}}, \delta\} - E\left\{S_{\text{mis}}(\eta)^{\otimes 2} \mid \mathbf{y}_{\text{obs}}, \delta\right\}.$$

Using (2.18), we have

$$S_{\text{mis}}(\eta) = S_{\text{com}}(\eta) - \bar{S}(\eta).$$

and, because $E\{S_{\text{com}}(\eta) \mid \mathbf{y}_{\text{obs}}, \delta\} = \bar{S}(\eta)$, we have

$$E\left\{S_{\text{mis}}(\eta)^{\otimes 2} \mid \mathbf{y}_{\text{obs}}, \delta\right\} = E\left\{S_{\text{com}}(\eta)^{\otimes 2} \mid \mathbf{y}_{\text{obs}}, \delta\right\} - \bar{S}(\eta)^{\otimes 2} \tag{2.27}$$

and the result follows. □

The score function $S_{\text{mis}}(\eta)$ defined in (2.25) is the score function associated with the conditional distribution $f(\mathbf{y}, \delta \mid \mathbf{y}_{\text{obs}}, \delta)$. The expected information derived from $S_{\text{mis}}(\eta)$ is often called the *missing information* and is defined by

$$\mathcal{I}_{\text{mis}}(\eta) = E\left\{-\frac{\partial}{\partial \eta'} S_{\text{mis}}(\eta)\right\}.$$

Also,

$$\mathcal{I}_{\text{mis}}(\eta) = E\left\{S_{\text{mis}}(\eta)^{\otimes 2}\right\}.$$

Using (2.27), the missing information satisfies

$$\mathcal{I}_{\text{mis}}(\eta) = \mathcal{I}_{\text{com}}(\eta) - \mathcal{I}_{\text{obs}}(\eta), \tag{2.28}$$

where

$$\mathcal{I}_{\text{com}}(\eta) = E\left\{-\frac{\partial}{\partial \eta'} S_{\text{com}}(\eta)\right\}$$

is the expected information associated with the complete-sample likelihood. The equality (2.28) is called the *missing information principle* by Orchard and Woodbury (1972). An alternative expression of the missing information principle is

$$V\{S_{\text{mis}}(\eta)\} = V\{S_{\text{com}}(\eta)\} - V\{\bar{S}(\eta)\} = E[V\{S_{\text{com}}(\eta) \mid \mathbf{y}_{\text{obs}}, \delta\}]. \tag{2.29}$$

Note that $V\{S_{\text{com}}(\eta)\} = \mathcal{I}_{\text{com}}(\eta)$ and $V\{S_{\text{obs}}(\eta)\} = \mathcal{I}_{\text{obs}}(\eta)$.

Remark 2.2. *An alternative proof of Theorem 2.7 can be given as follows. By Theorem 2.5, the observed information of $L_{\text{obs}}(\eta)$ can be expressed as*

$$I_{\text{obs}}(\eta) = -\frac{\partial}{\partial \eta'} \bar{S}(\eta)$$

where $\bar{S}(\eta) = E\{S_{\text{com}}(\eta) \mid \mathbf{y}_{\text{obs}}, \delta; \eta\}$. Thus, we have

$$\begin{aligned}
\frac{\partial}{\partial \eta'} \bar{S}(\eta) &= \frac{\partial}{\partial \eta'} \int S_{\text{com}}(\eta; \mathbf{y}) f(\mathbf{y}, \delta \mid \mathbf{y}_{\text{obs}}, \delta; \eta)\, d\mu(\mathbf{y}) \\
&= \int \left\{\frac{\partial}{\partial \eta'} S_{\text{com}}(\eta; \mathbf{y})\right\} f(\mathbf{y}, \delta \mid \mathbf{y}_{\text{obs}}, \delta; \eta)\, d\mu(\mathbf{y}) \\
&\quad + \int S_{\text{com}}(\eta; \mathbf{y}) \left\{\frac{\partial}{\partial \eta'} f(\mathbf{y}, \delta \mid \mathbf{y}_{\text{obs}}, \delta; \eta)\right\} d\mu(\mathbf{y}) \\
&= E\{\partial S_{\text{com}}(\eta)/\partial \eta' \mid \mathbf{y}_{\text{obs}}, \delta\} \\
&\quad + \int S_{\text{com}}(\eta; \mathbf{y}) \left\{\frac{\partial}{\partial \eta'} \log f(\mathbf{y}, \delta \mid \mathbf{y}_{\text{obs}}, \delta; \eta)\right\} f(\mathbf{y}, \delta \mid \mathbf{y}_{\text{obs}}, \delta; \eta)\, d\mu(\mathbf{y}).
\end{aligned}$$

The first term is equal to $-E\{I_{\text{com}}(\eta) \mid \mathbf{y}_{\text{obs}}, \delta\}$ and the second term is equal to

$$\begin{aligned}
E\{S_{\text{com}}(\eta) S_{\text{mis}}(\eta)' \mid \mathbf{y}_{\text{obs}}, \delta\} &= E\left[\{\bar{S}(\eta) + S_{\text{mis}}(\eta)\} S_{\text{mis}}(\eta)' \mid \mathbf{y}_{\text{obs}}, \delta\right] \\
&= E\{S_{\text{mis}}(\eta) S_{\text{mis}}(\eta)' \mid \mathbf{y}_{\text{obs}}, \delta\}
\end{aligned}$$

because

$$E\{\bar{S}(\eta) S_{\text{mis}}(\eta)' \mid \mathbf{y}_{\text{obs}}, \delta\} = E\left[\bar{S}(\eta)\{S_{\text{com}}(\eta) - \bar{S}(\eta)\}' \mid \mathbf{y}_{\text{obs}}, \delta\right] = \bar{S}(\eta)^{\otimes 2} - \bar{S}(\eta)^{\otimes 2} = \mathbf{0}.$$

Example 2.7. *Consider the following bivariate normal distribution:*

$$\begin{pmatrix} y_{1i} \\ y_{2i} \end{pmatrix} \sim N\left[\begin{pmatrix} \mu_1 \\ \mu_2 \end{pmatrix}, \begin{pmatrix} \sigma_{11} & \sigma_{12} \\ \sigma_{12} & \sigma_{22} \end{pmatrix}\right], \tag{2.30}$$

for $i = 1, 2, \cdots, n$. For simplicity, assume that σ_{11}, σ_{12} and σ_{22} are known constants and $\mu = (\mu_1, \mu_2)'$ is the parameter of interest. The complete sample score function for μ is

$$S_{\text{com}}(\mu) = \sum_{i=1}^{n} S_{\text{com}}^{(i)}(\mu) = \sum_{i=1}^{n} \begin{pmatrix} \sigma_{11} & \sigma_{12} \\ \sigma_{12} & \sigma_{22} \end{pmatrix}^{-1} \begin{pmatrix} y_{1i} - \mu_1 \\ y_{2i} - \mu_2 \end{pmatrix}. \tag{2.31}$$

The information matrix of μ based on the complete sample is

$$\mathcal{I}_{\text{com}}(\mu) = n \begin{pmatrix} \sigma_{11} & \sigma_{12} \\ \sigma_{12} & \sigma_{22} \end{pmatrix}^{-1}.$$

Suppose that there are some missing values in y_{1i} and y_{2i} and the original sample is partitioned into four sets:

$$\begin{aligned} H &= \text{both } y_1 \text{ and } y_2 \text{ are observed} \\ K &= \text{only } y_1 \text{ is observed} \\ L &= \text{only } y_2 \text{ is observed} \\ M &= \text{both } y_1 \text{ and } y_2 \text{ are missing.} \end{aligned}$$

Let n_H, n_K, n_L, n_M represent the size of H, K, L, M, respectively. Assume that the response mechanism does not depend on the value of (y_1, y_2) and so it is MAR. In this case, the observed score function of μ based on a single observation in set K is

$$\begin{aligned} E\left\{S_{\text{com}}^{(i)}(\mu) \mid y_{1i}, i \in K\right\} &= \begin{pmatrix} \sigma_{11} & \sigma_{12} \\ \sigma_{12} & \sigma_{22} \end{pmatrix}^{-1} \begin{pmatrix} y_{1i} - \mu_1 \\ E(y_{2i} \mid y_{1i}) - \mu_2 \end{pmatrix} \\ &= \begin{pmatrix} \sigma_{11}^{-1}(y_{1i} - \mu_1) \\ 0 \end{pmatrix}. \end{aligned}$$

Similarly, we have

$$E\left\{S_{\text{com}}^{(i)}(\mu) \mid y_{2i}, i \in L\right\} = \begin{pmatrix} 0 \\ \sigma_{22}^{-1}(y_{2i} - \mu_2) \end{pmatrix}.$$

Therefore, the expected information matrix of μ from the observed likelihood is

$$\mathcal{I}_{\text{obs}}(\mu) = n_H \begin{pmatrix} \sigma_{11} & \sigma_{12} \\ \sigma_{12} & \sigma_{22} \end{pmatrix}^{-1} + n_K \begin{pmatrix} \sigma_{11}^{-1} & 0 \\ 0 & 0 \end{pmatrix} + n_L \begin{pmatrix} 0 & 0 \\ 0 & \sigma_{22}^{-1} \end{pmatrix} \tag{2.32}$$

and the asymptotic variance of the MLE of μ can be obtained by the inverse of $\mathcal{I}_{\text{obs}}(\mu)$. In the special case of $n_L = n_M = 0$,

$$\{\mathcal{I}_{\text{obs}}(\mu)\}^{-1} = \left\{ n_H \begin{pmatrix} \sigma_{11} & \sigma_{12} \\ \sigma_{12} & \sigma_{22} \end{pmatrix}^{-1} + n_K \begin{pmatrix} \sigma_{11}^{-1} & 0 \\ 0 & 0 \end{pmatrix} \right\}^{-1}.$$

Using

$$(A + c\mathbf{b}\mathbf{b}')^{-1} = A^{-1} - A^{-1}\mathbf{b}\left(c^{-1} + \mathbf{b}'A^{-1}\mathbf{b}\right)^{-1}\mathbf{b}'A^{-1}$$

with

$$A = n_H \begin{pmatrix} \sigma_{11} & \sigma_{12} \\ \sigma_{12} & \sigma_{22} \end{pmatrix}^{-1},$$

$\mathbf{b} = (1,0)'$ *and* $c = n_K \sigma_{11}^{-1}$, *we have* $c^{-1} + \mathbf{b}'A^{-1}\mathbf{b} = (1/n_H + 1/n_K)\sigma_{11}$,

$$\{\mathcal{I}_{\text{obs}}(\mu)\}^{-1} = \frac{1}{n_H} \begin{pmatrix} \sigma_{11} & \sigma_{12} \\ \sigma_{12} & \sigma_{22} \end{pmatrix} + \left(\frac{1}{n} - \frac{1}{n_H}\right) \begin{pmatrix} \sigma_{11} & \sigma_{12} \\ \sigma_{12} & \sigma_{12}^2/\sigma_{11} \end{pmatrix}.$$

The asymptotic variance of the MLE of μ_1 is equal to σ_{11}/n and the asymptotic variance of the MLE of μ_2 is equal to $\sigma_{22}\rho^2/n + (1-\rho^2)\sigma_{22}/n_H = \sigma_{22}/n_H - \rho^2\sigma_{22}(1/n_H - 1/n)$, where $\rho = \sigma_{12}/\sqrt{\sigma_{11}\sigma_{22}}$. Thus, by incorporating the partial response, the asymptotic variance is reduced by $\rho^2\sigma_{22}(1/n_H - 1/n)$.

Exercises

1. Show that, for a p-dimensional multivariate normal distribution, $\mathbf{y} \sim N(\mu, \Sigma)$, where $\mu = (\mu_1, \cdots, \mu_p)$ and $\Sigma = (\sigma_{ij})$, the Fisher information of $\theta = (\mu', \sigma_{11}, \sigma_{12}, \cdots, \sigma_{pp})'$ is given by

$$\mathcal{I}(\theta) = \begin{pmatrix} \Sigma^{-1} & \mathbf{0} \\ \mathbf{0} & \frac{1}{2} tr\left(\Sigma^{-1} \frac{\partial \Sigma}{\partial \theta} \Sigma^{-1} \frac{\partial \Sigma}{\partial \theta}\right) \end{pmatrix}.$$

2. Show that for the general exponential family model with log-density of the form

$$\log f(x;\theta) = T(x)\eta(\theta) - A(\theta) + c(x)$$

we have

$$\mathcal{I}(\hat{\theta}) = I(\hat{\theta})$$

where $\hat{\theta}$ is the MLE of θ. If θ is the canonical parameter, then $\mathcal{I}(\theta) = I(\theta)$.

3. Prove (2.5).
4. Assume that $t_1, t_2, \cdots, t_n$ are IID with pdf $p_\theta(t) = \theta e^{-\theta t} I(t > 0)$. Instead of observing t_i, we observe (y_i, δ_i) where

$$y_i = \begin{cases} t_i & \text{if } \delta_i = 1 \\ c & \text{if } \delta_i = 0 \end{cases}$$

and

$$\delta_i = \begin{cases} 1 & \text{if } t_i \leq c \\ 0 & \text{if } t_i > c, \end{cases}$$

where c is the known censoring time. Answer the following questions:

(a) Obtain the observed likelihood and find the MLE for θ.

(b) Show that the Fisher information for the observed likelihood is $I(\theta) = \sum_{i=1}^{n} \delta_i / \theta^2$ and the expected information from the observed likelihood is $\mathcal{I}(\theta) = n\left(1 - e^{-\theta c}\right)/\theta^2$.

(c) For variance estimation, we have two candidates:

$$I(\hat{\theta}) = \sum_{i=1}^{n} \delta_i / \hat{\theta}^2$$

or

$$\mathcal{I}(\hat{\theta}) = n\left(1 - e^{-\hat{\theta} c}\right)/\hat{\theta}^2.$$

Discuss which one do you prefer. Why?

5. Let $f(y;\theta)$ be the density of a distribution and $\theta' = (\theta_1', \theta_2')'$, where θ_1 is a p_1-dimensional vector and θ_2 is a p_2-dimensional vector. Assume that the expected Fisher information matrix $\mathcal{I}(\theta)$ based on n observations can be written as

$$\mathcal{I}(\theta) = \begin{pmatrix} \mathcal{I}_{11} & \mathcal{I}_{12} \\ \mathcal{I}_{21} & \mathcal{I}_{22} \end{pmatrix} = \begin{pmatrix} E(S_1 S_1') & E(S_1 S_2') \\ E(S_2 S_1') & E(S_2 S_2') \end{pmatrix}$$

where $S_1(\theta) = \partial l(\theta)/\partial \theta_1$ and $S_2(\theta) = \partial l(\theta)/\partial \theta_2$. Thus, the asymptotic variance of $\hat{\theta}_{MLE}$ is $\{\mathcal{I}(\theta)\}^{-1}$.

(a) Derive the inverse of $\mathcal{I}(\theta)$ using the following steps.

i. Consider the following transformation:

$$\tilde{S}(\theta) \equiv \begin{pmatrix} \tilde{S}_1(\theta) \\ \tilde{S}_2(\theta) \end{pmatrix} = \begin{pmatrix} I & -\mathcal{I}_{12}\mathcal{I}_{22}^{-1} \\ 0 & I \end{pmatrix} \begin{bmatrix} S_1(\theta) \\ S_2(\theta) \end{bmatrix} = TS(\theta)$$

ii. Compute $E(\tilde{S}\tilde{S}')$. In particular, show that $E(\tilde{S}_1\tilde{S}_1') = \mathcal{I}_{11} - \mathcal{I}_{12}\mathcal{I}_{22}^{-1}\mathcal{I}_{21}$.

iii. Find the inverse of $\mathcal{I}(\theta) = E(SS') = T^{-1}E(\tilde{S}\tilde{S}')(T^{-1})'$.

(b) Show that the asymptotic variance of the MLE of θ_1 (when θ_2 is unknown) is $\left(\mathcal{I}_{11} - \mathcal{I}_{12}\mathcal{I}_{22}^{-1}\mathcal{I}_{21}\right)^{-1}$.

(c) Note that the asymptotic variance of the MLE of θ_1 when θ_2 is known is $\mathcal{I}_{11}^{-1}$. Compare the two variances. Which one is smaller and why?

6. Let $(x_1, y_1)', \cdots, (x_n, y_n)'$ be a random sample from a bivariate normal distribution with mean $(\mu_x, \mu_y)'$ and variance-covariance matrix $\Sigma = \begin{pmatrix} \sigma_x^2 & \sigma_{xy} \\ \sigma_{xy} & \sigma_y^2 \end{pmatrix}$. We are interested in estimating $\theta = \mu_y$.

(a) Assuming that parameters $\mu_x, \sigma_x^2, \sigma_{xy}, \sigma_y^2$ are known, find the maximum likelihood estimator (MLE) of θ and compute its variance.

(b) If the other parameters are also unknown, derive the MLE of θ. Is the estimator here or the estimator from in (a) more efficient? Explain.

7. Consider a bivariate variable (Y_1, Y_2) where

$$(Y_1, Y_2) = \begin{cases} (1,1) & \text{with probability } \pi_{11} \\ (1,0) & \text{with probability } \pi_{10} \\ (0,1) & \text{with probability } \pi_{01} \\ (0,0) & \text{with probability } \pi_{00} \end{cases}$$

with $\pi_{00} + \pi_{01} + \pi_{10} + \pi_{11} = 1$. To answer the following questions, it may be helpful to define $\pi_{1+} = P(Y_1 = 1)$, $\pi_{1|1} = P(Y_2 = 1 \mid Y_1 = 1)$ and $\pi_{1|0} = P(Y_2 = 1 \mid Y_1 = 0)$. Note that there is one-to-one correspondence between $\theta_1 = (\pi_{00}, \pi_{01}, \pi_{10})$ and $\theta_2 = (\pi_{1+}, \pi_{1|1}, \pi_{1|0})$. The realized sample observations are presented in Table 2.1.

Table 2.1 A 2 × 2 table with a supplemental margin for y_1

Set	y_1	y_2	Count
	1	1	100
H	1	0	50
	0	1	75
	0	0	75
K	1		40
	0		60

(a) Compute the observed likelihood and score functions in terms of θ_2.

(b) Obtain the maximum likelihood estimates for θ_1.

(c) Obtain the observed information matrix for θ_1.

8. Assume that $(x_i, y_i)'$, $i = 1, 2, \cdots, n$, are n independent realizations of the random variable $(X, Y)'$ whose distribution follows a canonical exponential family model with log-density of the form

$$\log f(x, y; \theta) = \theta' T(x, y) - A(\theta) + c(x, y).$$

(a) Show that the complete-sample score function for θ is

$$S_c(\theta) = \sum_{i=1}^{n} \{T(x_i, y_i) - E(T)\}$$

and the complete-sample Fisher information can be estimated by

$$\sum_{i=1}^{n} (T_i - \bar{T}_n)(T_i - \bar{T}_n)'$$

where $T_i = T(x_i, y_i)$ and $\bar{T}_n = n^{-1}\sum_{i=1}^{n} T_i$.

(b) Suppose that x_i are always observed and y_i are subject to missing. Assume that y_i is observed for the first r observations and the remaining $n-r$ elements are missing in y. Show that the expected information associated with the observed likelihood is equal to

$$\mathcal{I}_{\text{obs}}(\theta) = nV(T) - (n-r)E\{V(T \mid X)\}.$$

Discuss the missing information principle in this setup.

9. Consider the following regression model

$$y_i = \beta_0 + \beta_1 x_i + e_i$$

where $x_i \sim N(\mu_x, \sigma_x^2)$, $e_i \sim N(0, \sigma_e^2)$, and e_i is independent of x_i. In addition, a surrogate measurement of x_i, denoted by z_i, is available with distribution

$$z_i = x_i + u_i$$

with $u_i \sim N(0, \sigma_u^2)$. This is often called *measurement error model*. The surrogate variable z_i is conditionally independent of y_i given x_i. That is, we have $f(z \mid x, y) = f(z \mid x)$. Assume that $\mu_x, \sigma_x^2, \sigma_u^2$ are known. Throughout the sample, we observe w_i and y_i. We are interested in estimating $\theta = (\beta_0, \beta_1, \sigma_e^2)$. Answer the following questions.

(a) Under no measurement error ($\sigma_u^2 \equiv 0$), obtain the complete-data likelihood function and its score functions.

(b) Under the existence of measurement error, obtain the observed likelihood function.

(c) Use Bayes' theorem or other techniques, derive the conditional distribution of x_i given z_i and y_i.

(d) Compute the mean score functions for θ.

10. Under the setup of Example 2.7, use

$$(A+D)^{-1} = A^{-1} - A^{-1}\left(D^{-1} + A^{-1}\right)^{-1} A^{-1},$$

where A and D are invertible matrices of the same dimension, to compute the inverse of $\mathcal{I}_{\text{obs}}(\mu)$ in (2.32) in terms of $\sigma_{11}, \sigma_{12}, \sigma_{22}$ and n_H, n_K, n_L. Discuss the efficiency gain of the MLEs of μ_1 and μ_2 compared to the case of using only set H.

Chapter 3

Computation

3.1 Introduction

Maximum likelihood estimation (MLE) plays a central role in statistical inference. The actual computation to obtain maximum likelihood estimators can be challenging in many situations, specially in missing data problems. We first review some of the popular methods of computing maximum likelihood estimators and some issues associated with the computation.

In principle, we are interested in finding the solution

$$\hat{\theta} = \arg\max_{\theta} L(\theta).$$

The MLE satisfies the score equation given by

$$S(\theta) \equiv \frac{\partial}{\partial\theta} \log L(\theta) = 0 \tag{3.1}$$

which is generally a system of nonlinear equations. In most cases, the MLE can be obtained by solving the score equation (3.1).

To discuss the computational approaches of solving the score equation (3.1), we first review some methods of solving $g(\theta) = 0$ for θ. If $g(\theta)$ is a scalar function of θ, then the solution to $g(\theta) = 0$ can be easily obtained by the bisection method, which is based on the intermediate value theorem: If g is continuous for all θ in the interval $g(\theta_1) g(\theta_2) < 0$, then a root of $g(\theta)$ lies in the interval (θ_1, θ_2). To describe the bisection method, let $[a_0, b_0]$ be an interval in the domain of $g(\theta)$ such that $g(a_0) g(b_0) < 0$. Hence, the solution θ^* to $g(\theta) = 0$ lies in $[a_0, b_0]$. The bisection method can find the root θ^* for $g(\theta) = 0$ by the following iterative steps:

[Step 1] Set $x_t = (a_t + b_t)/2$.

[Step 2] Evaluate the signs of $g(a_t)g(x_t)$ and $g(x_t)g(b_t)$. If $g(a_t)g(x_t) < 0$ then set $a_{t+1} = a_t$ and $b_{t+1} = x_t$. If $g(b_t)g(x_t) < 0$ then set $a_{t+1} = x_t$ and $b_{t+1} = b_t$.

[Step 3] Stop if the convergence criterion is met. Otherwise, increase t by one and go to step 1.

For the convergence criterion, we can use $|a_{t+1} - b_{t+1}| < \varepsilon$ for a sufficiently small $\varepsilon > 0$. The bisection method is easy to execute and does not require computing the partial derivatives. The bisection method is an example of the bracketing method, which can be roughly described as finding a root within a sequence of nested intervals of decreasing length.

Also popular is Newton's method, or the Newton–Raphson method. It uses a linear approximation of $g(\theta)$ at $\theta^{(t)}$

$$g(\theta) \cong g(\theta^{(t)}) + \left\{\partial g(\theta^{(t)})/\partial\theta'\right\} \left(\theta - \theta^{(t)}\right).$$

We can next make a mental transformation to change $g(\theta)$ to 0, and θ to $\theta^{(t+1)}$. Thus, Newton's method for finding the solution $\hat{\theta}$ to $g(\theta) = 0$ can be described as

$$\theta^{(t+1)} = \theta^{(t)} - \left\{\partial g(\theta^{(t)})/\partial\theta'\right\}^{-1} g\left(\theta^{(t)}\right). \tag{3.2}$$

If $g(\theta)$ is equal to $S(\theta)$, the score function for θ, then (3.2) can be written as

$$\theta^{(t+1)} = \theta^{(t)} + \left[I\left(\theta^{(t)}\right)\right]^{-1} S\left(\theta^{(t)}\right). \tag{3.3}$$

This is the formula for the scoring method. The scoring method can be modified to guarantee that the sequence $\{\theta^{(t)}; t = 1, \cdots\}$ increases the likelihood:

$$\theta^{(t+1)} = \theta^{(t)} + \alpha \left[I\left(\theta^{(t)}\right)\right]^{-1} S\left(\theta^{(t)}\right)$$

for $\alpha \in (0,1]$. If $L(\hat{\theta}^{(t+1)}) < L(\hat{\theta}^{(t)})$, then use $\alpha = \alpha/2$ and compute $\theta^{(t+1)}$ again. Such modification is often called the *ascent method.*

In the scoring method, the behavior of $I(\theta^{(t)})$ can be problematic if $\theta^{(t)}$ is far from the MLE $\hat{\theta}$. Thus, instead of using the observed Fisher information $I(\theta)$ in (3.3), we can use the expected Fisher information to get

$$\theta^{(t+1)} = \theta^{(t)} + \left[\mathcal{I}\left(\theta^{(t)}\right)\right]^{-1} S\left(\theta^{(t)}\right). \tag{3.4}$$

This algorithm is called the *Fisher scoring method.* Generally speaking, the Fisher scoring method can expect rapid computational improvement at the beginning, while Newton's method works better for refinement near the end (Givens and Hoeting, 2005).

Example 3.1. *Consider the logistic regression model*

$$y_i \overset{i.i.d.}{\sim} Bernoulli\,(p_i)$$

with

$$logit\,(p_i) = \ln\left(\frac{p_i}{1-p_i}\right) = \mathbf{x}_i'\beta.$$

The log-likelihood function for β is

$$\begin{aligned} l(\beta) &= \sum_{i=1}^{n}\{y_i \ln(p_i) + (1-y_i)\ln(1-p_i)\} \\ &= \sum_{i=1}^{n}\{y_i(\mathbf{x}_i'\beta) - \ln(1+\exp(\mathbf{x}_i'\beta))\} \end{aligned}$$

The score function is given by

$$S(\beta) = \frac{\partial}{\partial\beta} l(\beta) = \sum_{i=1}^{n}\{y_i - p_i(\beta)\}\mathbf{x}_i \tag{3.5}$$

and the Hessian is given by

$$I(\beta) = -\frac{\partial}{\partial\beta'} S(\beta) = \sum_{i=1}^{n} p_i(\beta)\{1-p_i(\beta)\}\mathbf{x}_i\mathbf{x}_i'.$$

Because the Hessian does not depend on the y_i's, the expected Fisher information evaluated at $\hat{\theta}$ (MLE) is equal to the observed Fisher information. Hence, the Fisher scoring method is equal to the scoring method. The Fisher scoring method can be expressed as

$$\beta^{(t+1)} = \beta^{(t)} + \left[\sum_{i=1}^{n} p_i^{(t)}(1-p_i^{(t)})\mathbf{x}_i\mathbf{x}_i'\right]^{-1} \sum_{i=1}^{n}(y_i - p_i^{(t)})\mathbf{x}_i$$

where $p_i^{(t)} = p_i(\beta^{(t)})$.

We now discuss the convergence properties of Newton's method. For simplicity, we consider only a scalar parameter θ and scalar equations, $g(\theta) = 0$. The updating equation for Newton's method can be written as

$$\theta^{(t+1)} = \theta^{(t)} - \frac{g(\theta^{(t)})}{g'(\theta^{(t)})}. \tag{3.6}$$

The convergence of Newton's method depends on the shape of g and the starting value in the iteration. To discuss the convergence properties of Newton's method for $g(\theta) = 0$, suppose that g is twice differentiable with continuous second-order derivatives satisfying $g''(\theta^*) \neq 0$, where θ^* is a solution to $g(\theta) = 0$. A Taylor expansion of $g(\theta^*) = 0$ around $\theta^{(t)}$ leads to

$$0 = g(\theta^*) \quad = \quad g(\theta^{(t)}) + g'(\theta^{(t)})\left(\theta^* - \theta^{(t)}\right) + 0.5g''(q)\left(\theta^* - \theta^{(t)}\right)^2$$

where q is between θ^* and $\theta^{(t)}$. Multiplying both sides of the above equation by $1/g'\left(\theta^{(t)}\right)$ and using (3.6), we have

$$\varepsilon^{(t+1)} = \left(\varepsilon^{(t)}\right)^2 \frac{g''(q)}{2g'\left(\theta^{(t)}\right)}, \tag{3.7}$$

where $\varepsilon^{(t)} = \theta^{(t)} - \theta^*$. Now consider a neighborhood of θ^*, $B_\delta(\theta^*) = \{\theta; |\theta - \theta^*| < \delta\}$ for $\delta > 0$. Define

$$c(\delta) = \max_{\theta_1, \theta_2 \in B_\delta(\theta^*)} \left| \frac{g''(\theta_1)}{2g'(\theta_2)} \right|.$$

Since $c(\delta) \to |g''(\theta^*)/\{2g'(\theta^*)\}|$, as $\delta \to 0$, it follows that $\delta c(\delta) \to 0$ as $\delta \to 0$ and so $\delta c(\delta) < 1$ for some $\delta > 0$. For $\theta^{(t)} \in B_\delta(\theta^*)$, by (3.7),

$$\left|c(\delta)\varepsilon^{(t+1)}\right| \leq \left\{c(\delta)\varepsilon^{(t)}\right\}^2$$

which implies

$$\left|c(\delta)\varepsilon^{(t+1)}\right| \leq \left\{c(\delta)\varepsilon^{(1)}\right\}^{2^t}. \tag{3.8}$$

Thus, if the initial value $\theta^{(1)}$ is chosen to satisfy $\left|\varepsilon^{(1)}\right| < \delta$, then (3.8) implies

$$\left|\varepsilon^{(t+1)}\right| \leq \frac{\{c(\delta)\delta\}^{2^t}}{c(\delta)}$$

which converges to zero as $t \to \infty$, because $|c(\delta)\delta| < 1$. Therefore, if g'' is continuous and θ^* is a root of $g(\theta) = 0$, then there exists a neighborhood of x^* for which Newton's method converges to θ^* if $\theta^{(0)}$ is in that neighborhood. We formalize the above result in the following theorem.

Theorem 3.1. *Let $g : (a,b) \to R$ be a differentiable function obeying the following conditions:*

1. *g is twice differentiable with continuous g''.*
2. *$|g'|$ is bounded away from 0 on (a,b), and $\inf_{\theta \in (a,b)} |g'(\theta)| > 0$.*
3. *There exists a point $\theta^* \in (a,b)$ such that $g(\theta^*) = 0$.*

Then there is $\delta > 0$ such that whenever the initial value $\theta^{(0)} \in (\theta^* - \delta, \theta^* + \delta)$, the sequence $\{\theta^{(t)}\}$ of iterations produced by Newton's method lies in $(\theta^* - \delta, \theta^* + \delta)$, and $\theta^{(t)} \to \theta^*$ as $t \to \infty$.

Theorem 3.1 is a local convergence theorem. It establishes the existence of an interval $(\theta^* - \delta, \theta^* + \delta)$ in which Newton's method converges but gives very little information about that interval. In application one often wants a global convergence theorem in the sense that the iterative scheme converges regardless of the starting point. The simplest version of the global convergence theorem is as follows: If a convex function g has a root and is twice continuously differentiable, then

Newton's method converges to the root from any starting point. For details, see Allen and Issacson (1998).

We now discuss the order of convergence of the iterative solutions.

Definition 3.1. *A sequence $\{\theta^{(t)}\}$ that converges to θ^* is of order p if*

$$\lim_{t\to\infty} \|\theta^{(t)} - \theta^*\| = 0$$

and

$$\lim_{t\to\infty} \frac{\|\theta^{(t+1)} - \theta^*\|}{\|\theta^{(t)} - \theta^*\|^p} = c$$

for some constant $c \neq 0$.

For Newton's method, by (3.7), we have

$$\frac{\theta^{(t+1)} - \theta^*}{\left(\theta^{(t)} - \theta^*\right)^2} = \frac{g''(q)}{2g'\left(\theta^{(t)}\right)}.$$

Then, if Newton's method converges, q also converges to θ^* and

$$\lim_{t\to\infty} \frac{\|\theta^{(t+1)} - \theta^*\|}{\|\theta^{(t)} - \theta^*\|^2} = \left|\frac{g''(\theta^*)}{2g'(\theta^*)}\right| \neq 0.$$

Thus, its convergence is of quadratic order, i.e. $p = 2$. A quadratic order convergence is quite fast. This is regarded as the major strength of Newton's method. Another advantage of Newton's method is that the one-step estimator with a $\sqrt{n}$-consistent initial point is asymptotically efficient. See Chapter 6 of Lehmann (1983).

Remark 3.1. *Newton's method requires computing the Hessian and also updating the Hessian for each iteration, which can be at a huge computational cost, specially when the dimension of θ is large. Instead of (3.2), an alternative iterative method of obtaining $\theta^{(t)}$, called the* Quasi-Newton method, *can be expressed as*

$$\theta^{(t+1)} = \theta^{(t)} - \left(M^{(t)}\right)^{-1} S(\theta^{(t)}) \tag{3.9}$$

where $M^{(t)}$ is a $p \times p$ matrix approximating the Hessian evaluated at $\theta^{(t)}$. The choice of $M^{(t)} = -\mathcal{I}(\theta^{(t)})$ leads to the Fisher scoring method in (3.3). The secant method finds $M^{(t)}$ by solving

$$S(\theta^{(t)}) - S(\theta^{(t-1)}) = \left(M^{(t)}\right)\left(\theta^{(t)} - \theta^{(t-1)}\right)$$

or even set $M^{(t)}$ to be M, where M is sufficiently close to the Hessian matrix at the solution, $-I(\theta^*)$. In this case, the full quadratic convergence of the Newton's method is lost. It can be shown that under some reasonable assumptions, the convergence rate is superlinear in the sense that $\|\theta^{(t+1)} - \theta^*\| < h_t\|\theta^{(t)} - \theta^*\|$ for some h_t that converges to zero as $t \to \infty$. In the IID case, $S(\theta) = \sum_{i=1}^n S_i(\theta)$, Redner and Walker (1984) and Meilijson (1989) have suggested using $M^{(t)} = -\hat{H}(\theta^{(t)})$ in (3.9) where

$$\hat{H}(\theta) = \sum_{i=1}^n \left\{S_i(\theta) - n^{-1}\sum_{j=1}^n S_j(\theta)\right\}^{\otimes 2}. \tag{3.10}$$

Note that $\hat{H}(\theta)$ is consistent for $\mathcal{I}(\theta) = V_\theta\{S(\theta)\}$ and $\hat{H}(\theta)$ is very easy to compute. The convergence rate is known to be linear.

We now discuss direct computation under the existence of missing data. Under the setup of Section 2.4, the MLE that maximizes the observed likelihood can be obtained as a solution to the observed score equation given by

$$S_{\text{obs}}(\theta) = 0 \tag{3.11}$$

where $S_{\text{obs}}(\theta) = \partial \ln L_{\text{obs}}(\theta)/\partial\theta$. The following example illustrates the direct computation method with missing data.

Example 3.2. *Consider the following bivariate normal distribution*

$$\begin{pmatrix} X_i \\ Y_i \end{pmatrix} \sim N\left[\begin{pmatrix} \mu_x \\ \mu_y \end{pmatrix}, \begin{pmatrix} \sigma_{xx} & \sigma_{xy} \\ \sigma_{xy} & \sigma_{yy} \end{pmatrix}\right], \tag{3.12}$$

where $\sigma_{xx}, \sigma_{xy}, \sigma_{yy}$ are known constants. Suppose that we have complete response in the first $r(< n)$ units $\{(x_i, y_i); i = 1, 2, \cdots, r\}$ and $n - r$ partial responses from the remaining $n - r$ units $\{x_i; i = r+1, r+2, \cdots, n\}$. In this case, under MAR, the observed score function for $\mu = (\mu_x, \mu_y)'$ can be written as

$$\bar{S}(\mu) = \sum_{i=1}^{r} \begin{pmatrix} \sigma_{xx} & \sigma_{xy} \\ \sigma_{xy} & \sigma_{yy} \end{pmatrix}^{-1} \begin{pmatrix} x_i - \mu_x \\ y_i - \mu_y \end{pmatrix} + \sum_{i=r+1}^{n} \begin{pmatrix} \sigma_{xx} & \sigma_{xy} \\ \sigma_{xy} & \sigma_{yy} \end{pmatrix}^{-1} \begin{pmatrix} x_i - \mu_x \\ E(y_i \mid x_i) - \mu_y \end{pmatrix},$$

where $E(y_i \mid x_i) = \mu_y + (\sigma_{xy}/\sigma_{xx})(x_i - \mu_x)$. The solution to the observed score equation is

$$\hat{\mu}_x = \bar{x}_n \equiv \frac{1}{n}\sum_{i=1}^{n} x_i$$

and

$$\hat{\mu}_y = \bar{y}_r + \frac{\sigma_{xy}}{\sigma_{xx}}(\bar{x}_n - \bar{x}_r)$$

where $(\bar{x}_r, \bar{y}_r) = r^{-1}\sum_{i=1}^{r}(x_i, y_i)$. The variance of $\hat{\mu}_y$ is equal to $\sigma_{yy}\rho^2/n + (1-\rho^2)\sigma_{yy}/r$, which is consistent with the findings of Example 2.7.

In the above example, the mean score equations are linear and so the computation is straightforward. In general, the mean score equations are nonlinear and we often rely on iterative computation methods. The Fisher-scoring method applied to (3.11) can be written as

$$\hat{\theta}^{(t+1)} = \hat{\theta}^{(t)} + \left\{\mathcal{I}_{\text{obs}}\left(\hat{\theta}^{(t)}\right)\right\}^{-1} S_{\text{obs}}\left(\hat{\theta}^{(t)}\right). \tag{3.13}$$

Under the IID setup, we can use the empirical Fisher information $\hat{H}(\theta)$ in (3.10) as an estimator of the expected Fisher information $\mathcal{I}_{\text{obs}}(\theta)$.

Example 3.3. *Consider the following normal-theory mixed effect model*

$$y_{ij} = \mathbf{x}_{ij}'\beta + u_i + e_{ij}, \quad u_i \overset{i.i.d.}{\sim} N(0, \sigma_u^2), \; e_{ij} \overset{i.i.d.}{\sim} N(0, \sigma_e^2)$$

for $i = 1, \cdots, m; j = 1, \cdots, n_i$, and e_{ij} are independent of u_i. We observe $(\mathbf{x}_{ij}, y_{ij})$ for each unit j in cluster i. In this case, the cluster-specific effect u_i can be treated as missing data. The complete sample likelihood for $\theta = (\beta, \sigma_u^2, \sigma_e^2)$ is

$$L_{\text{com}}(\theta) = \prod_i \left[\prod_{j=1}^{n_i}\left\{\frac{1}{\sigma_e}\phi\left(\frac{y_{ij} - \mathbf{x}_{ij}'\beta - u_i}{\sigma_e}\right)\right\}\frac{1}{\sigma_u}\phi\left(\frac{u_i}{\sigma_u}\right)\right]$$

where $\phi(x) = (2\pi)^{-1/2}\exp(-x^2/2)$ is the probability density function of the standard normal distribution. The complete-sample score functions are then

$$
\begin{aligned}
S_{\text{com},1}(\theta) &\equiv \partial \log\{L_{\text{com}}(\theta)\}/\partial\beta = \sum_i\sum_j (y_{ij} - \mathbf{x}_{ij}'\beta - u_i)\,\mathbf{x}_{ij}/\sigma_e^2 \\
S_{\text{com},2}(\theta) &\equiv \partial \log\{L_{\text{com}}(\theta)\}/\partial\sigma_u^2 = \frac{1}{2\sigma_u^4}\sum_i (u_i^2 - \sigma_u^2) \\
S_{\text{com},3}(\theta) &\equiv \partial \log\{L_{\text{com}}(\theta)\}/\partial\sigma_e^2 = \frac{1}{2\sigma_e^4}\sum_i\sum_j (e_{ij}^2 - \sigma_e^2),
\end{aligned}
$$

where $e_{ij} = y_{ij} - \mathbf{x}_{ij}'\beta - u_i$. Using

$$
u_i \mid \mathbf{x}_i, \mathbf{y}_i \sim N\left(\tau_i(\bar{y}_i - \bar{\mathbf{x}}_i'\beta), \sigma_u^2(1-\tau_i)\right), \tag{3.14}
$$

where $\tau_i = \sigma_u^2/(\sigma_u^2 + \sigma_e^2/n_i)$, the observed score function can be computed by taking the conditional expectation of the complete-sample score functions with respect to the conditional distribution in (3.14). For example, the observed score equation for β is given by

$$
\bar{S}_1(\beta) \equiv \sum_i\sum_j \{y_{ij} - \mathbf{x}_{ij}'\beta - \tau_i(\bar{y}_i - \bar{\mathbf{x}}_i'\beta)\}\,\mathbf{x}_{ij}/\sigma_e^2 = 0.
$$

The resulting $\hat{\beta}$ is obtained by regressing $y_{ij} - \tau_i\bar{y}_i$ on $(\mathbf{x}_{ij} - \tau_i\bar{\mathbf{x}}_i)$. Fuller and Battese (1973) obtained the same result from the estimated generalized least square method.

3.2 Factoring likelihood approach

If the missing data is MAR and the responses are monotone in the sense that the set of respondents for one variable is a proper subset of the set of respondents for another variable, then the computation for the MLE can be simplified using the factoring likelihood approach. To simplify the presentation, consider a bivariate random variable (X,Y) with density $f(x,y;\theta)$. Assume that x_i are completely observed and y_i are subject to missingness. Without loss of generality, assume that y_i are observed only for the first $r < n$ elements.

In this case, assuming MAR, the observed likelihood for θ can be written

$$
L_{\text{obs}}(\theta) = \prod_{i=1}^{r} f(x_i, y_i;\theta) \prod_{i=r+1}^{n} \int f(x_i, y_i;\theta)\,dy_i. \tag{3.15}
$$

Let $f_X(x;\theta_1)$ be the marginal density of X and $f_{Y|X}(y \mid x;\theta_2)$ be the density for the conditional distribution of Y given $X = x$. Since $f(x_i,y_i;\theta) = f_X(x_i;\theta_1) f_{Y|X}(y_i \mid x_i;\theta_2)$, we can rearrange (3.15) as

$$
L_{\text{obs}}(\theta) = \prod_{i=1}^{n} f_X(x_i;\theta_1) \prod_{i=1}^{r} f_{Y|X}(y_i \mid x_i;\theta_2). \tag{3.16}
$$

When the observed likelihood is written as (3.16), note that we can write

$$
L_{\text{obs}}(\theta) = L_1(\theta_1) \times L_2(\theta_2), \tag{3.17}
$$

and the MLEs for each parameter can be obtained by separately maximizing the corresponding likelihood. If parameters θ_1 and θ_2 satisfies (3.17), then the two parameters are called *orthogonal*. If the two parameters, θ_1 and θ_2 are orthogonal, the information matrix for $\theta = (\theta_1,\theta_2)$ becomes a block-diagonal and the MLE of θ_1 is independent of the MLE of θ_2, at least asymptotically (Cox and Reid, 1987).

Example 3.4. *Consider the same setup as in Example 3.2, except that σ_{xx}, σ_{xy}, and σ_{yy} are also unknown parameters. There are now five parameters to identify the distribution. The observed likelihood for $\theta = (\mu_x, \mu_y, \sigma_{xx}, \sigma_{xy}, \sigma_{yy})'$ is*

$$L_{\text{obs}}(\theta) = \prod_{i=1}^{r} f(x_i, y_i; \mu_x, \mu_y, \sigma_{xx}, \sigma_{xy}, \sigma_{yy}) \times \prod_{i=r+1}^{n} f(x_i; \mu_x, \sigma_{xx})$$

Finding the MLE of θ would require an iterative computation method. An alternative parametrization is

$$\begin{aligned} X_i &\sim N(\mu_x, \sigma_{xx}) \\ Y_i \mid X_i = x &\sim N(\beta_0 + \beta_1 x, \sigma_{ee}) \end{aligned} \tag{3.18}$$

where $\beta_1 = \sigma_{xy}/\sigma_{xx}$, $\beta_0 = \mu_y - \beta_1 \mu_x$, and $\sigma_{ee} = \sigma_{yy} - \sigma_{xy}^2/\sigma_{xx}$. Under this new parametrization,

$$\begin{aligned} L_{\text{obs}}(\theta) &= \prod_{i=1}^{n} f(x_i; \mu_x, \sigma_{xx}) \times \prod_{i=1}^{r} f(y_i \mid x_i; \beta_0, \beta_1, \sigma_{ee}) \\ &= L_1(\mu_x, \sigma_{xx}) \times L_2(\beta_0, \beta_1, \sigma_{ee}) \end{aligned}$$

and the two parameters, $\theta_1 = (\mu_x, \sigma_{xx})$ and $\theta_2 = (\beta_0, \beta_1, \sigma_{ee})$, are orthogonal in the sense of satisfying (3.17). The MLEs are obtained by maximizing each component of $L_{\text{obs}}(\theta)$ separately, which is given by

$$\begin{aligned} \hat{\mu}_x &= \bar{x}_n \\ \hat{\sigma}_{xx} &= S_{xxn} \end{aligned}$$

and

$$\begin{aligned} \hat{\beta}_1 &= S_{xyr}/S_{xxr} \\ \hat{\beta}_0 &= \bar{y}_r - \hat{\beta}_1 \bar{x}_r \\ \hat{\sigma}_{ee} &= S_{yyr} - S_{xyr}^2/S_{xxr}, \end{aligned}$$

where the subscript r denotes that the statistics are computed from the r complete respondents only, while the subscript n denotes that the statistics are computed from the whole sample of size n. Thus, the MLEs for the original parametrization are

$$\begin{aligned} \hat{\mu}_y &= \hat{\beta}_0 + \hat{\beta}_1 \hat{\mu}_x = \bar{y}_r + \hat{\beta}_1 (\hat{\mu}_x - \bar{x}_r) \\ \hat{\sigma}_{yy} &= S_{yyr} + \hat{\beta}_1^2 (\hat{\sigma}_{xx} - S_{xxr}) \\ \hat{\sigma}_{xy} &= S_{xyr} \frac{\hat{\sigma}_{xx}}{S_{xxr}}. \end{aligned} \tag{3.19}$$

Furthermore, we can compute

$$\hat{\rho} = r_{xy} \times \left(\frac{\hat{\sigma}_{xx}}{S_{xxr}} \right)^{1/2} \left(\frac{\hat{\sigma}_{yy}}{S_{yyr}} \right)^{-1/2},$$

where $r_{xy} = S_{xyr}/(S_{xxr}S_{yyr})^{1/2}$. The MLE of μ_y, $\hat{\mu}_y$ in (3.19), is called the regression estimator *and is very popular in sample surveys.*

Example 3.5. *We consider the setup of Exercise 7 in Chapter 2. Suppose that we have complete response in the first $r(< n)$ units $\{(y_{1i}, y_{2i}); i = 1, 2, \cdots, r\}$ and $n - r$ partial responses from the remaining $n - r$ units $\{y_{1i}; i = r+1, r+2, \cdots, n\}$. The observed likelihood can be written as*

$$L_{\text{obs}}(\theta_2) = \prod_{i=1}^{n} \pi_{1+}^{y_{1i}} (1 - \pi_{1+})^{1 - y_{1i}} \times \prod_{i=1}^{r} \left\{ \pi_{1|1}^{y_{2i}} (1 - \pi_{1|1})^{1 - y_{2i}} \right\}^{y_{1i}} \left\{ \pi_{1|0}^{y_{2i}} (1 - \pi_{1|0})^{1 - y_{2i}} \right\}^{1 - y_{1i}}.$$

Because we can write

$$L_{\text{obs}}(\pi_{1+},\pi_{1|1},\pi_{1|0}) = L_1(\pi_{1+})L_2(\pi_{1|1})L_3(\pi_{1|0})$$

for some $L_1(\cdot)$, $L_2(\cdot)$, *and* $L_3(\cdot)$, *we can obtain the MLE by separately maximizing each likelihood component. Thus, we have*

$$\begin{aligned} \hat{\pi}_{1+} &= \frac{1}{n}\sum_{i=1}^{n} y_{1i} \\ \hat{\pi}_{1|1} &= \frac{\sum_{i=1}^{r} y_{1i}y_{2i}}{\sum_{i=1}^{r} y_{1i}} \\ \hat{\pi}_{1|0} &= \frac{\sum_{i=1}^{r}(1-y_{1i})y_{2i}}{\sum_{i=1}^{r}(1-y_{1i})}. \end{aligned}$$

The MLE for π_{ij} *can then be obtained by* $\hat{\pi}_{ij} = \hat{\pi}_{i+}\hat{\pi}_{j|i}$ *for* $i=0,1$ *and* $j=0,1$.

The factoring likelihood approach was first proposed by Anderson (1957), and further discussed by Rubin (1974). It is particularly useful for the monotone missing pattern, where we can relabel the variable in such a way that the set of respondents for each variable is monotonely decreasing:

$$R_1 \supset R_2 \supset \cdots \supset R_p$$

where R_i denotes the set of respondents for Y_i after relabeling. In this case, under MAR, the observed likelihood can be written as

$$L_{\text{obs}}(\theta) = \prod_{i\in R_1} f(y_{1i};\theta_1) \times \prod_{i\in R_2} f(y_{2i} \mid y_{1i};\theta_2) \times \cdots \times \prod_{i\in R_p} f(y_{pi} \mid y_{p-1,i},\cdots,y_{1,i};\theta_p)$$

and the parameters $\theta_1,\cdots,\theta_p$ are orthogonal. The MLE for each component of the parameters can be obtained by maximizing each component of the observed likelihood.

We now consider some possible extensions to the nonmonotone missing data, discussed by Kim and Shin (2012). In the case of nonmonotone missing data, we have the following steps to apply the factoring likelihood approach:

[Step 1] Partition the original sample into several disjoint sets according to the missing pattern.

[Step 2] Compute the MLEs for the identified parameters separately in each partition of the sample.

[Step 3] Combine the estimators to get a set of final estimates in a generalized least squares (GLS) form.

To simplify the presentation, we describe the proposed method in the bivariate normal setup with a nonmonotone missing pattern. The joint distribution of $(x,y)'$ is parameterized by the five parameters using model (3.12) or (3.18). For the convenience of the factoring method, we use the parametrization in (3.18) and let $\theta = (\beta_0,\beta_1,\sigma_{ee},\mu_x,\sigma_{xx})'$.

In Step 1, we partition the sample into several disjoint sets according to the pattern of missingness. In the case of a nonmonotone missing pattern with two variables, we have $3 = 2^2-1$ types of respondents that contain information about the parameters. The first set H has both x and y observed, the second set K has x observed but y missing, and the third set L has y observed but x missing. See Table 3.1. Let n_H, n_K, n_L be the sample sizes of the sets H,K,L, respectively. The case of both x and y missing can be safely removed from the sample.

In Step 2, we obtain the parameter estimates in each set: For set H, we have the five parameters $\eta_H = (\beta_0,\beta_1,\sigma_{ee},\mu_x,\sigma_{xx})'$ of the conditional distribution of y given x and the marginal distribution of x, with MLEs $\hat{\eta}_H = (\hat{\beta}_{0,H},\hat{\beta}_{1,H},\hat{\sigma}_{ee,H},\hat{\mu}_{x,H},\hat{\sigma}_{xx,H})'$. For set K, the MLEs $\hat{\eta}_K = (\hat{\mu}_{x,K},\hat{\sigma}_{xx,K})'$ are obtained for $\eta_K = (\mu_x,\sigma_{xx})'$, the parameters of the marginal distribution of x. For set L, the MLEs $\hat{\eta}_L = (\hat{\mu}_{y,L},\hat{\sigma}_{yy,L})'$ are obtained for $\eta_L = (\mu_y,\sigma_{yy})'$, where $\mu_y = \beta_0+\beta_1\mu_x$ and $\sigma_{yy} = \sigma_{ee}+\beta_1^2\sigma_{xx}$.

Table 3.1 An illustration of the missing data structure under a bivariate normal distribution

Set	x	y	Sample Size	Estimable parameters
H	Observed	Observed	n_H	$\mu_x, \mu_y, \sigma_{xx}, \sigma_{xy}, \sigma_{yy}$
K	Observed	Missing	n_K	μ_x, σ_{xx}
L	Missing	Observed	n_L	μ_y, σ_{yy}

In Step 3, we use the GLS method to combine the three estimators $\hat{\eta}_H, \hat{\eta}_K, \hat{\eta}_L$ to get a final estimator for the parameter θ. Let $\hat{\eta} = (\hat{\eta}_H', \hat{\eta}_K', \hat{\eta}_L')'$. Then

$$\hat{\eta} = \left(\hat{\beta}_{0,H}, \hat{\beta}_{1,H}, \hat{\sigma}_{ee,H}, \hat{\mu}_{x,H}, \hat{\sigma}_{xx,H}, \hat{\mu}_{x,K}, \hat{\sigma}_{xx,K}, \hat{\mu}_{y,L}, \hat{\sigma}_{yy,L}\right)'. \tag{3.20}$$

The expected value of this estimator is

$$\eta(\theta) = \left(\beta_0, \beta_1, \sigma_{ee}, \mu_x, \sigma_{xx}, \mu_x, \sigma_{xx}, \beta_0 + \beta_1\mu_x, \sigma_{ee} + \beta_1^2\sigma_{xx}\right)' \tag{3.21}$$

and the asymptotic covariance matrix is

$$\mathbf{V} = diag\left\{\frac{\Sigma_{yy.x}}{n_H}, \frac{2\sigma_{ee}^2}{n_H}, \frac{\sigma_{xx}}{n_H}, \frac{2\sigma_{xx}^2}{n_H}, \frac{\sigma_{xx}}{n_K}, \frac{2\sigma_{xx}^2}{n_K}, \frac{\sigma_{yy}}{n_L}, \frac{2\sigma_{yy}^2}{n_L}\right\}, \tag{3.22}$$

where

$$\Sigma_{yy.x} = \begin{pmatrix} \sigma_{ee}\left(1 + \sigma_{xx}^{-1}\mu_x^2\right) & -\sigma_{xx}^{-1}\sigma_{ee}\mu_x \\ -\sigma_{xx}^{-1}\sigma_{ee}\mu_x & \sigma_{xx}^{-1}\sigma_{ee} \end{pmatrix}.$$

Note that

$$\Sigma_{yy.x} = \{E[(1,x)(1,x)']\}^{-1}\sigma_{ee} = \begin{pmatrix} 1 & \mu_x \\ \mu_x & \sigma_{xx} + \mu_x^2 \end{pmatrix}^{-1}\sigma_{ee}.$$

Deriving the asymptotic covariance matrix of the first five estimates in (3.20) is straightforward. Relevant reference can be found, for example, in Subsection 7.2.2 of Little and Rubin (2002). We have a block-diagonal structure of $\mathbf{V}$ in (3.22) because $\hat{\mu}_{1K}$ and $\hat{\sigma}_{11K}$ are independent due to normality, and observations between different sets are independent due to the IID assumptions.

Note that the nine elements in $\eta(\theta)$ are related to each other because they are all functions of the five elements of vector θ. The information contained in the four extra equations has not yet been utilized in constructing the estimators $\hat{\eta}_H, \hat{\eta}_K, \hat{\eta}_L$. The information can be employed to construct a fully efficient estimator of θ by combining $\hat{\eta}_H, \hat{\eta}_K, \hat{\eta}_L$ through a GLS regression of $\hat{\eta} = (\hat{\eta}_H', \hat{\eta}_K', \hat{\eta}_L')'$ on θ as follows:

$$\hat{\eta} - \eta(\hat{\theta}_S) = (\partial\eta/\partial\theta')(\theta - \hat{\theta}_S) + \text{ error},$$

where $\hat{\theta}_S$ is an initial estimator.

The expected value and variance of $\hat{\eta}$ in (3.21) and (3.22) can furnish the specification of a nonlinear model of the five parameters in θ. Using a Taylor series expansion on the nonlinear model, a step of the Gauss–Newton method can be formulated as

$$\mathbf{e}_\eta = \mathbf{X}\left(\theta - \hat{\theta}_S\right) + \mathbf{u}, \tag{3.23}$$

where $\mathbf{e}_\eta = \hat{\eta} - \eta\left(\hat{\theta}_S\right)$, $\eta\left(\hat{\theta}_S\right)$ is the vector (3.21) evaluated at $\hat{\theta}_S$,

$$\mathbf{X} \equiv \frac{\partial\eta}{\partial\theta'} = \begin{pmatrix} 1 & 0 & 0 & 0 & 0 & 0 & 0 & 1 & 0 \\ 0 & 1 & 0 & 0 & 0 & 0 & 0 & \mu_x & 2\beta_1\sigma_{xx} \\ 0 & 0 & 1 & 0 & 0 & 0 & 0 & 0 & 1 \\ 0 & 0 & 0 & 1 & 0 & 1 & 0 & \beta_1 & 0 \\ 0 & 0 & 0 & 0 & 1 & 0 & 1 & 0 & \beta_1^2 \end{pmatrix}', \tag{3.24}$$

and, approximately,

$$\mathbf{u} \sim (\mathbf{0}, \mathbf{V}),$$

where $\mathbf{V}$ is the covariance matrix defined in (3.22). Relations among parameters η, θ, $\mathbf{X}$, and $\mathbf{V}$ are summarized in Table 3.2.

Table 3.2 Summary for the bivariate normal case

x	y	Data Set	Size	Estimable parameters	Asymptotic variance
O	M	K	n_K	$\eta_K = \theta_1$	$W_K = diag(\sigma_{xx}, 2\sigma_{xx}^2)$
O	O	H	n_H	$\eta_H = (\theta_1, \theta_2)'$	$W_H = diag(W_K, \Sigma_{yy.x}, 2\sigma_{ee}^2)$
M	O	L	n_L	$\eta_L = (\mu_y, \sigma_{yy})'$	$W_L = diag(\sigma_{yy}, 2\sigma_{yy}^2)$

O: observed, M: missing, $\theta_1 = (\mu_x, \sigma_{xx})'$, $\theta_2 = (\beta_0, \beta_1, \sigma_{ee})'$, $\eta = (\eta_H', \eta_K', \eta_L')'$, $\theta = \eta_H$, $V = diag(W_H/n_H, W_K/n_K, W_L/n_L)$, $\Sigma_{ee} = \{E[(1,x)(1,x)']\}^{-1}\sigma_{ee}$, $\mathbf{X} = \partial\eta/\partial\theta'$, $\mu_y = \beta_0 + \beta_1\mu_x$, $\sigma_{yy} = \sigma_{ee} + \beta_1^2\sigma_{xx}$

The procedure can be carried out iteratively until convergence. Given the current value $\hat{\theta}^{(t)}$, the solution of the Gauss–Newton method can be obtained iteratively as

$$\hat{\theta}^{(t+1)} = \hat{\theta}^{(t)} + \left(\mathbf{X}_{(t)}'\hat{\mathbf{V}}_{(t)}^{-1}\mathbf{X}_{(t)}\right)^{-1}\mathbf{X}_{(t)}'\hat{\mathbf{V}}_{(t)}^{-1}\left\{\hat{\eta} - \eta\left(\hat{\theta}^{(t)}\right)\right\}, \tag{3.25}$$

where $\mathbf{X}_{(t)}$ and $\hat{\mathbf{V}}_{(t)}$ are evaluated from $\mathbf{X}$ in (3.24) and $\mathbf{V}$ in (3.22), respectively, using the current value $\hat{\theta}^{(t)}$. The covariance matrix of the estimator in (3.25) can be estimated by

$$\mathbf{C} = \left(\mathbf{X}_{(t)}'\hat{\mathbf{V}}_{(t)}^{-1}\mathbf{X}_{(t)}\right)^{-1}, \tag{3.26}$$

when the iteration is stopped at the t-th iteration. Gauss–Newton method for the estimation of non-linear models can be found, for example, in Seber and Wild (1989).

Remark 3.2. *The factoring likelihood method also provides a useful tool for efficiency comparison between different estimators that ignore some part of partial response. In the above bivariate normal example, suppose that we wish to compare the following four types of the estimates:* $\hat{\theta}_H$, $\hat{\theta}_{HK}$, $\hat{\theta}_{HL}$, $\hat{\theta}_{HKL}$ *which are the MLEs obtained from data set* H, $H \cup K$, $H \cup L$, $H \cup K \cup L$, *respectively. The Gauss-Newton estimator (3.25) is asymptotically equal to* $\hat{\theta}_{HKL}$. *Write*

$$\mathbf{X}' = [\mathbf{X}_H'\ \mathbf{X}_K'\ \mathbf{X}_L']$$

where $\mathbf{X}_H'$ *is the left* 5×5 *submatrix of* $\mathbf{X}'$, $\mathbf{X}_K'$ *is the* 5×2 *submatrix in the middle of* $\mathbf{X}'$, *and* $\mathbf{X}_L'$ *is the* 5×2 *submatrix in the right side of* $\mathbf{X}'$. *Similarly, we can decompose* $\hat{\eta}' = (\hat{\eta}_H', \hat{\eta}_K', \hat{\eta}_L')$ *and* $\mathbf{V} = diag\{\mathbf{V}_H, \mathbf{V}_K, \mathbf{V}_L\}$.

Note that the asymptotic variance of $\hat{\theta}_H$ *is* $\left(\mathbf{X}_H'\hat{\mathbf{V}}_H^{-1}\mathbf{X}_H\right)^{-1}$. *Similarly, we have*

$$Var\left(\hat{\theta}_{HL}\right) = \left(\mathbf{X}_H'\hat{\mathbf{V}}_H^{-1}\mathbf{X}_H + \mathbf{X}_L'\hat{\mathbf{V}}_L^{-1}\mathbf{X}_L\right)^{-1},$$

$$Var\left(\hat{\theta}_{HK}\right) = \left(\mathbf{X}_H'\hat{\mathbf{V}}_H^{-1}\mathbf{X}_H + \mathbf{X}_K'\hat{\mathbf{V}}_K^{-1}\mathbf{X}_K\right)^{-1},$$

and

$$Var\left(\hat{\theta}_{HKL}\right) = \left(\mathbf{X}_H'\hat{\mathbf{V}}_H^{-1}\mathbf{X}_H + \mathbf{X}_K'\hat{\mathbf{V}}_K^{-1}\mathbf{X}_K + \mathbf{X}_L'\hat{\mathbf{V}}_L^{-1}\mathbf{X}_L\right)^{-1}.$$

Using the matrix algebra such as

$$\left(\mathbf{X}_H'\hat{\mathbf{V}}_H^{-1}\mathbf{X}_H + \mathbf{X}_L'\hat{\mathbf{V}}_L^{-1}\mathbf{X}_L\right)^{-1} = \left(\mathbf{X}_H'\hat{\mathbf{V}}_H^{-1}\mathbf{X}_H\right)^{-1}$$
$$- \left(\mathbf{X}_H'\hat{\mathbf{V}}_H^{-1}\mathbf{X}_H\right)^{-1}\mathbf{X}_L'\left[\hat{\mathbf{V}}_L + \mathbf{X}_L\left(\mathbf{X}_H'\hat{\mathbf{V}}_H^{-1}\mathbf{X}_H\right)^{-1}\mathbf{X}_L'\right]^{-1}\mathbf{X}_L\left(\mathbf{X}_H'\hat{\mathbf{V}}_H^{-1}\mathbf{X}_H\right)^{-1},$$

we can derive expressions for the variances of the estimators.

For estimates of the slope parameter, the asymptotic variances are

$$
\begin{aligned}
Var(\hat{\beta}_{21.1,HK}) &= \frac{\sigma_{22.1}}{\sigma_{11} n_H} = Var(\hat{\beta}_{21.1,H}) \\
Var(\hat{\beta}_{21.1,HL}) &= \frac{\sigma_{22.1}}{\sigma_{11} n_H} \left\{1 - 2p_L \rho^2 \left(1-\rho^2\right)\right\} \\
Var\left(\hat{\beta}_{21\cdot 1,HKL}\right) &= \frac{\sigma_{22.1}}{\sigma_{11} n_H} \left\{1 - \frac{2p_L \rho^2 \left(1-\rho^2\right)}{1 - p_L p_K \rho^4}\right\},
\end{aligned}
$$

where $\rho^2 = \sigma_{12}^2/(\sigma_{11}\sigma_{22})$, $p_K = n_K/(n_H + n_K)$ *and* $p_L = n_L/(n_H + n_L)$. *Thus, we have*

$$
Var(\hat{\beta}_{21.1,H}) = Var(\hat{\beta}_{21.1,HK}) \geq Var(\hat{\beta}_{21.1,HL}) \geq Var(\hat{\beta}_{21.1,HKL}). \tag{3.27}
$$

Here strict inequalities generally hold except for special trivial cases. Note that the asymptotic variance of $\hat{\beta}_{21\cdot 1,HK}$ *is the same as the variance of* $\hat{\beta}_{21\cdot 1,H}$, *which implies that there is no gain of efficiency by adding set K (missing* y_2*) to H. On the other hand, by comparing* $Var(\hat{\beta}_{21\cdot 1,HL})$ *with* $Var(\hat{\beta}_{21\cdot 1,H})$, *we observe an efficiency gain by adding a set L (missing* y_1*) to H. It is interesting to observe that even though adding K (the data set with missing* y_2*) to H does not improve the efficiency of the regression parameter estimate, i.e.,* $Var(\hat{\beta}_{21.1,H}) = Var(\hat{\beta}_{21.1,HK})$, *adding K to (H, L) does improve the efficiency, i.e.,* $Var(\hat{\beta}_{21.1,HL}) > Var(\hat{\beta}_{21.1,HKL})$.

Example 3.6. *For a numerical example, we consider the data set originally presented by Little (1982) and also discussed in Little and Rubin (2002). Table 3.3 presents a partially classified data in a* 2×2 *table with supplemental margins for both the classification variables. For the orthogonal parametrization, we use* $\eta_H = \left(\pi_{1|1}, \pi_{1|2}, \pi_{+1}\right)'$ *where* $\pi_{1|1} = P(y_2 = 1 \mid y_1 = 1)$, $\pi_{1|2} = P(y_2 = 1 \mid y_1 = 2)$, $\pi_{1+} = P(y_1 = 1)$. *We also set* $\theta = \eta_H$. *Note that the validity of the proposed method does not depend on the choice of the parametrization. This parametrization makes the computation of the information matrix simple.*

Table 3.3 A 2 × 2 table with supplemental margins for both variables

Set	y_1	y_2	Count
	1	1	100
H	1	2	50
	2	1	75
	2	2	75
K	1		30
	2		60
L		1	28
		2	60

From the data in Table 3.3, the five observations for the three parameters are

$$
\begin{aligned}
\hat{\eta} &= \left(\hat{\pi}_{1|1,H}, \hat{\pi}_{1|2,H}, \hat{\pi}_{1+,H}, \hat{\pi}_{1+,K}, \hat{\pi}_{+1,L}\right)' \\
&= (100/150, 75/150, 150/300, 30/90, 28/88)'
\end{aligned}
$$

with the expectations

$$
\eta(\theta) = \left(\pi_{1|1}, \pi_{1|2}, \pi_{1+}, \pi_{1+}, \pi_{1|1}\pi_{1+} + \pi_{1|2} - \pi_{1|2}\pi_{1+}\right)'
$$

and the variance-covariance matrix

$$
\mathbf{V} = diag\left\{\frac{\pi_{1|1}\left(1-\pi_{1|1}\right)}{n_H \pi_{1+}}, \frac{\pi_{1|2}\left(1-\pi_{1|2}\right)}{n_H(1-\pi_{1+})}, \frac{\pi_{+1}\left(1-\pi_{+1}\right)}{n_H}, \frac{\pi_{1+}\left(1-\pi_{1+}\right)}{n_K}, \frac{\pi_{+1}\left(1-\pi_{+1}\right)}{n_L}\right\},
$$

where $\pi_{+1} = P(y_2 = 1)$. The Gauss-Newton method as described in (3.25) can be used to solve the nonlinear model of the three parameters, where the initial estimator of θ is $\hat{\theta}_S = (100/150, 75/150, 180/390)'$ and the X matrix is

$$\mathbf{X} = \begin{pmatrix} 1 & 0 & 0 & 0 & \pi_{+1} \\ 0 & 1 & 0 & 0 & 1-\pi_{+1} \\ 0 & 0 & 1 & 1 & \pi_{1|1}-\pi_{1|2} \end{pmatrix}'.$$

The resulting one-step estimates are $\hat{\pi}_{11} = 0.28, \hat{\pi}_{12} = 0.18, \hat{\pi}_{21} = 0.24$, and $\hat{\pi}_{22} = 0.31$. The standard errors of the estimated values are computed by (3.26) and are 0.0205, 0.0174, 0.0195, 0.0211 *for $\hat{\pi}_{11}, \hat{\pi}_{12}, \hat{\pi}_{21}, \hat{\pi}_{22}$, respectively.*

The proposed method can be shown to be algebraically equivalent to the Fisher scoring method in (3.13), but avoids the burden of obtaining the observed likelihood. Instead, the MLEs separately computed from each partition of the marginal likelihoods and the full likelihoods are combined in a natural way.

3.3 EM algorithm

When finding the MLE from the observed likelihood $L_{\text{obs}}(\eta)$, where $\eta = (\theta, \phi)$, we often encounter two problems associated with Newton's method:

1. Computing the second order partial derivatives of the log-likelihood can be cumbersome.
2. The likelihood does not always increase for each iteration.

The Expectation-Maximization (EM) algorithm, proposed by Dempster et al. (1977), is an iterative algorithm of finding the MLE without computing the second order partial derivatives. To describe the EM algorithm, suppose that we are interested in finding the MLE $\hat{\eta}$ that maximizes the observed likelihood $L_{\text{obs}}(\eta)$. Let $\eta^{(t)}$ be the current value of the parameter estimate of η. The EM algorithm can be defined iteratively by carrying out the following E-step and M-steps:

[E-step] Compute

$$Q\left(\eta \mid \eta^{(t)}\right) = E\left\{\ln f(\mathbf{y}, \delta; \eta) \mid \mathbf{y}_{\text{obs}}, \delta; \eta^{(t)}\right\} \tag{3.28}$$

[M-step] Find $\eta^{(t+1)}$ that maximizes $Q\left(\eta \mid \eta^{(t)}\right)$.

One of the most important properties of the EM algorithm is that its step never decreases the likelihood:

$$L_{\text{obs}}\left(\eta^{(t+1)}\right) \geq L_{\text{obs}}\left(\eta^{(t)}\right). \tag{3.29}$$

This makes EM a numerically stable procedure as it climbs the likelihood surface. Newton's method does not necessarily satisfy (3.29). The following theorem presents this result formally.

Theorem 3.2. *If $Q(\eta^{(t+1)} \mid \eta^{(t)}) \geq Q(\eta^{(t)} \mid \eta^{(t)})$, then (3.29) is satisfied.*

Proof. By the definition of the observed likelihood,

$$\ln L_{\text{obs}}(\eta) = \ln f(\mathbf{y}, \delta; \eta) - \ln f(\mathbf{y}, \delta \mid \mathbf{y}_{\text{obs}}, \delta; \eta).$$

Taking the conditional expectation of the above equation given $(\mathbf{y}_{\text{obs}}, \delta)$ evaluated at $\eta^{(t)}$, we have

$$\ln L_{\text{obs}}(\eta) = Q\left(\eta \mid \eta^{(t)}\right) - H\left(\eta \mid \eta^{(t)}\right), \tag{3.30}$$

where

$$H\left(\eta \mid \eta^{(t)}\right) = E\left\{\ln f(\mathbf{y}, \delta \mid \mathbf{y}_{\text{obs}}, \delta; \eta) \mid \mathbf{y}_{\text{obs}}, \delta, \eta^{(t)}\right\}. \tag{3.31}$$

Using Lemma 2.1, $H(\eta^{(t)} \mid \eta^{(t)}) \geq H(\eta \mid \eta^{(t)})$ for all $\eta \neq \eta^{(t)}$. Therefore, for any $\eta^{(t+1)}$ that satisfies $Q(\eta^{(t+1)} \mid \eta^{(t)}) \geq Q(\eta^{(t)} \mid \eta^{(t)})$, we have $H(\eta^{(t)} \mid \eta^{(t)}) \geq H(\eta^{(t+1)} \mid \eta^{(t)})$ and the result follows. □

Remark 3.3. *An alternative proof of Theorem 3.2 is made as follows. Writing*

$$\begin{aligned}
\ln L_{\text{obs}}(\eta) &= \ln \int_{\mathcal{R}(y_{\text{obs}},\delta)} f(\mathbf{y},\delta;\eta)d\mu(\mathbf{y}) \\
&= \ln \int_{\mathcal{R}(y_{\text{obs}},\delta)} \frac{f(\mathbf{y},\delta;\eta)}{f(\mathbf{y},\delta;\eta^{(t)})} f\left(\mathbf{y},\delta \mid \mathbf{y}_{\text{obs}},\delta;\eta^{(t)}\right) L_{\text{obs}}(\eta^{(t)})d\mu(\mathbf{y}) \\
&= \ln E\left\{\frac{f(\mathbf{y},\delta;\eta)}{f(\mathbf{y},\delta;\eta^{(t)})} \mid \mathbf{y}_{\text{obs}},\delta;\eta^{(t)}\right\} + \ln L_{\text{obs}}(\eta^{(t)}),
\end{aligned}$$

we have

$$\begin{aligned}
\ln L_{\text{obs}}(\eta) - \ln L_{\text{obs}}(\eta^{(t)}) &= \ln E\left\{\frac{f(\mathbf{y},\delta;\eta)}{f(\mathbf{y},\delta;\eta^{(t)})} \mid \mathbf{y}_{\text{obs}},\delta;\eta^{(t)}\right\} \\
&\geq E\left[\ln\left\{\frac{f(\mathbf{y},\delta;\eta)}{f(\mathbf{y},\delta;\eta^{(t)})}\right\} \mid \mathbf{y}_{\text{obs}},\delta;\eta^{(t)}\right] \\
&= Q(\eta \mid \eta^{(t)}) - Q(\eta^{(t)} \mid \eta^{(t)}),
\end{aligned}$$

where the above inequality follows from Lemma 2.1. Therefore, $Q(\eta^{(t+1)} \mid \eta^{(t)}) \geq Q(\eta^{(t)} \mid \eta^{(t)})$ implies (3.29).

We now discuss the convergence of the EM sequence $\{\eta^{(t)}\}$. By (3.29), the sequence $\{L_{\text{obs}}(\eta^{(t)})\}$ is monotone increasing and it is bounded above if the MLE exists. Thus, the sequence of $L_{\text{obs}}(\eta^{(t)})$ converges to some value L^*. In most cases, L^* is a stationary value in the sense that $L^* = L_{\text{obs}}(\eta^*)$ for some η^* at which $\partial L_{\text{obs}}(\eta)/\partial\eta = 0$. Under fairly weak conditions, such as $Q(\eta \mid \gamma)$ in (3.28) satisfies

$$\partial Q(\eta \mid \gamma)/\partial\eta \text{ is continuous in } \eta \text{ and } \gamma, \tag{3.32}$$

the EM sequence $\{\eta^{(t)}\}$ converges to a stationary point η^*. In particular, if $L_{\text{obs}}(\eta)$ is unimodal in Ω with η^* being the only stationary point, then, under (3.32), any EM sequence $\{\eta^{(t)}\}$ converges to the unique maximizer η^* of $L_{\text{obs}}(\eta)$. Further convergence details can be found in Wu (1983) and McLachlan and Krishnan (2008).

Remark 3.4. *Expression (3.30) gives further insight on the observed information matrix associated with the observed likelihood. Note that (3.30) leads to*

$$0 = \frac{\partial}{\partial\eta_0} Q(\eta \mid \eta_0) - \frac{\partial}{\partial\eta_0} H(\eta \mid \eta_0) \tag{3.33}$$

and

$$I_{\text{obs}}(\eta) = -\frac{\partial^2}{\partial\eta\partial\eta'} \log L_{\text{obs}}(\eta) = -\frac{\partial^2}{\partial\eta\partial\eta'} Q(\eta \mid \eta_0) + \frac{\partial^2}{\partial\eta\partial\eta'} H(\eta \mid \eta_0) \tag{3.34}$$

for any η_0. By applying (2.2) to $S_{\text{mis}}(\eta_0)$, we have

$$E\{S_{\text{mis}}(\eta_0) \mid \mathbf{y}_{\text{obs}},\delta;\eta_0\} = 0 \tag{3.35}$$

and, by the definition of $S_{\text{mis}}(\eta)$ in (2.25),

$$\begin{aligned}
\frac{\partial^2}{\partial\eta\partial\eta'} H(\eta \mid \eta_0)|_{\eta=\eta_0} &= \frac{\partial}{\partial\eta'} E\{S_{\text{mis}}(\eta) \mid \mathbf{y}_{\text{obs}},\delta;\eta_0\}|_{\eta=\eta_0} \\
&= -\frac{\partial}{\partial\eta_0'} E\{S_{\text{mis}}(\eta) \mid \mathbf{y}_{\text{obs}},\delta;\eta_0\}|_{\eta=\eta_0} \\
&= -\frac{\partial^2}{\partial\eta_0\partial\eta'} H(\eta \mid \eta_0)|_{\eta=\eta_0} \\
&= -\frac{\partial^2}{\partial\eta_0\partial\eta'} Q(\eta \mid \eta_0)|_{\eta=\eta_0}
\end{aligned}$$

where the first and the third equalities follow from the definition of $H(\eta \mid \eta_o)$ in (3.31), the second equality follows from (3.35), and the last equality follows from (3.33). Therefore, we have

$$I_{\text{obs}}(\eta_0) = -\left\{ \frac{\partial^2}{\partial \eta \partial \eta'} Q(\eta \mid \eta_0) + \frac{\partial^2}{\partial \eta_0 \partial \eta'} Q(\eta \mid \eta_0) \right\} |_{\eta=\eta_0}, \tag{3.36}$$

which is another way of computing the observed information using only $Q(\eta \mid \eta_0)$ in the EM algorithm. The formula (3.36) was first derived by Oakes (1999) and is sometimes called Oakes' formula. *Because we can write*

$$\frac{\partial}{\partial \eta} Q(\eta \mid \eta_0) = E\left\{ S_{\text{com}}(\eta) \mid \mathbf{y}_{\text{obs}}, \delta; \eta_0 \right\},$$

result (3.36) is equivalent to

$$I_{\text{obs}}(\eta_0) = -\left[\frac{\partial}{\partial \eta'} E\left\{ S_{\text{com}}(\eta) \mid \mathbf{y}_{\text{obs}}, \delta; \eta_0 \right\} + \frac{\partial}{\partial \eta_0'} E\left\{ S_{\text{com}}(\eta) \mid \mathbf{y}_{\text{obs}}, \delta; \eta_0 \right\} \right] |_{\eta=\eta_0} = -\frac{\partial}{\partial \eta'} \bar{S}(\eta)|_{\eta=\eta_0},$$

which was discussed in Remark 2.2.

Note that

$$\begin{aligned}
\frac{\partial}{\partial \eta_0'} E\left\{ S_{\text{mis}}(\eta) \mid \mathbf{y}_{\text{obs}}, \delta; \eta_0 \right\} &= \frac{\partial}{\partial \eta_0'} \int S_{\text{mis}}(\eta) f(\mathbf{y}, \delta \mid \mathbf{y}_{\text{obs}}, \delta; \eta_0) d\mu(\mathbf{y}) \\
&= \int S_{\text{mis}}(\eta) \left\{ \frac{\partial}{\partial \eta_0'} f(\mathbf{y}, \delta \mid \mathbf{y}_{\text{obs}}, \delta; \eta_0) \right\} d\mu(\mathbf{y}) \\
&= \int S_{\text{mis}}(\eta) S_{\text{mis}}(\eta_0)' f(\mathbf{y}, \delta \mid \mathbf{y}_{\text{obs}}, \delta; \eta_0) d\mu(\mathbf{y}) \\
&= E\left\{ S_{\text{mis}}(\eta) S_{\text{mis}}(\eta_0)' \mid \mathbf{y}_{\text{obs}}, \delta; \eta_0 \right\}
\end{aligned} \tag{3.37}$$

and so

$$\frac{\partial^2}{\partial \eta \partial \eta'} H(\eta \mid \eta_0)|_{\eta=\eta_0} = \frac{\partial}{\partial \eta_0'} E\left\{ S_{\text{mis}}(\eta) \mid \mathbf{y}_{\text{obs}}, \delta; \eta_0 \right\} |_{\eta=\eta_0} = E\left\{ S_{\text{mis}}(\eta)^{\otimes 2} \mid \mathbf{y}_{\text{obs}}, \delta; \eta_0 \right\}.$$

Thus, by (2.27), we have

$$\frac{\partial^2}{\partial \eta_0 \partial \eta'} Q(\eta \mid \eta_0)|_{\eta=\eta_0} = E\left\{ S_{\text{com}}(\eta_0)^{\otimes 2} \mid \mathbf{y}_{\text{obs}}, \delta; \eta_0 \right\} - \bar{S}(\eta_0)^{\otimes 2}$$

and (3.36) is equal to (2.22). This proves the equivalence between Oakes' formula and Louis' formula.

The EM algorithm can be expressed as

$$\eta^{(t+1)} = \arg\max_{\eta \in \Omega} Q(\eta \mid \eta^{(t)}) \tag{3.38}$$

which is often obtained by

$$\eta^{(t+1)} \leftarrow \text{solution to } \bar{S}(\eta \mid \eta^{(t)}) = 0 \tag{3.39}$$

where

$$\bar{S}(\eta \mid \eta^{(t)}) = \frac{\partial}{\partial \eta} Q(\eta \mid \eta^{(t)}) = E\left\{ S_{\text{com}}(\eta) \mid \mathbf{y}_{\text{obs}}, \delta, \eta^{(t)} \right\}.$$

Equivalently,

$$\bar{S}(\eta^{(t+1)} \mid \eta^{(t)}) = 0. \tag{3.40}$$

Now let η^* be the (unique) limit point of $\eta^{(t)}$. Applying a Taylor expansion of (3.40) with respect to $(\eta^{(t+1)}, \eta^{(t)})$ around its limiting point (η^*, η^*), we have

$$\begin{aligned} 0 &= \bar{S}(\eta^{(t+1)} \mid \eta^{(t)}) \cong \bar{S}(\eta^* \mid \eta^*) \\ &\quad + \{\bar{S}_1(\eta^* \mid \eta^*)\} \left(\eta^{(t+1)} - \eta^*\right) + \{\bar{S}_2(\eta^* \mid \eta^*)\} \left(\eta^{(t)} - \eta^*\right) \end{aligned} \tag{3.41}$$

where

$$\begin{aligned} \bar{S}_1(\eta^{(t+1)} \mid \eta^{(t)}) &= \partial \bar{S}(\eta^{(t+1)} \mid \eta^{(t)}) / \partial (\eta^{(t+1)})' \\ \bar{S}_2(\eta^{(t+1)} \mid \eta^{(t)}) &= \partial \bar{S}(\eta^{(t+1)} \mid \eta^{(t)}) / \partial (\eta^{(t)})'. \end{aligned}$$

Note that, by the definition of the mean score function, we have $\bar{S}(\eta^* \mid \eta^*) = 0$. Also, it can be shown that

$$\bar{S}_1(\eta^* \mid \eta^*) = -\mathcal{I}_{\text{com}} \tag{3.42}$$

and

$$p\lim \bar{S}_2(\eta^* \mid \eta^*) = \mathcal{I}_{\text{mis}}. \tag{3.43}$$

Thus, since $\bar{S}(\eta^* \mid \eta^*) = 0$, (3.41) implies

$$\eta^{(t+1)} - \eta^* \cong \mathcal{I}_{\text{com}}^{-1} \mathcal{I}_{\text{mis}} \left(\eta^{(t)} - \eta^*\right) \tag{3.44}$$

and so

$$\eta^{(t+1)} - \eta^{(t)} \cong \mathcal{J}_{\text{mis}} \left(\eta^{(t)} - \eta^{(t-1)}\right).$$

That is, the convergence rate for the EM sequence is linear. The matrix $\mathcal{J}_{\text{mis}} = \mathcal{I}_{\text{com}}^{-1} \mathcal{I}_{\text{mis}}$ is called the *fraction of missing information*. The fraction of missing information may vary across different components of $\eta^{(t)}$, suggesting that certain components of $\eta^{(t)}$ may approach η^* rapidly while other components may require many iterations. Roughly speaking, the rate of convergence of a vector sequence $\eta^{(t)}$ from the EM algorithm is given by the largest eigenvalue of the matrix $\mathcal{J}_{\text{mis}}$.

Example 3.7. *Consider the following exponential distribution.*

$$z_i \sim f(z; \theta) = \theta exp(-\theta z), \; z > 0.$$

Instead of observing z_i, we observe $\{(y_i, \delta_i); i = 1, 2, \cdots, n\}$ where

$$\begin{aligned} y_i &= min\{z_i, c_i\} \\ \delta_i &= \begin{cases} 1 & if\, z_i \le c_i \\ 0 & if\, z_i > c_i. \end{cases} \end{aligned}$$

The score function for θ is

$$S(\theta) = \sum_{i=1}^{n} \left(\frac{1}{\theta} - z_i\right).$$

Now, we need to evaluate

$$E(z_i \mid y_i, \delta_i) = \begin{cases} y_i & if\, \delta_i = 1 \\ E_\theta(z_i \mid z_i > c_i) & if\, \delta_i = 0 \end{cases}$$

For given θ,

$$E_\theta(z_i \mid z_i > c_i) = c_i + 1/\theta$$

by a property of the exponential distribution. Thus,

$$\hat{\theta}_{MLE} = \left[\frac{1}{r} \sum_{i=1}^{n} \{\delta_i y_i + (1 - \delta_i) c_i\}\right]^{-1},$$

where $r = \sum_{i=1}^{n} \delta_i$.

Example 3.8. *Consider the following random variable*

$$Y_i = (1-\delta_i) Z_{1i} + \delta_i Z_{2i}, \quad i = 1, 2, \cdots, n$$

where $Z_{1i} \sim N(\mu_1, \sigma_1^2)$, $Z_{2i} \sim N(\mu_2, \sigma_2^2)$, and $\delta_i \sim Bernoulli(\pi)$. Suppose that we do not observe Z_{1i}, Z_{i2}, δ_i but only observe Y_i in the sample. The parameter of interest is $\theta = (\mu_1, \mu_2, \sigma_1^2, \sigma_2^2, \pi)$. We treat δ_i as a latent variable *(i.e., a variable that is always missing) and consider the following complete sample likelihood:*

$$L_{\text{com}}(\theta) = \prod_{i=1}^{n} pdf(y_i, \delta_i \mid \theta)$$

where

$$pdf(y, \delta \mid \theta) = \left[\phi\left(y \mid \mu_1, \sigma_1^2\right)\right]^{1-\delta} \left[\phi\left(y \mid \mu_2, \sigma_2^2\right)\right]^{\delta} \pi^{\delta} (1-\pi)^{1-\delta}$$

and $\phi(y \mid \mu, \sigma^2)$ is the density of a $N(\mu, \sigma^2)$ distribution. Hence,

$$\begin{aligned} \ln L_{\text{com}}(\theta) &= \sum_{i=1}^{n} \left\{(1-\delta_i) \ln \phi\left(y_i \mid \mu_1, \sigma_1^2\right) + \delta_i \ln \phi\left(y_i \mid \mu_2, \sigma_2^2\right)\right\} \\ &\quad + \sum_{i=1}^{n} \left\{\delta_i \ln(\pi_i) + (1-\delta_i) \ln(1-\pi_i)\right\}. \end{aligned}$$

To apply the EM algorithm, the E-step can be expressed as

$$\begin{aligned} Q\left(\theta \mid \theta^{(t)}\right) &= \sum_{i=1}^{n} \left\{\left(1-w_i^{(t)}\right) \ln \phi\left(y_i \mid \mu_1, \sigma_1^2\right) + w_i^{(t)} \ln \phi\left(y_i \mid \mu_2, \sigma_2^2\right)\right\} \\ &\quad + \sum_{i=1}^{n} \left\{w_i^{(t)} \ln(\pi_i) + \left(1-w_i^{(t)}\right) \ln(1-\pi_i)\right\}, \end{aligned}$$

where $w_i^{(t)} = E\left(\delta_i \mid y_i, \theta^{(t)}\right)$ with

$$E(\delta_i \mid y_i, \theta) = \frac{\pi_i \phi\left(y_i \mid \mu_2, \sigma_2^2\right)}{(1-\pi_i)\phi\left(y_i \mid \mu_1, \sigma_1^2\right) + \pi_i \phi\left(y_i \mid \mu_2, \sigma_2^2\right)}.$$

The M-step for updating θ is

$$\frac{\partial}{\partial \theta} Q\left(\theta \mid \theta^{(t)}\right) = 0,$$

which can be written as

$$\begin{aligned} \mu_j^{(t+1)} &= \sum_{i=1}^{n} w_{ij}^{(t)} y_i \Big/ \sum_{i=1}^{n} w_{ij}^{(t)} \\ \sigma_j^{2(t+1)} &= \sum_{i=1}^{n} w_{ij}^{(t)} \left(y_i - \mu_j^{(t+1)}\right)^2 \Big/ \sum_{i=1}^{n} w_{ij}^{(t)} \\ \pi^{(t+1)} &= \sum_{i=1}^{n} w_i^{(t)} / n \end{aligned}$$

for $j = 1, 2$, where $w_{i1}^{(t)} = 1 - w_i^{(t)}$ and $w_{i2}^{(t)} = w_i^{(t)}$.

Remark 3.5. *Consider the exponential family of distributions of the form*

$$f(\mathbf{y}; \theta) = b(\mathbf{y}) exp\left\{\theta' \mathbf{T}(\mathbf{y}) - A(\theta)\right\}. \tag{3.45}$$

Under MAR, the E-step of the EM algorithm is

$$Q\left(\theta \mid \theta^{(t)}\right) = constant + \theta' E\left\{\mathbf{T}(\mathbf{y}) \mid \mathbf{y}_{\text{obs}}, \theta^{(t)}\right\} - A(\theta) \tag{3.46}$$

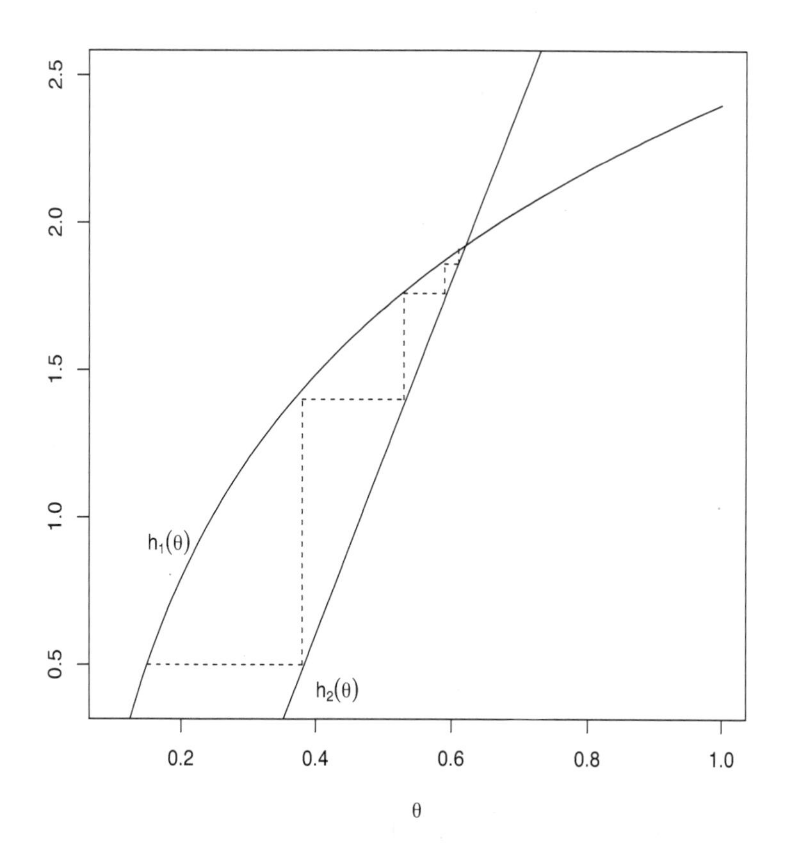

Figure 3.1 Illustration of EM algorithm for exponential family.

and the M-step is

$$\frac{\partial}{\partial\theta}Q\left(\theta \mid \theta^{(t)}\right)=0 \quad \Longleftrightarrow \quad E\left\{\mathbf{T}(\mathbf{y}) \mid \mathbf{y}_{\mathrm{obs}};\theta^{(t)}\right\}=\frac{\partial}{\partial\theta}A(\theta).$$

Because $\int f(\mathbf{y};\theta)\,d\mathbf{y}=1$, *we have*

$$\frac{\partial}{\partial\theta}A(\theta)=E\left\{\mathbf{T}(\mathbf{y});\theta\right\}.$$

Therefore, the M-step reduces to finding $\theta^{(t+1)}$ *as a solution to*

$$E\left\{\mathbf{T}(\mathbf{y}) \mid \mathbf{y}_{\mathrm{obs}},\theta^{(t)}\right\}=E\left\{\mathbf{T}(\mathbf{y});\theta\right\}. \tag{3.47}$$

Navidi (1997) used a graphical illustration for the EM algorithm finding the solution to (3.47). Let $h_1(\theta)=E\{\mathbf{T}(\mathbf{y}) \mid \mathbf{y}_{\mathrm{obs}},\theta\}$ *and* $h_2(\theta)=E\{\mathbf{T}(\mathbf{y});\theta\}$*. The EM algorithm finds the solution to* $h_1(\theta)=h_2(\theta)$ *iteratively by obtaining* $\theta^{(t+1)}$ *which solves* $h_1(\theta^{(t)})=h_2(\theta)$ *for* θ*, as illustrated in Figure 3.5.*

Note that, by (3.46),

$$\frac{\partial^2}{\partial\theta\partial\theta'}Q(\theta \mid \theta_0) = -A''(\theta)$$

and

$$\frac{\partial^2}{\partial\theta\partial\theta_0}Q(\theta \mid \theta_0) = \frac{\partial}{\partial\theta_0}E\{T(\mathbf{y}) \mid \mathbf{y}_{\text{obs}};\theta_0\},$$

so by (3.36),

$$I_{\text{obs}}(\theta) = A''(\theta) - \frac{\partial}{\partial\theta}E\{T(\mathbf{y}) \mid \mathbf{y}_{\text{obs}};\theta\},$$

which can also be obtained by taking the partial derivatives of

$$-S_{\text{obs}}(\theta) = A'(\theta) - E\{T(\mathbf{y}) \mid \mathbf{y}_{\text{obs}},\theta\}$$

with respect to θ.

Example 3.9. *Consider the bivariate normal model in (3.12). Assume that both* x_i *and* y_i *are subject to missingness under MAR. Let* $\delta_i^{(x)}$ *be the response indicator function for* x_i *and* $\delta_i^{(y)}$ *be the response indicator function for* y_i. *The sufficient statistic for* $\theta = (\mu_x, \mu_y, \sigma_{xx}, \sigma_{xy}, \sigma_{yy})$ *is*

$$T = \left(\sum_{i=1}^n x_i, \sum_{i=1}^n y_i, \sum_{i=1}^n x_i^2, \sum_{i=1}^n x_i y_i, \sum_{i=1}^n y_i^2\right),$$

and because the distribution belongs to the exponential family, the EM algorithm reduces to solving

$$\sum_{i=1}^n E\left\{(x_i, y_i, x_i^2, x_i y_i, y_i^2) \mid \delta_i^{(x)}, \delta_i^{(y)}, \delta_i^{(x)} x_i, \delta_i^{(y)} y_i; \theta^{(t)}\right\} = \sum_{i=1}^n E\{(x_i, y_i, x_i^2, x_i y_i, y_i^2); \theta\}$$

for θ. *The above conditional expectation can be obtained using the usual conditional expectation under normality. For example, if* $\delta_i^{(x)} = 1$ *and* $\delta_i^{(y)} = 0$, *the conditional expectation is equal to*

$$\begin{aligned} & E\left\{(x_i, y_i, x_i^2, x_i y_i, y_i^2) \mid \delta_i^{(x)}, \delta_i^{(y)}, \delta_i^{(x)} x_i, \delta_i^{(y)} y_i; \theta\right\} \\ = & \left(x_i, \beta_0 + \beta_1 x_i, x_i^2, x_i(\beta_0 + \beta_1 x_i), (\beta_0 + \beta_1 x_i)^2 + \sigma_{ee}\right), \end{aligned}$$

where $\beta_1 = \sigma_{xy}/\sigma_{xx}$, $\beta_0 = \mu_y - \beta_1 \mu_x$, *and* $\sigma_{ee} = \sigma_{yy} - \beta_1 \sigma_{xy}$.

If $\mathbf{y}$ is a categorical variable that takes values in set S_y, then the E-step can be easily computed by a weighted summation

$$E\left\{\ln f(\mathbf{y}, \delta; \eta) \mid \mathbf{y}_{\text{obs}}, \delta; \eta^{(t)}\right\} = \sum_{\mathbf{y}\in S_y} P\left(\mathbf{y} \mid \mathbf{y}_{\text{obs}}, \delta, \eta^{(t)}\right) \ln f(\mathbf{y}, \delta; \eta) \tag{3.48}$$

where the summation is over all possible values of $\mathbf{y}$ and $P(\mathbf{y} \mid \mathbf{y}_{\text{obs}}, \delta, \eta^{(t)})$ is the conditional probability of taking $\mathbf{y}$ given $\mathbf{y}_{\text{obs}}$ and δ evaluated at $\eta^{(t)}$. The conditional probability $P\left(\mathbf{y} \mid \mathbf{y}_{\text{obs}}, \delta; \eta^{(t)}\right)$ can be treated as the weight assigned for the categorical variable $\mathbf{y}$. That is, if $S(\eta) = \sum_{i=1}^n S(\eta; \mathbf{y}_i, \delta_i)$ is the score function for η, then the EM algorithm using (3.48) can be obtained by solving

$$\sum_{i=1}^n \sum_{\mathbf{y}\in S_y} P\left(\mathbf{y}_i = \mathbf{y} \mid \mathbf{y}_{i,\text{obs}}, \delta_i; \eta^{(t)}\right) S(\eta; \mathbf{y}, \delta_i) = 0$$

for η to get $\eta^{(t+1)}$. Ibrahim (1990) called this approach *EM by weighting*.

Example 3.10. *We now return to the example in Example 3.6. The parameters of interest are* $\pi_{ij} = P(Y_1 = i,\, Y_2 = j)$, $i = 1,2, j = 1,2$. *The sufficient statistics for the parameters are* $n_{ij}, i = 1,2; j = 1,2$, *where* n_{ij} *is the sample size for the set with* $Y_1 = i$ *and* $Y_2 = j$. *Let* $n_{i+} = n_{i1} + n_{i2}$ *and* $n_{+j} = n_{1j} + n_{2j}$. *Let* $n_{ij,H}$ *be the number of elements in set H with* $Y_1 = i$ *and* $Y_2 = j$. *We define* $n_{ij,K}$ *and* $n_{ij,L}$ *in a similar fashion. The E-step computes the conditional expectation of the sufficient statistics. This gives*

$$n_{ij}^{(t)} = E\left(n_{ij} \mid data, \pi_{ij}^{(t)}\right) = n_{ij,H} + n_{i+,K}\frac{\pi_{ij}^{(t)}}{\pi_{i+}^{(t)}} + n_{+j,L}\frac{\pi_{ij}^{(t)}}{\pi_{+j}^{(t)}},$$

for $i = 1,2; j = 1,2$. *In the M-step, the parameters are updated by* $\pi_{ij}^{(t+1)} = n_{ij}^{(t)}/n$.

Example 3.11. *We now return to the example in Example 2.5. In the E-step, the conditional expectations of the score functions are computed by*

$$\bar{S}_1\left(\beta \mid \beta^{(t)}, \phi^{(t)}\right) = \sum_{\delta_i=1} \{y_i - p_i(\beta)\}\mathbf{x}_i + \sum_{\delta_i=0}\sum_{j=0}^{1} w_{ij(t)}\{j - p_i(\beta)\}\mathbf{x}_i,$$

where

$$\begin{aligned} w_{ij(t)} &= Pr\left(Y_i = j \mid \mathbf{x}_i, \delta_i = 0; \beta^{(t)}, \phi^{(t)}\right) \\ &= \frac{Pr\left(Y_i = j \mid \mathbf{x}_i; \beta^{(t)}\right) Pr\left(\delta_i = 0 \mid \mathbf{x}_i, j; \phi^{(t)}\right)}{\sum_{y=0}^{1} Pr\left(Y_i = y \mid \mathbf{x}_i; \beta^{(t)}\right) Pr\left(\delta_i = 0 \mid \mathbf{x}_i, y; \phi^{(t)}\right)} \end{aligned}$$

and

$$\bar{S}_2\left(\phi \mid \beta^{(t)}, \phi^{(t)}\right) = \sum_{\delta_i=1} \{\delta_i - \pi(\mathbf{x}_i, y_i; \phi)\}(\mathbf{x}_i', y_i)' + \sum_{\delta_i=0}\sum_{j=0}^{1} w_{ij(t)}\{\delta_i - \pi_i(\mathbf{x}_i, j; \phi)\}(\mathbf{x}_i', j)'.$$

In the M-step, the parameter estimates are updated by solving

$$\left[\bar{S}_1\left(\beta \mid \beta^{(t)}, \phi^{(t)}\right), \bar{S}_2\left(\phi \mid \beta^{(t)}, \phi^{(t)}\right)\right] = (0,0)$$

for β *and* ϕ.

Example 3.12. *Suppose that we have a random sample from a t-distribution with mean* μ *and scale parameter* σ *with* ν *degrees of freedom such that* $x_i = \mu + \sigma e_i$ *with* $e_i \overset{indep}{\sim} t(\nu)$. *Note that we can write*

$$e_i = u_i/\sqrt{w_i}$$

where

$$x_i \mid w_i \sim N\left(\mu, \sigma^2/w_i\right), \quad w_i \sim \chi^2_\nu/\nu. \tag{3.49}$$

In this case, suppose that we are interested in estimating $\theta = (\mu, \sigma)$ *by the maximum likelihood method. To apply the EM algorithm, we can treat* w_i *in (3.49) as a latent variable.*

By Bayes' formula, we have

$$\begin{aligned} f(w_i \mid x_i) &\propto f(w_i) f(x_i \mid w_i) \\ &\propto (w_i\nu)^{\frac{\nu}{2}-1}\exp\left(-\frac{w_i\nu}{2}\right) \times \left(\sigma^2/w_i\right)^{-1/2}\exp\left\{-\frac{w_i}{2}\left(\frac{x_i-\mu}{\sigma}\right)^2\right\} \\ &\sim Gamma\left[\frac{\nu+1}{2}, 2\left\{\nu + \left(\frac{x_i-\mu}{\sigma}\right)^2\right\}^{-1}\right] \end{aligned}$$

Thus, the E-step of EM algorithm can be written as

$$E(w_i \mid x_i, \theta^{(t)}) = \frac{\nu+1}{\nu + \left(d_i^{(t)}\right)^2},$$

where $d_i^{(t)} = (x_i - \mu^{(t)})/\sigma^{(t)}$. *The M-step is then*

$$\begin{aligned} \mu^{(t+1)} &= \frac{\sum_{i=1}^n w_i^{(t)} x_i}{\sum_{i=1}^n w_i^{(t)}} \\ \sigma^{2(t+1)} &= \frac{1}{n}\sum_{i=1}^n w_i^{(t)} \left(x_i - \mu^{(t+1)}\right)^2 \end{aligned}$$

where $w_i^{(t)} = E(w_i \mid x_i, \theta^{(t)})$.

Next, we briefly discuss how to speed up the convergence of the EM algorithm. Let $\tilde{\theta}^{(t)}$ be the sequence from the usual EM algorithm and $\hat{\theta}^{(t)}$ be the sequence from the Newton method. That is,

$$\hat{\theta}^{(t+1)} = \hat{\theta}^{(t)} + \left\{\mathcal{I}_{\text{obs}}(\hat{\theta}^{(t)})\right\}^{-1} S_{\text{obs}}(\hat{\theta}^{(t)})$$

and

$$\tilde{\theta}^{(t+1)} = \tilde{\theta}^{(t)} + \left\{\bar{I}_{\text{com}}(\tilde{\theta}^{(t)})\right\}^{-1} S_{\text{obs}}(\tilde{\theta}^{(t)})$$

where $\bar{I}_{\text{com}}(\theta) = -E\left\{\dot{S}_{\text{com}}(\theta) \mid \mathbf{y}_{\text{obs}}, \delta\right\}$ and $\dot{S}_{\text{com}}(\theta) = \partial S_{\text{com}}(\theta)/\partial\theta'$. Thus, we have

$$\hat{\theta}^{(t+1)} - \hat{\theta}^{(t)} = \left\{\mathcal{I}_{\text{obs}}(\hat{\theta}^{(t)})\right\}^{-1} \bar{I}_{\text{com}}(\tilde{\theta}^{(t)}) \left\{\tilde{\theta}^{(t+1)} - \tilde{\theta}^{(t)}\right\}.$$

Because of the missing information principle, we can write $\mathcal{I}_{\text{obs}}(\theta) = E\{\bar{I}_{\text{com}}(\tilde{\theta})\} - \mathcal{I}_{\text{mis}}(\theta)$ and

$$\hat{\theta}^{(t+1)} - \hat{\theta}^{(t)} = \left\{I - \mathcal{J}_{\text{mis}}(\hat{\theta}^{(t)})\right\}^{-1} \left\{\tilde{\theta}^{(t+1)} - \tilde{\theta}^{(t)}\right\}. \tag{3.50}$$

where $\mathcal{J}_{\text{mis}}(\theta) = \mathcal{I}_{\text{mis}}(\theta)\{\mathcal{I}_{\text{com}}(\theta)\}^{-1}$ is the fraction of missing information. Setting $\tilde{\theta}^{(t)} = \hat{\theta}^{(t)}$, the procedure in (3.50) can be used to speed up the convergence of the EM algorithm. Such acceleration algorithm is commonly known as *Aitken acceleration* and was also proposed by Louis (1982).

3.4 Monte Carlo computation

In many situations, the analytic computation methods are not directly applicable because the integral associated with the conditional expectation in the mean score equation is not necessarily of a closed form. In this case, the Monte Carlo method is often used to approximately compute the expectation.

To explain the Monte Carlo method, suppose that we are interested in computing $\theta = E\{h(X)\}$ where X is a random variable with known density $f(x)$. In this case, a Monte Carlo approximation of θ is

$$\hat{\theta}_m = \frac{1}{m}\sum_{i=1}^m h(x_i) \tag{3.51}$$

where $x_1, \cdots, x_m$ are m realizations of random variable X. By the law of large numbers, $\hat{\theta}_m$ in (3.51) converges in probability to θ as $m \to \infty$. Since the variance of $\hat{\theta}_m$ is $m^{-1}\sigma_h^2$ where $\sigma_h^2 = V\{h(X)\}$, we have $\hat{\theta}_m - \theta = O_p(m^{-1/2})$ if σ_h^2 is finite, where $X_n = O_p(1)$ denotes that X_n is bounded in probability. The Monte Carlo method is attractive because the computation is simple, and it is directly applicable when X is a random vector.

There are two main issues in the Monte Carlo computation. The first is how to generate samples from a target distribution f. This type of problems, i.e., simulation problems, can be critical when the target density does not follow a familiar parametric density. The second issue is how to reduce the variance of the Monte Carlo estimator for a given Monte Carlo sample size m. Variance reduction techniques in Monte Carlo sampling are not covered here and can be found, for example, in Givens and Hoeting (2005).

Probably the simplest method of simulation is based on the probability integral transformation. For any continuous distribution function F, if $U \sim \text{Unif}(0,1)$, then $X = F^{-1}(U)$ has cumulative distribution function equal to F. The probability integral transformation approach is not directly applicable to multivariate distribution.

Rejection sampling is another popular technique for simulation. Given a density of interest f, suppose that there exist a density g and a constant M such that

$$f(x) \leq Mg(x) \tag{3.52}$$

on the support of f. The rejection sampling method proceeds as follows:

1. Sample $Y \sim g$ and $U \sim U(0,1)$, where $U(0,1)$ denotes the uniform (0,1) distribution.
2. Reject Y if
$$U > \frac{f(Y)}{Mg(Y)}. \tag{3.53}$$
In this case, do not record the value of Y as an element in the target random sample and return to step 1.
3. Otherwise, keep the value of Y. Set $X = Y$, and consider X to be an element of the target random sample.

In the rejection sampling method,

$$\begin{aligned} P(X \leq y) &= P\left\{Y \leq y \mid U \leq \frac{f(Y)}{Mg(Y)}\right\} \\ &= \frac{\int_{-\infty}^{y} \int_{0}^{f(x)/Mg(x)} du g(x) dx}{\int_{-\infty}^{\infty} \int_{0}^{f(x)/Mg(x)} du g(x) dx} \\ &= \frac{\int_{-\infty}^{y} f(x) dx}{\int_{-\infty}^{\infty} f(x) dx}. \end{aligned}$$

Note that the rejection sampling method is applicable when the density f is known up to a multiplicative factor, because the above equality still follows even if $f(x) \propto f_1(x)$ with $f_1(x) < Mg(x)$ and the decision rule (3.53) uses $f_1(x)$ instead of $f(x)$. Gilks and Wild (1992) proposed an adaptive rejection algorithm that computes the envelops function $g(x)$ automatically.

Importance sampling is also a very popular Monte Carlo approximation method. Writing

$$\theta \equiv \int h(x) f(x) dx = \int h(x) \frac{f(x)}{g(x)} g(x) dx$$

for some density $g(x)$, we can approximate θ by

$$\hat{\theta} = \sum_{i=1}^{m} w_i h(X_i)$$

where

$$w_i = \frac{f(X_i)/g(X_i)}{\sum_{j=1}^{m} f(X_j)/g(X_j)}$$

and $X_1, \cdots, X_m$ are IID from a distribution with density $g(x)$. We require that the support of g include the entire support of f. Furthermore, g should have heavier tails than f, or at least the ratio $f(x)/g(x)$ should never grow too large. The distribution g that generates the Monte Carlo sample is called the *proposal distribution* and the distribution f is called the *target distribution*.

Markov Chain Monte Carlo (MCMC) is also a frequently used Monte Carlo computation tool. A sequence of random variable $\{X_t; t = 0, 1, 2, \cdots\}$ is called a *Markov chain* if the distribution of each element depends only on the previous one. That is, it satisfies

$$P(X_t \mid X_0, X_1, \cdots, X_{t-1}) = P(X_t \mid X_{t-1})$$

for each $t = 1, 2, \cdots$. Markov Chain Monte Carlo (MCMC) refers to a body of methods for generating pseudorandom draws from probability distributions via Markov chains.

To explain the idea of the MCMC methods, let Z be a generic random vector with density $f(Z)$, which is difficult to simulate from directly. In this case, we can construct a Markov chain $\left\{Z^{(t)}; t = 1, 2 \cdots\right\}$ with f as its stationary distribution so that

$$P\left(Z^{(t)}\right) \to f \text{ as } t \to \infty$$

or

$$\frac{1}{N}\sum_{t=1}^{N} h\left(Z^{(t)}\right) \to E_f[h(Z)] = \int h(z) f(z)\, dz \tag{3.54}$$

as $N \to \infty$. A Markov chain that satisfies (3.54) is called *ergodic*. Note that the random variables $Z^{(1)}, Z^{(2)}, \cdots, Z^{(N)}$ are not IID but still (3.54) can be satisfied.

Gibbs sampling, introduced by Geman and Geman (1984), is one of the MCMC methods. Given $Z^{(t)} = (Z_1^{(t)}, Z_2^{(t)}, \cdots, Z_J^{(t)})$, Gibbs sampling draws $Z^{(t+1)}$ by sampling from the full conditionals of f,

$$\begin{aligned}
Z_1^{(t+1)} &\sim P\left(Z_1 \mid Z_2^{(t)}, Z_3^{(t)}, \cdots, Z_J^{(t)}\right) \\
Z_2^{(t+1)} &\sim P\left(Z_2 \mid Z_1^{(t+1)}, Z_3^{(t)}, \cdots, Z_J^{(t)}\right) \\
&\ \vdots \\
Z_J^{(t+1)} &\sim P\left(Z_J \mid Z_1^{(t+1)}, Z_2^{(t+1)}, \cdots, Z_{J-1}^{(t+1)}\right).
\end{aligned}$$

Under mild regularity conditions, $P(Z^{(t)}) \to f$ as $t \to \infty$.

Example 3.13. *Suppose $Z = (Z_1, Z_2)'$ is bivariate normal,*

$$\begin{pmatrix} Z_1 \\ Z_2 \end{pmatrix} \sim N\left[\begin{pmatrix} 0 \\ 0 \end{pmatrix}, \begin{pmatrix} 1 & \rho \\ \rho & 1 \end{pmatrix}\right].$$

The Gibbs sampler would be

$$\begin{aligned}
Z_1 &\sim N(\rho Z_2, 1 - \rho^2) \\
Z_2 &\sim N(\rho Z_1, 1 - \rho^2).
\end{aligned}$$

After a suitably large "burn-in period" we would find that

$$\begin{aligned}
Z_1^{(t+1)}, Z_1^{(t+2)}, \cdots, Z_1^{(t+n)} &\sim N(0, 1) \\
Z_2^{(t+1)}, Z_2^{(t+2)}, \cdots, Z_2^{(t+n)} &\sim N(0, 1).
\end{aligned}$$

Note that, if $\rho \neq 0$ the samples are dependent.

Another very popular MCMC method is the *Metropolis–Hastings (MH) algorithm*, proposed by Metropolis et al. (1953) and Hastings (1970). To explain the MH algorithm, let $f(Z)$ be a known distribution on $\mathbb{R}^k$ except for the normalizing constant. The aim is to generate $Z \sim f$. The method starts with the selection of the initial value $Z^{(0)}$ with the requirement that $f(Z^{(0)}) > 0$. Given $Z^{(t)}$, the MH algorithm generating $Z^{(t+1)}$ can be described as follows:

1. Sample a candidate value Z^* from a proposal distribution $g(Z \mid Z^{(t)})$.
2. Compute the ratio
$$R\left(Z^*, Z^{(t)}\right) = \frac{g(Z^{(t)} \mid Z^*)}{g\left(Z^* \mid Z^{(t)}\right)} \frac{f(Z^*)}{f\left(Z^{(t)}\right)}.$$
3. Sample a value $Z^{(t+1)}$ as
$$Z^{(t+1)} = \begin{cases} Z^* & \text{with probability } \rho\left(Z^{(t)}, Z^*\right) \\ Z^{(t)} & \text{with probability } 1 - \rho\left(Z^{(t)}, Z^*\right) \end{cases}$$
where $\rho(x,y) = \min\{R(x,y), 1\}$.

A rigorous justification for the ergodicity (3.54) of the MH method is beyond the scope of this book and can be found in Robert and Casella (1999). In the independent chain where $g(Z^* \mid Z^{(t)}) = g(Z^*)$, the Metropolis–Hastings ratio is
$$R\left(Z^*, Z^{(t)}\right) = \frac{f(Z^*)/g(Z^*)}{f\left(Z^{(t)}\right)/g\left(Z^{(t)}\right)},$$
which is the ratio of the importance weight for Z^* over the importance weight for $Z^{(t)}$. Thus, the Metropolis–Hastings ratio $R(Z^*, Z^{(t)})$ is also called the importance ratio. Roughly speaking, the basic idea of the MH algorithm is

1. From the current position x, move to y according to $g(y \mid x)$,
2. Stay at y with probability $\min\{f(y)/f(x), 1\}$.

Example 3.14. *Suppose that* $Y_1, \cdots, Y_n \overset{i.i.d.}{\sim} N(\theta, 1)$ *and the prior distribution for* θ *is Cauchy (0,1) with density*
$$\pi(\theta) = \frac{1}{\pi(1+\theta^2)}. \tag{3.55}$$
The posterior distribution can be derived as
$$\begin{aligned} \pi(\theta \mid y) &\propto exp\left\{-\frac{\sum_{i=1}^n (y_i - \theta)^2}{2}\right\} \times \frac{1}{1+\theta^2} \\ &\propto exp\left\{-\frac{n(\theta - \bar{y})^2}{2}\right\} \times \frac{1}{1+\theta^2}. \end{aligned}$$
If we want to generate $\theta \sim \pi(\theta \mid y)$, *we can apply the MH algorithm as follows:*

1. *Generate* θ^* *from Cauchy (0,1).*
2. *Given* $y_1, \cdots, y_n$, *compute the importance ratio*
$$R\left(\theta^*, \theta^{(t)}\right) = \frac{f(\theta^*)}{f\left(\theta^{(t)}\right)},$$
where $f(\theta) = \exp\left\{-n(\theta - \bar{y})^2/2\right\}$ *and* $\pi(\theta)$ *is defined in (3.55).*
3. *Accept* θ^* *as* $\theta^{(t+1)}$ *with probability* $\rho(\theta^{(t)}, \theta^*) = \min\left\{R\left(\theta^{(t)}, \theta^*\right), 1\right\}$.

3.5 Monte Carlo EM

In the EM algorithm, the E-step involves computing

$$Q\left(\eta \mid \eta^{(t)}\right) = E\left\{\ln f(\mathbf{y},\delta;\eta) \mid \mathbf{y}_{\text{obs}},\delta;\eta^{(t)}\right\},$$

which involves integration. To avoid the computational challenges in the E-step of the EM algorithm, Wei and Tanner (1990) proposed the Monte Carlo EM (MCEM) method based on approximating the Q function by a Monte Carlo sampling. That is,

$$Q\left(\eta \mid \eta^{(t)}\right) \cong \frac{1}{m}\sum_{j=1}^{m} \ln f\left(\mathbf{y}^{*(j)},\delta;\eta\right),$$

where $\mathbf{y}^{*(j)} = (\mathbf{y}_{\text{obs}}, \mathbf{y}_{mis}^{*(j)})$ and $\mathbf{y}_{mis}^{*(1)},\cdots,\mathbf{y}_{mis}^{*(m)} \overset{i.i.d.}{\sim} h\left(\mathbf{y}_{mis} \mid \mathbf{y}_{\text{obs}},\delta;\eta^{(t)}\right)$. That is, using the Monte Carlo approximation for the E-step, whereas the M-step remains the same. If the density $h(\mathbf{y}_{mis} \mid \mathbf{y}_{\text{obs}},\delta;\eta)$ is a nonstandard distribution, then, by the Bayes' formula,

$$h(\mathbf{y}_{mis} \mid \mathbf{y}_{\text{obs}},\delta;\hat{\eta}) = \frac{f(\mathbf{y};\hat{\theta})P(\delta \mid \mathbf{y};\hat{\phi})}{\int f(\mathbf{y};\hat{\theta})P(\delta \mid \mathbf{y};\hat{\phi})d\mathbf{y}_{mis}}. \tag{3.56}$$

Thus, given the parameter estimate $\hat{\psi} = (\hat{\theta},\hat{\phi})$, the Metropolis–Hastings algorithm can be implemented as follows:

[Step 1] For a given t-th value $\mathbf{y}^{(t)} = \left(\mathbf{y}_{\text{obs}},\mathbf{y}_{mis}^{(t)}\right)$, generate $\mathbf{y}^* = (\mathbf{y}_{\text{obs}},\mathbf{y}_{mis}^*)$ from $f(\mathbf{y};\hat{\theta})$ and compute the importance ratio

$$R(\mathbf{y}^*,\mathbf{y}^{(t)}) = \frac{P(\delta \mid \mathbf{y}^*;\hat{\phi})}{P(\delta \mid \mathbf{y}^{(t)};\hat{\phi})}.$$

[Step 2] If $R(\mathbf{y}^*,\mathbf{y}^{*(t)}) > 1$, then $\mathbf{y}^{(t+1)} = \mathbf{y}^*$. Otherwise, choose

$$\mathbf{y}^{(t+1)} = \begin{cases} \mathbf{y}^* & \text{with probability } R(\mathbf{y}^*,\mathbf{y}^{(t)}) \\ \mathbf{y}^{(t)} & \text{with probability } 1-R(\mathbf{y}^*,\mathbf{y}^{(t)}). \end{cases}$$

Instead of the Metropolis–Hastings algorithm, because $P(\delta \mid \mathbf{y};\hat{\phi})$ is bounded by 1, we can also use the rejection algorithm to generate Monte Carlo samples from the conditional distribution in (3.56). That is, we first generate $\mathbf{y}^* = (\mathbf{y}_{\text{obs}},\mathbf{y}_{mis}^*)$ from $f(\mathbf{y};\hat{\theta})$ and then accept it with probability $P(\delta \mid \mathbf{y}^*;\hat{\phi})$.

Example 3.15. *Consider the conditional distribution*

$$y_i \sim f(y_i \mid x_i;\theta).$$

Assume that x_i is always observed but we observe y_i only when $\delta_i = 1$ where $\delta_i \sim$ Bernoulli$\{\pi_i(\phi)\}$ and

$$\pi_i(\phi) = \pi(x_i,y_i;\phi) = \frac{\exp(\phi_0+\phi_1 x_i+\phi_2 y_i)}{1+\exp(\phi_0+\phi_1 x_i+\phi_2 y_i)}. \tag{3.57}$$

To implement the MCEM method, we need to generate samples from

$$f(y_i \mid x_i,\delta_i=0;\hat{\theta},\hat{\phi}) = \frac{f(y_i \mid x_i;\hat{\theta})\{1-\pi(x_i,y_i;\hat{\phi})\}}{\int f(y_i \mid x_i;\hat{\theta})\{1-\pi(x_i,y_i;\hat{\phi})\}dy_i}.$$

We can use the following rejection method

[Step 1] Generate y_i^* from $f(y_i \mid x_i;\hat{\theta})$.

[Step 2] Using y_i^* generated from Step 1, compute

$$\pi_i^*(\hat{\phi}) = \frac{\exp\left(\hat{\phi}_0 + \hat{\phi}_1 x_i + \hat{\phi}_2 y_i^*\right)}{1+\exp\left(\hat{\phi}_0 + \hat{\phi}_1 x_i + \hat{\phi}_2 y_i^*\right)}. \tag{3.58}$$

Accept y_i^* with probability $1-\pi_i^*(\hat{\phi})$.

Using the above two steps, we can create m values of y_i, denoted by $y_i^{*(1)}, \cdots, y_i^{*(m)}$, and the M-step can be implemented by solving

$$\sum_{i=1}^{n}\sum_{j=1}^{m} S\left(\theta; x_i, y_i^{*(j)}\right) = 0$$

and

$$\sum_{i=1}^{n}\sum_{j=1}^{m}\left\{\delta_i - \pi(\phi; x_i, y_i^{*(j)})\right\}\left(1, x_i, y_i^{*(j)}\right) = \mathbf{0},$$

where $S(\theta; x_i, y_i) = \partial \log f(y_i \mid x_i; \theta)/\partial\theta$.

Example 3.16. *In Example 3.15, instead of having y_i missing when $\delta_i = 0$, suppose that a grouped version of y_i, denoted by $\tilde{y}_i = T(y_i)$, is observed. For example, in some surveys, total income is reported in an interval measure for various reasons. Also, age is often reported as groups. In this case, we observe x_i, δ_i, and $\delta_i y_i + (1-\delta_i)\tilde{y}_i$, for $n = 1, 2, \cdots, n$. The mapping T from y_i to $\tilde{y}_i$ is deterministic in the sense that, if y_i is known, $\tilde{y}_i$ is uniquely determined. To implement the MCEM method, we need to generate samples from*

$$f\left(y_i \mid x_i, \tilde{y}_i, \delta_i = 0; \hat{\theta}, \hat{\phi}\right) = \frac{f(y_i \mid x_i; \hat{\theta})\{1-\pi(x_i, y_i; \hat{\phi})\} I\{y_i \in \mathcal{R}(\tilde{y}_i)\}}{\int_{\mathcal{R}(\tilde{y}_i)} f(y_i \mid x_i; \hat{\theta})\{1-\pi(x_i, y_i; \hat{\phi})\} dy_i}, \tag{3.59}$$

where $\mathcal{R}(\tilde{y}_i) = \{y; T(y) = \tilde{y}_i\}$. Thus, the rejection method for generating samples from (3.59) is described as follows:

[Step 1] Generate y_i^* from $f(y_i \mid x_i; \hat{\theta})$.

[Step 2] If $y_i^* \in \mathcal{R}(\tilde{y}_i)$, then accept y_i^* with probability $1-\pi_i^*(\hat{\phi})$ where $\pi_i^*(\hat{\phi})$ is computed by (3.58). Otherwise, go to Step 1.

The observed information matrix can be computed by Louis' formula (2.22) using Monte Carlo computation. Using the alternative formula (2.23), the Monte Carlo approximation of the observed information matrix for the observed likelihood can be computed by

$$\hat{I}_{obs}^*(\eta) = -\frac{1}{m}\sum_{j=1}^{m}\frac{\partial}{\partial\eta'} S_{\text{com}}(\eta; \mathbf{y}^{*(j)}, \delta) - \frac{1}{m}\sum_{j=1}^{m}\left\{S_{\text{com}}(\eta; \mathbf{y}^{*(j)}, \delta) - \bar{S}_{\text{I,m}}^*(\eta)\right\}^{\otimes 2} \tag{3.60}$$

where $\bar{S}_{\text{I,m}}^*(\eta) = m^{-1}\sum_{j=1}^{m} S_{\text{com}}(\eta; \mathbf{y}^{*(j)}, \delta)$.

Example 3.17. *Suppose that we are interested in estimating parameters in the conditional distribution $f(y \mid x; \theta)$. Instead of observing (x_i, y_i), suppose that we observe (z_i, y_i) in the sample, where z is conditionally independent of y conditional on x and $Cov(x, z) \neq 0$. Such variable z is called* instrumental variable *of x. We assume that we select a random sample, called* validation sample *or* calibration sample, *from the original sample and observe x_i in addition to (z_i, y_i). Thus, we can obtain a consistent estimator for the conditional distribution $g(x \mid z)$ from the validation sample. Such setup is often called* external calibration *in measurement error model literature (Guo and Little, 2011) because the target distribution for x_i is, by Bayes' theorem and the conditional independence of z and y given x,*

$$f(x_i \mid y_i, z_i) \propto f(y_i \mid x_i) g(x_i \mid z_i). \tag{3.61}$$

Thus, we may rely on the Monte Carlo method to compute the E-step of the EM algorithm.

To apply the MCEM approach, we can generate x_i^* *from* $\hat{g}(x_i \mid z_i)$ *and then accept with probability proportional to* $f(y_i \mid x_i^*; \hat{\theta})$. *If MH algorithm is to be used, the probability of accepting* x_i^* *is computed by*

$$\rho(x_i^*, x_i^{(t-1)}) = min\left\{ \frac{f(y_i \mid x_i^*; \hat{\theta})}{f(y_i \mid x_i^{(t-1)}; \hat{\theta})}, 1 \right\}.$$

Once m Monte Carlo samples are generated from $f(x_i \mid y_i, z_i)$, *the parameters are updated by solving*

$$\sum_{i \in V} S(\theta; x_i, y_i) + \sum_{i \in V^c} \sum_{j=1}^{m} S(\theta; x_i^{*(j)}, y_i) = 0$$

for θ, *where V is the set of sample indices for the validation sample and* $S(\theta; x, y) = \partial \ln f(y \mid x; \theta)/\partial \theta$.

Example 3.18. *We now consider the parameter estimation problem in generalized linear mixed models (GLMM). Let* y_{ij} *be a binary random variable (that takes values 0 or 1) with probability* $p_{ij} = P(y_{ij} = 1 \mid x_{ij}, a_i)$ *and assume that*

$$logit(p_{ij}) = x_{ij}'\beta + a_i$$

where x_{ij} *is a p-dimension covariate associated with the j-th repetition of unit i,* β *is the parameter of interest that can represent the treatment effect due to* $\mathbf{x}$, *and* a_i *represents the random effect associate with unit i. We assume that* a_i *are IID* $N(0, \sigma_a^2)$. *In this case, the latent variable, the variable that is always missing, is* a_i, *and the observed likelihood is*

$$L_{obs}(\beta, \sigma^2) = \prod_i \int \left\{ \prod_j p(x_{ij}, a_i; \beta)^{y_{ij}} [1 - p(x_{ij}, a_i; \beta)]^{1-y_{ij}} \frac{1}{\sigma_a} \phi\left(\frac{a_i}{\sigma_a}\right) \right\} da_i$$

where $\phi(\cdot)$ *is the pdf of the standard normal distribution. To apply the MCEM approach, we generate* a_i^* *from*

$$f\left(a_i \mid \mathbf{x}_i, \mathbf{y}_i; \hat{\beta}, \hat{\sigma}\right) \propto f_1(\mathbf{y}_i \mid \mathbf{x}_i, a_i; \hat{\beta}) f_2(a_i; \hat{\sigma}_a), \tag{3.62}$$

where

$$f_1\left(\mathbf{y}_i \mid \mathbf{x}_i, a_i; \hat{\beta}\right) = \prod_j p(x_{ij}, a_i; \beta)^{y_{ij}} [1 - p(x_{ij}, a_i; \beta)]^{1-y_{ij}}$$

and $f_2(a_i; \sigma) = \phi(a_i/\sigma)/\sigma$. *To generate* a_i^* *from (3.62) using the rejection sampling method, we first generate* a_i^* *from* $f_2(a_i; \hat{\sigma})$ *and then accept it with probability proportional to* $f_1(\mathbf{y}_i \mid \mathbf{x}_i, a_i^*; \hat{\beta})$. *We can also use the MH algorithm to generate samples from (3.62). To use the MH algorithm, we choose the proposal density* $f_2(a_i; \hat{\sigma})$. *Then, we accept* a_i^* *generated from* $f_2(a_i; \hat{\sigma}_a)$ *as* $a^{(t)}$ *with probability*

$$\rho\left(a_i^*, a_i^{(t-1)}\right) = min\left\{ \frac{f_1(\mathbf{y}_i \mid \mathbf{x}_i, a_i^*; \hat{\beta})}{f_1(\mathbf{y}_i \mid \mathbf{x}_i, a_i^{(t-1)}; \hat{\beta})}, 1 \right\}.$$

Otherwise, we choose $a^{(t)} = a^{(t-1)}$.

In the MCEM method, the convergence of the EM sequence of $\{\theta^{(t)}\}$ is hard to check. Booth and Hobert (1999) discuss some convergence criteria for MCEM.

3.6 Data augmentation

In this section, we consider an alternative method of generating Monte Carlo samples in the analysis of missing data. Assume that the complete data $\mathbf{y}$ can be written as $\mathbf{y} = (\mathbf{y}_{\text{obs}}, \mathbf{y}_{\text{mis}})$, with $\mathbf{y}_{\text{mis}}$ representing the missing part of $\mathbf{y}$. To make the presentation simple, assume an ignorable missing mechanism. Let $f(\mathbf{y};\theta)$ be the joint density of the complete data $\mathbf{y}$ with parameter θ. Ideally, the Monte Carlo samples are generated from the conditional distribution of the missing values given the observation evaluated at true parameter value θ_0

$$\mathbf{y}_{\text{mis}}^* \sim f(\mathbf{y}_{\text{mis}} \mid \mathbf{y}_{\text{obs}};\theta_0).$$

Since the true value is unknown, the MCEM method updates the parameter estimate $\hat{\theta}$ iteratively. In the MCEM method, the parameter estimates are updated deterministically in the sense that the best parameter estimate is chosen based on the current evaluation of the conditional expectation in the MCEM algorithm. Alternatively, one can consider a stochastic update of the parameter estimates by treating the parameter as a random variable as we do in Bayesian analysis. By treating θ as a random variable, we can generate the Monte Carlo samples by averaging over the parameter values. That is,

$$\mathbf{y}_{mis}^* \sim f_1(\mathbf{y}_{\text{mis}} \mid \mathbf{y}_{\text{obs}}) = \int f(\mathbf{y}_{mis} \mid \mathbf{y}_{\text{obs}},\theta)\, p_{obs}(\theta \mid \mathbf{y}_{\text{obs}})d\theta. \tag{3.63}$$

Here, $f_1(\mathbf{y}_{\text{mis}} \mid \mathbf{y}_{\text{obs}})$ is often called the *posterior predictive distribution* and $p_{obs}(\theta \mid \mathbf{y}_{\text{obs}})$ is the posterior distribution of θ given the observation $\mathbf{y}_{\text{obs}}$ and is called the *observed posterior distribution.* One can generate Monte Carlo samples from (3.63) by the following steps:

1. Generate θ^* from $p_{obs}(\theta \mid \mathbf{y}_{\text{obs}})$.
2. Given θ^*, generate $\mathbf{y}_{\text{mis}}^*$ from $f(\mathbf{y}_{\text{mis}} \mid \mathbf{y}_{\text{obs}},\theta^*)$.

Often it is hard to generate samples from the observed posterior distribution $p_{obs}(\theta \mid \mathbf{y}_{\text{obs}})$ directly. Instead, one can consider generating them from

$$\theta^* \sim p_{\text{obs}}(\theta \mid \mathbf{y}_{\text{obs}}) = \int p(\theta \mid \mathbf{y}_{\text{obs}},\mathbf{y}_{\text{mis}})\, f_1(\mathbf{y}_{\text{mis}} \mid \mathbf{y}_{\text{obs}})\, d\mathbf{y}_{\text{mis}}, \tag{3.64}$$

where $p(\theta \mid \mathbf{y}_{\text{obs}},\mathbf{y}_{\text{mis}}) = p(\theta \mid \mathbf{y})$ is the conditional distribution of θ given $\mathbf{y}$ and $f_1(\mathbf{y}_{mis} \mid \mathbf{y}_{\text{obs}})$ is the posterior predictive distribution defined in (3.63). The conditional distribution $p(\theta \mid \mathbf{y})$ is often called the *posterior distribution* (under complete response) in Bayesian literature.

To generate samples from the posterior predictive distributions (3.63) and the observed posterior distribution (3.64), we can use the following iterative method:

[Imputation Step (I-step)] Given the parameter value θ^*, generate $\mathbf{y}_{\text{mis}}^*$ from $f(\mathbf{y}_{\text{mis}} \mid \mathbf{y}_{\text{obs}},\theta^*)$.

[Posterior Step (P-step)] Given the imputed value $\mathbf{y}^* = (\mathbf{y}_{\text{obs}},\mathbf{y}_{\text{mis}}^*)$, generate θ^* from $p(\theta \mid \mathbf{y}^*)$.

The convergence to a stable posterior predictive distribution is very hard to check and often requires tremendous computation time. See Tanner and Wong (1987) for more details.

Example 3.19. *Let $y_1,y_2,\cdots,y_n$ be n independent realizations of a random variable Y following a Bernoulli(θ) distribution, where $\theta = P(Y=1)$ and the prior distribution for θ is Beta(α,β). In this case, the posterior distribution under complete data is*

$$\theta \mid (y_1,y_2,\cdots,y_n) \sim Beta\left(\alpha+\sum_{i=1}^{n} y_i,\beta+n-\sum_{i=1}^{n} y_i\right).$$

Now, suppose that the first r elements are observed in y and the remaining $n-r$ elements are missing in y. Then the observed posterior becomes

$$\theta \mid (y_1,y_2,\cdots,y_r) \sim Beta\left(\alpha+\sum_{i=1}^{r} y_i,\beta+r-\sum_{i=1}^{r} y_i\right). \tag{3.65}$$

We can derive (3.65) using data augmentation. In the I-step, given the current parameter value

$\theta^{*(t)}$, we generate $y_i^{*(t)}$ from a Bernoulli $(\theta^{*(t)})$ distribution for $i = r+1, \cdots, n$. For $i = 1, 2, \cdots, r$, we set $y_i^{*(t)} = y_i$. In the P-step, the parameters are generated by

$$\theta^{*(t+1)} \mid (y_1^{*(t)}, \cdots, y_n^{*(t)}) \sim Beta\left(\alpha + \sum_{i=1}^{n} y_i^{*(t)}, \beta + n - \sum_{i=1}^{n} y_i^{*(t)}\right).$$

Note that, for $i > r$,

$$\begin{aligned} E\left(y_i^{*(t+1)} \mid Y^{*(t)}\right) &= E\left(\theta^{*(t)} \mid Y^{*(t)}\right) \\ &= \frac{\alpha + \sum_{i=1}^{r} y_i + \sum_{i=r+1}^{n} y_i^{*(t)}}{\alpha + \beta + n} \\ &= \frac{\alpha + \sum_{i=1}^{r} y_i}{\alpha + \beta + r} + \lambda \left(\frac{\sum_{i=r+1}^{n} y_i^{*(t)}}{n-r} - \frac{\alpha + \sum_{i=1}^{r} y_i}{\alpha + \beta + r}\right) \end{aligned}$$

where $Y^{*(t)} = (y_1^{*(t)}, \cdots, y_n^{*(t)})$ and $\lambda = (n-r)/(\alpha + \beta + n)$. Writing

$$E\left(y_i^{*(t+1)} \mid Y^{*(t)}\right) = a_0 + \lambda(a^{(t)} - a_0)$$

where $a_0 = (\alpha + \sum_{i=1}^{r} y_i)/(\alpha + \beta + r)$ and $a^{(t)} = (\sum_{i=r+1}^{n} y_i^{*(t)})/(n-r)$, we can obtain

$$E\left(y_i^{*(t+1)} \mid Y^{*(1)}\right) = a_0 + \lambda^t \left(a^{(1)} - a_0\right).$$

Thus, as $\lambda < 1$,

$$\lim_{t \to \infty} E\left(y_i^{*(t+1)} \mid y_1, y_2 \cdots, y_r\right) = a_0 = \frac{\alpha + \sum_{i=1}^{r} y_i}{\alpha + \beta + r},$$

which can also be obtained directly from (3.65).

Example 3.20. *Assume that the scalar outcome variable Y_i follows the model*

$$\begin{aligned} y_i &= \mathbf{x}_i'\beta + e_i, \\ e_i &\overset{i.i.d.}{\sim} N(0, \sigma^2). \end{aligned} \tag{3.66}$$

The p-dimensional $\mathbf{x}_i$'s are observed on the complete sample and are assumed to be fixed. We assume that the first r units are the respondents. Let $\mathbf{y}_r = (y_1, y_2, \cdots, y_r)'$ and $X_r = (\mathbf{x}_1, \mathbf{x}_2, \cdots, \mathbf{x}_r)'$. Also, let $\mathbf{y}_{n-r} = (y_{r+1}, y_{r+2}, \cdots, y_n)'$ and $X_{n-r} = (\mathbf{x}_{r+1}, \mathbf{x}_{r+2}, \cdots, \mathbf{x}_n)'$.

The suggested Bayesian imputation method for model (3.66) is as follows:

[Posterior Step] Draw

$$\sigma^{*2} \mid \mathbf{y}_r \overset{i.i.d.}{\sim} (r-p)\hat{\sigma}_r^2/\chi^2_{r-p}, \tag{3.67}$$

and

$$\beta^* \mid (\mathbf{y}_r, \sigma^*) \overset{i.i.d.}{\sim} N\left(\hat{\beta}_r, (X_r'X_r)^{-1}\sigma^{*2}\right), \tag{3.68}$$

where $\hat{\sigma}_r^2 = (r-p)^{-1}\mathbf{y}_r'\left[I - X_r(X_r'X_r)^{-1}X_r'\right]\mathbf{y}_r$ and $\hat{\beta}_r = (X_r'X_r)^{-1}X_r'\mathbf{y}_r$.

[Imputation Step] For each missing unit $j = r+1, \cdots, n$, draw

$$e_j^* \mid (\beta^*, \sigma^*) \overset{i.i.d.}{\sim} N(0, \sigma^{*2}).$$

Then, $y_j^* = \mathbf{x}_j'\beta^* + e_j^*$ is the imputed value associated with unit j.

The above procedure assumes a constant prior for $(\beta, \log\sigma)$ and an ignorable response mechanism in the sense of Rubin (1976). Because of the monotone missing pattern, there is no need to iterate the procedures. The posterior distributions, (3.67) and (3.68), follow from known distributions.

We can also apply a Gibbs' sampling approach to generate the imputed values:

[P-step] Given the current imputed data, $\mathbf{y}_n^{*(t)} = \left(y_1, \cdots, y_r, y_{r+1}^{*(t)}, \cdots, y_n^{*(t)}\right)'$, update the parameters by generating from the posterior distribution

$$\sigma^{*(t)2} \mid \mathbf{y}_n^{*(t)} \stackrel{i.i.d.}{\sim} (n-p)\,\hat{\sigma}_n^{*(t)2} / \chi^2_{n-p}$$

and

$$\beta^{*(t)} \mid \left(\mathbf{y}_n^{*(t)}, \sigma^{*(t)}\right) \stackrel{i.i.d.}{\sim} N\left(\hat{\beta}_n^{(t)}, \left(X_n'X_n\right)^{-1} \sigma^{*(t)2}\right),$$

where $\hat{\sigma}_n^{*(t)2} = (n-p)^{-1} \mathbf{y}_n^{*(t)'} \left[I - X_n\left(X_n'X_n\right)^{-1} X_n'\right] \mathbf{y}_n^{*(t)}$ *and* $\hat{\beta}_n^{(t)} = \left(X_n'X_n\right)^{-1} X_n'\mathbf{y}_n^{*(t)}$.

[I-step] Draw

$$e_j^{*(t)} \mid \left(\beta^{*(t)}, \sigma^{*(t)}\right) \stackrel{i.i.d.}{\sim} N\left(0, \sigma^{*(t)2}\right).$$

Then, the imputed value for unit j is

$$y_j^{*(t)} = \mathbf{x}_j'\beta^{*(t)} + e_j^{*(t)}.$$

There are two main uses of data augmentation. First, data augmentation is used for parameter simulation. That is, it can be used to generate parameters from posterior distributions. The simulated parameters can be used to make an inference. The second use of data augmentation is to generate missing values from the posterior predictive distribution. Multiple imputation is a tool for making an inference using data from data augmentation. More details about multiple imputation inference will be covered in Section 4.5.

Exercises

1. Let $\bar{S}(\eta \mid \eta_0) = E\left\{S_{\text{com}}(\eta) \mid \mathbf{y}_{\text{obs}}, \delta; \eta_0\right\}$.
   (a) Prove that
   $$\frac{\partial}{\partial \eta_0'} \bar{S}(\eta \mid \eta_0)|_{\eta=\eta_0} = E\left\{S_{\text{com}}(\eta_0) S_{\text{mis}}(\eta_0)' \mid \mathbf{y}_{\text{obs}}, \delta; \eta_0\right\}.$$
   (b) Prove that $E\left\{S(\eta_0) S_{\text{mis}}(\eta_0)' \mid \mathbf{y}_{\text{obs}}, \delta; \eta_0\right\} = E\left\{S_{\text{mis}}(\eta_0) S_{\text{mis}}(\eta_0)' \mid \mathbf{y}_{\text{obs}}, \delta; \eta_0\right\}$.
   (c) Prove (3.43).
2. Under the setup of Remark 3.2, prove the following:
   (a) Equations involving the marginal parameters:
   $$\begin{aligned} Var(\hat{\mu}_{1,HK}) &= \frac{\sigma_{11}}{n_H}\left(1 - p_K\right), \\ Var(\hat{\mu}_{1,HL}) &= \frac{\sigma_{11}}{n_H}\left(1 - p_L\rho^2\right), \\ Var(\hat{\mu}_{1,HKL}) &= \frac{\sigma_{11}}{n_H}\left\{\left(1 - p_K\right) - \frac{p_L\rho^2\left(1-p_K\right)^2}{\left(1 - p_L p_K \rho^2\right)}\right\}, \end{aligned}$$

and

$$
\begin{aligned}
Var(\hat{\sigma}_{11,HK}) &= \frac{2\sigma_{11}^2}{n_H}(1-p_K), \\
Var(\hat{\sigma}_{11,HL}) &= \frac{2\sigma_{11}^2}{n_H}\left(1-p_L\rho^4\right), \\
Var(\hat{\sigma}_{11,HKL}) &= \frac{2\sigma_{11}^2}{n_H}\left\{(1-p_K)-\frac{p_L\rho^4(1-p_K)^2}{1-p_Lp_K\rho^4}\right\}.
\end{aligned}
$$

(b) If $n_K = n_L$, then

$$Var(\hat{\mu}_{1,H}) \geq Var(\hat{\mu}_{1,HL}) \geq Var(\hat{\mu}_{1,HK}) \geq Var(\hat{\mu}_{1,HKL})$$

and

$$Var(\hat{\sigma}_{11,H}) \geq Var(\hat{\sigma}_{11,HL}) \geq Var(\hat{\sigma}_{11,HK}) \geq Var(\hat{\sigma}_{11,HKL}).$$

Give some intuitive explanation on why $\hat{\mu}_{1,HK}$ is more efficient than $\hat{\mu}_{1,HL}$. Discuss when equalities hold.

3. Consider the partially classified categorical data in Table 3.3. Using the EM algorithm, find the maximum likelihood estimates of the parameters and compute standard errors.
4. Consider the data in Table 3.4 which is adapted from Table 1.4-2 of Bishop et al. (1975). Table 3.4 gives the data for a 2^3 table of three categorical variables (Y_1 = Clinic, Y_2 = Parental care, Y_3 = Survival), with one supplemental margin for Y_2 and Y_3 and another supplemental margin for Y_1 and Y_3. In this setup, Y_i are all dichotomous, taking either 0 or 1, and 8 parameters can be defined as $\pi_{ijk} = Pr(Y_1 = i, Y_2 = j, Y_3 = k),\ i = 0,1; j = 0,1; k = 0,1.$
For the orthogonal parametrization, we use

$$\eta = \left(\pi_{1|11}, \pi_{1|10}, \pi_{1|01}, \pi_{1|00}, \pi_{+1|1}, \pi_{+1|0}, \pi_{++1}\right)'$$

where $\pi_{i|jk} = Pr(y_1 = i \mid y_2 = j, y_3 = k)$, $\pi_{+j|k} = Pr(y_2 = j \mid y_3 = k)$, $\pi_{++k} = Pr(y_3 = k)$.

Table 3.4 A 2^3 table with supplemental margins

Set	y_1	y_2	y_3	Count
	1	1	1	293
	1	0	1	176
	0	1	1	23
H	0	0	1	197
	1	1	0	4
	1	0	0	3
	0	1	0	2
	0	0	0	17
	1		1	100
K	0		1	82
	1		0	5
	0		0	6
		1	1	90
L		0	1	150
		1	0	5
		0	0	10

(a) Use the factoring likelihood approach to estimate parameter η and estimated standard errors.

(b) Use the EM algorithm to estimate parameter η and use the Louis formula to compute the estimated standard errors.

5. A survey of households in several communities in north-central Iowa was conducted to determine people's views of the community in which they lived. We consider two variables, "Age of Respondent" and "Number of Years Residing in Community." An initial mailing was made to 1,023 households. After two additional mailings a total of 787 eligible units responded. The respondents were divided into seven categories on the basis of age. The age categories and the number of responses to each mailing are given in Table 3.5, which was originally presented in Drew and Fuller (1980).

Table 3.5 Responses by age for a community study

Age	First mailing	Second mailing	Third mailing	No response after 3 attempts
15-24	28	17	11	
25-34	63	26	16	
35-44	73	32	23	
45-54	97	36	12	236
55-64	97	32	15	
65-74	72	26	13	
75+	47	28	23	

We assume that the population is partitioned into $K = 7$ age categories. We assume that a proportion $1-\gamma$ of the population is composed of hard-core nonrespondents who will never answer the survey. We assume that the fraction of hard-core nonrespondents is the same in each age category. Let f_k be the population proportion in category k such that $\sum_{k=1}^{K} f_k = 1$. Let p_k be the conditional probability of response for a unit in category k. Thus, the probability of response at the r-th call for an individual in the k-th category is

$$\pi_{rk} = \gamma(1-p_k)^{r-1} p_k f_k, r = 1,2,3$$

Also, the probability that an individual in category k will not have responded after $R = 3$ calls is

$$\pi_0 = (1-\gamma) + \gamma \sum_{k=1}^{K} (1-p_k)^3 f_k.$$

The joint distribution of $n_{rk}(k = 1,\cdots,7; r = 1,2,3)$, and n_0 follows from a multinomial distribution with parameter $\pi_{rk}(k = 1,\cdots,7; r = 1,2,3)$, and π_0.

(a) Compute the observed likelihood from the data in Table 3.5 and find the maximum likelihood estimates of the parameters.

(b) Use the EM algorithm to estimate the parameters by applying the following steps:

i. Write $\pi_0 = \sum_{k=1}^{K} \pi_{0k}$, where $\pi_{0k} = \pi_{0k,M} + \pi_{0k,R}$, $\pi_{0k,M} = (1-\gamma)f_k$, and $\pi_{0k,R} = \gamma(1-p_k)^R f_k$. Let $n_{0k} = n_{0k,M} + n_{0k,R}$ be the decomposition of $n_0 = n\pi_0$ corresponding to $\pi_{0k} = \pi_{0k,M} + \pi_{0k,R}$. Find the full-sample likelihood function assuming that $n_{0k,M}$, $n_{0k,R}$, $n_{1k},\cdots n_{Rk}$ are available for $k = 1,\cdots,K$ and follow a multinomial distribution.

ii. Construct the EM algorithm from the full-sample likelihood and compute the MLE of the parameters.

6. Consider bivariate random variable (X,Y) with joint density $f_1(x;\alpha)f_2(y \mid x;\beta)$, where $f_1(x;\alpha)$ is the marginal density of X and $f_2(y \mid x;\beta)$ is the conditional density of Y given $X = x$. Assume that

f_1 is a normal distribution with parameter $\alpha = (\mu_x, \sigma_x^2)$ and f_2 are the exponential distribution with mean $\mu_y(x)$, where $1/\mu_y(x) = \beta_0 + \beta_1 x$. Suppose that we have n observations of (x_i, y_i) from the distribution of (X, Y) but some of the items are subject to missingness. We are interested in estimating $\beta = (\beta_0, \beta_1)$ from the data. Assume that the missing data mechanism is ignorable.

(a) Discuss how to obtain the maximum likelihood estimate of β if x_i are always observed and y_i are subject to missingness.

(b) Describe the EM algorithm when y_i are always observed and x_i are subject to missingness.

7. Let $X_1, \cdots, X_n$ be independently and identically distributed random variables with probability density function (pdf)

$$f_X(x;\theta) = \begin{cases} \theta e^{-\theta x} & \text{if } x > 0 \\ 0 & \text{otherwise,} \end{cases}$$

and $\theta > 0$. $Y_1, \cdots, Y_n$ are independently derived by the following steps:

[Step 1] Generate X_i from $f_X(x;\theta)$ above.

[Step 2] Generate U_i from $\exp(1)$ distribution, independently from X_i, where $\exp(1)$ is the exponential distribution with mean 1.

[Step 3] If $U_i \le X_i$, then set $Y_i = X_i$. Otherwise, go to Step 1.

Find the joint density of $Y_1, \cdots, Y_n$ using the Bayes' formula. Find the MLE of θ based on the observations $y_1, \cdots, y_n$. Discuss how to estimate the variance of the MLE.

8. The EM algorithm in Example 3.12 can be further extended to the problem of robust regression where the error distribution in the regression model is assumed to follow from a t-distribution. Suppose that the model for robust regression can be written as

$$y_i = \beta_0 + \beta_1 x_i + \sigma e_i$$

where $e_i \overset{indep}{\sim} t(\nu)$ with known ν. Similarly to Example 3.12, we can write $e_i = u_i/\sqrt{w_i}$ where

$$y_i \mid (x_i, w_i) \sim N\left(\beta_0 + \beta_1 x_i, \sigma^2/w_i\right), \quad w_i \sim \chi_\nu^2/\nu.$$

Here, (x_i, y_i) are always observed and w_i are always missing. Answer the following questions.

(a) Show that the conditional distribution of w_i given x_i and y_i follows from Gamma (α, β) for some α and β. Find the constants α and β.

(b) Show that the E-step can be expressed as

$$E(w_i \mid x_i, y_i, \theta^{(t)}) = \frac{\nu + 1}{\nu + \left(d_i^{(t)}\right)^2},$$

where $d_i^{(t)} = (y_i - \beta_0^{(t)} - \beta_1^{(t)} x_i)/\sigma^{(t)}$.

(c) Show that the M-step can be written as

$$\begin{aligned}
\left(\mu_x^{(t)}, \mu_y^{(t)}\right) &= \sum_{i=1}^n w_i^{(t)} (x_i, y_i) / (\sum_{i=1}^n w_i^{(t)}) \\
\beta_0^{(t+1)} &= \mu_y^{(t)} - \hat{\beta}_1^{(t+1)} \mu_x^{(t)} \\
\beta_1^{(t+1)} &= \frac{\sum_{i=1}^n w_i^{(t)} (x_i - \mu_x^{(t)})(y_i - \mu_y^{(t)})}{\sum_{i=1}^n w_i^{(t)} (x_i - \mu_x^{(t)})^2} \\
\sigma^{2(t+1)} &= \frac{1}{n} \sum_{i=1}^n w_i^{(t)} \left(y_i - \beta_0^{(t)} - \beta_1^{(t)} x_i\right)^2
\end{aligned}$$

where $w_i^{(t)} = E(w_i \mid x_i, y_i, \theta^{(t)})$.

9. In repeated measures experiments we typically model the vectors of observations $\mathbf{y}_1, \mathbf{y}_2, \cdots, \mathbf{y}_n$ as

$$\mathbf{y}_i \mid b_i \sim N\left(\mathbf{x}_i'\beta + \mathbf{z}_i b_i, \sigma_e^2 I_{n_i}\right).$$

That is, each $\mathbf{y}_i$ is a vector of length n_i measured from subject i; $\mathbf{x}_i$ and $\mathbf{z}_i$ are known design matrices; β is a p-dimensional vector of parameters; b_i is a latent variable representing the random effect associated with individual i. We never observe b_i and simply assume that b_i' are IID with $N(0, \sigma_b^2)$. We want to derive the EM algorithm to estimate the unknown parameters β, σ_e^2, and σ_b^2.

(a) What is the log-likelihood based on the observed data? (Hint: What is the marginal distribution of $\mathbf{y}_i$?)

(b) Consider $(\mathbf{y}_i, b_i)$ for $i = 1, 2, \cdots, n$ as the "complete data." Write down the log-likelihood of the complete data. Use that and show that the E-step of the EM algorithm involves computing $\hat{b}_i \equiv E(b_i \mid \mathbf{y}_i)$ and $\widehat{b_i^2} \equiv E(b_i^2 \mid \mathbf{y}_i)$.

(c) Given the current values of the unknown parameters, show that

$$\hat{b}_i = \left(\mathbf{z}_i'\mathbf{z}_i + \frac{\sigma_e^2}{\sigma_b^2}\right)^{-1} \mathbf{z}_i'(\mathbf{y}_i - \mathbf{x}_i\beta)$$

$$\widehat{b_i^2} = \left(\hat{b}_i\right)^2 + \sigma_e^2\left(\mathbf{z}_i'\mathbf{z}_i + \frac{\sigma_e^2}{\sigma_b^2}\right)^{-1}.$$

(d) Obtain the M-step updates for the unknown parameters β, σ_e^2, and σ_b^2.

10. Consider the problem of estimating parameters in the regression model

$$y_i = \beta_0 + \beta_1 x_{1i} + \beta_2 x_{2i} + e_i$$

where $e_i \sim N(0, \sigma_e^2)$ and

$$\begin{pmatrix} x_{1i} \\ x_{2i} \end{pmatrix} \sim N\left[\begin{pmatrix} \mu_1 \\ \mu_2 \end{pmatrix}, \begin{pmatrix} \sigma_{11} & \sigma_{12} \\ \sigma_{12} & \sigma_{22} \end{pmatrix}\right].$$

Assume, for simplicity, that the parameters $(\mu_1, \mu_2, \sigma_{11}, \sigma_{12}, \sigma_{22})$ are known. Instead of observing (x_{1i}, x_{2i}, y_i), suppose that we observe (x_{1i}, z_i, y_i) in the sample such that $f(y \mid x_1, x_2, z) = f(y \mid x_1, x_2)$ and $Cov(z, x_2) \neq 0$. Assume that the conditional distribution $g(x_2 \mid z)$ is known. In this case, discuss how to implement the EM algorithm to estimate $\theta = (\beta_0, \beta_1, \sigma_e^2)$.

11. Consider the problem of estimating parameter θ in the conditional distribution of y conditional on x, given by $f(y \mid x; \theta)$. Instead of observing (x_i, y_i) throughout the sample, we observe (x_i, y_i) for $\delta_i = 1$ and observe y_i for $\delta_i = 0$. Thus, the covariate in the regression model is subject to missingness. Let $g(x; \alpha)$ be the marginal distribution of x_i in the original sample. Assume further that $P(\delta_i = 1 \mid x_i, y_i)$ does not depend on x_i. Thus, it is MAR. For simplicity, assume that $f(y \mid x; \theta)$ follows from a logistic regression model as in Example 3.1.

(a) Devise an EM algorithm for computing the MLE of θ when α is unknown.

(b) Devise an EM algorithm for computing the MLE of θ when α is known. Discuss the efficiency gain of the MLE of θ over the situation when α is unknown.

12. Consider a simple random effect logistic regression model with dichotomous observation y_{ij} and continuous covariate x_{ij} $(i = 1, 2, \cdots, n; j = 1, 2, \cdots, m)$ with conditional probability

$$Pr(y_{ij} = 1 \mid x_{ij}, u_i, \beta) = \frac{\exp(\beta x_{ij} + u_i)}{1 + \exp(\beta x_{ij} + u_i)},$$

where $u_i \sim N(0, \sigma^2)$ is an unobserved random effect. The vector of random effects $(u_1, \cdots, u_n)$ corresponds to the missing data.

(a) Use the rejection sampling idea to construct a Monte Carlo EM algorithm for obtaining the ML estimates of β and σ^2.

(b) Discuss how to use the Metropolis–Hastings algorithm to construct a MCMC EM algorithm for obtaining the ML estimates of β and σ^2.

(c) Create artificial data from the model with $\sigma^2 = 1$, $\beta = 1$, and $x_{ij} \sim N(5,1)$ with $n = 100$ and $m = 10$. Compare the above two methods numerically using the artificial data.

Chapter 4

Imputation

4.1 Introduction

Imputation can be viewed as a Monte Carlo approximation of the conditional expectation in Chapter 3. Unlike the usual Monte Carlo approximation method, the size of Monte Carlo sample (or the imputation size), denoted by m, is not necessarily large. The main motivation for imputation is to provide a completed data set so that the resulting point estimates are consistent among different users. For example, suppose that the parameter of interest is $\mu_g = E\{g(Y)\}$ so that the complete sample estimator of μ_g is

$$\hat{\mu}_{g,n} = \frac{1}{n}\sum_{i=1}^{n} g(y_i),$$

where $y_1, \cdots, y_n$ are n independent realizations of a random variable Y with density $f(y)$. If the imputed data set is provided with the values of y_i imputed for $\delta_i = 0$, then the imputed estimator for μ_g can be computed by

$$\hat{\mu}_{g,I} = \frac{1}{n}\sum_{i=1}^{n} \{\delta_i g(y_i) + (1-\delta_i) g(y_i^*)\}.$$

Here, the implicit assumption is that

$$E\{g(y_i) \mid \delta_i = 0\} = E\{g(y_i^*) \mid \delta_i = 0\} \tag{4.1}$$

holds at least approximately for each unit i with $\delta_i = 0$. A sufficient condition for (4.1) is to generate y_i^* from the conditional distribution of y_i given $\delta_i = 0$. If the imputed values are not provided then the user needs to obtain a prediction for $g(y_i)$ given $\delta_i = 0$, which can be difficult when little is known about the response mechanism. Further, without imputation, the resulting estimates for μ_g can be different when different users estimate the same parameter under different models. Sometimes it is desirable to avoid this type of inconsistency.

The following example illustrates the statistical issues associated with imputed data.

Example 4.1. *Let $(x,y)'$ be a vector of bivariate random variables. Assume that x_i are always observed and y_i are subject to missingness in the sample, and the probability of missingness does not depend on the value of y_i. That is, it is missing at random. In this case, a consistent estimator of $\theta = E(Y)$ based on a single imputation can be computed by*

$$\hat{\theta}_I = \frac{1}{n}\sum_{i=1}^{n} \{\delta_i y_i + (1-\delta_i) y_i^*\} \tag{4.2}$$

where y_i^* satisfies

$$E(y_i^* \mid \delta_i = 0) = E(y_i \mid \delta_i = 0). \tag{4.3}$$

If we have a model for $E(y_i \mid x_i)$, such as

$$y_i \sim N(\beta_0 + \beta_1 x_i, \sigma_e^2),$$

then we can use

$$y_i^* = \hat{\beta}_0 + \hat{\beta}_1 x_i + e_i^*,$$

where $E(\hat{\beta}_0, \hat{\beta}_1 \mid \mathbf{x}_n, \delta_n) = (\beta_0, \beta_1)$, $E(e_i^* \mid \mathbf{x}_n, \delta_n) = 0$, *and* $(\hat{\beta}_0, \hat{\beta}_1)$ *is the maximum likelihood estimator of* (β_0, β_1). *If* $e_i^* \equiv 0$, *the imputation method is called the* (deterministic) regression imputation. *The regression imputation with*

$$\left(\hat{\beta}_0, \hat{\beta}_1\right) = \left(\bar{y}_r - \hat{\beta}_1 \bar{x}_r, S_{xxr}^{-1} S_{xyr}\right)$$

satisfies

$$E\left(\hat{\theta}_I - \theta\right) = 0$$

and

$$V\left(\hat{\theta}_I\right) = \frac{1}{n}\sigma_y^2 + \left(\frac{1}{r} - \frac{1}{n}\right)\sigma_e^2 = \frac{\sigma_y^2}{r}\left\{1 - \left(1 - \frac{r}{n}\right)\rho^2\right\}. \tag{4.4}$$

If e_i^* *are randomly generated from a distribution with mean* 0 *and variance* $\hat{\sigma}_e^2$, *which is the MLE of* σ_e^2, *then*

$$V\left(\hat{\theta}_I\right) = \frac{1}{n}\sigma_y^2 + \left(\frac{1}{r} - \frac{1}{n}\right)\sigma_e^2 + \frac{n-r}{n^2}\sigma_e^2,$$

where the third term represents the additional variance due to the stochastic imputation.

Deterministic imputation is unbiased for estimating the mean but may not be unbiased for estimating the proportion. For example, if $\theta = Pr(Y < c) = E\{I(Y < c)\}$, the imputed estimator

$$\hat{\theta} = n^{-1} \sum_{i=1}^{n} \{\delta_i I(y_i < c) + (1 - \delta_i) I(y_i^* < c)\}$$

is unbiased if $E\{I(Y < c)\} = E\{I(Y^* < c)\}$, which holds only when the marginal distribution of y^* is the same as the marginal distribution of y. In general, under deterministic imputation, we have $E(y) = E(y^*)$ but $V(y) > V(y^*)$. For the (deterministic) regression imputation, $V(y^*) = \sigma_y^2(1 - \rho^2) < \sigma_y^2 = V(y)$. Stochastic regression imputation provides an approximate unbiased estimator of the proportions.

Inference with imputed data is challenging because the synthetic observations in the imputed data, $\{\delta_i y_i + (1 - \delta_i) y_i^*; i = 1, 2, \cdots, n\}$, are no longer independent even though the original observations are. For example, suppose that only the first r elements are observed and the imputed values for the remaining $n - r$ elements are all equal to $\bar{y}_r = \sum_{i=1} \delta_i y_i / r$. In this case, assuming MAR, the correlation between the two imputed values is equal to one. Generally, imputation increases the correlation between the imputed values and thus increases the variance of the resulting imputed estimator.

Note that in Example 4.1, the imputed values under the regression estimation is a function of $\hat{\beta} = (\hat{\beta}_0, \hat{\beta}_1)$. Thus, we can express the imputed estimator as $\hat{\theta}_I = \hat{\theta}_I(\hat{\beta})$ to emphasize the fact that it is a function of the estimated parameter $\hat{\beta}$. Here, β can be called a *nuisance parameter* in the sense that we are not directly interested in estimating β but we need to account for the effect of estimating β when making inferences about θ using $\hat{\theta}_I = \hat{\theta}_I(\hat{\beta})$.

4.2 Basic theory for imputation

In this section, we develop basic theories for estimating $\eta = (\theta, \phi)$ from the imputed data, where θ is the parameter in $f(\mathbf{y}; \theta)$ and ϕ is the parameter in $P(\delta \mid \mathbf{y}; \phi)$. In other words, θ is related to the complete-sample data and ϕ is related to the response mechanism. Suppose that $m \geq 1$ imputed values, say $\mathbf{y}_{\text{mis}}^{*(1)}, \cdots, \mathbf{y}_{\text{mis}}^{*(m)}$, are generated from $h(\mathbf{y}_{\text{mis}} \mid \mathbf{y}_{\text{obs}}, \delta; \hat{\eta}_p)$, which is the conditional distribution defined in (3.56) and $\hat{\eta}_p$ is the preliminary estimator of parameter η. Using the m imputed values, we can obtain the imputed score equation as

$$\bar{S}_{\text{I,m}}^*(\eta \mid \hat{\eta}_p) \equiv \frac{1}{m} \sum_{j=1}^{m} S_{\text{com}}\left(\eta; \mathbf{y}^{*(j)}, \delta\right) = 0, \tag{4.5}$$

where $\mathbf{y}^{*(j)} = (\mathbf{y}_{\text{obs}}, \mathbf{y}_{\text{mis}}^{*(j)})$. Let $\hat{\eta}_{\text{I,m}}^*$ be the solution to (4.5). Note that $\hat{\eta}_{\text{I,m}}^*$ is the one-step update of $\hat{\eta}_p$ using the imputed score equation based on $\bar{S}_{\text{I,m}}^*(\eta \mid \hat{\eta}_p)$ in (4.5). We are now interested in the asymptotic properties of $\hat{\eta}_{\text{I,m}}^*$.

Before discussing the asymptotic properties, we first present the following result without proof. For the regularity conditions in Lemma 4.1, see Zhou and Kim (2012, Lemma 1).

Lemma 4.1. *Let $\hat{\theta}$ be the solution to $\hat{U}(\theta) = 0$, where $\hat{U}(\theta)$ is a function of complete observations $\mathbf{y}_1, \cdots, \mathbf{y}_n$ and parameter θ. Let θ_0 be the solution to $E\{\hat{U}(\theta)\} = 0$. Then, under some regularity conditions,*

$$\hat{\theta} - \theta_0 \cong -\left[E\{\dot{U}(\theta_0)\}\right]^{-1} \hat{U}(\theta_0),$$

where $\dot{U}(\theta) = \partial \hat{U}(\theta)/\partial \theta'$ and the notation $A_n \cong B_n$ means that $B_n^{-1} A_n = 1 + R_n$ for some R_n which converges to zero in probability.

By Lemma 4.1 and by the definition of $\mathcal{I}_{\text{obs}}$, the solution $\hat{\eta}_{MLE}$ to $S_{\text{obs}}(\eta) = 0$ satisfies

$$\hat{\eta}_{MLE} - \eta_0 \cong \mathcal{I}_{\text{obs}}^{-1} S_{\text{obs}}(\eta_0). \tag{4.6}$$

To discuss the asymptotic properties of $\hat{\eta}_{\text{I,m}}^*$, we first consider the asymptotic properties of $\hat{\eta}_{\text{I},\infty}^* = p\lim_{m\to\infty} \hat{\eta}_{\text{I,m}}^*$ which solves

$$\bar{S}(\eta \mid \hat{\eta}_p) \equiv E\{S_{\text{com}}(\eta) \mid \mathbf{y}_{\text{obs}}, \delta; \hat{\eta}_p\} = 0, \tag{4.7}$$

where

$$\begin{aligned} E\{S_{\text{com}}(\eta) \mid \mathbf{y}_{\text{obs}}, \delta; \hat{\eta}_p\} &= p \lim_{m\to\infty} \frac{1}{m} \sum_{j=1}^{m} S_{\text{com}}\left(\eta; \mathbf{y}^{*(j)}, \delta\right) \\ &= \int S_{\text{com}}(\eta; \mathbf{y}, \delta) h(\mathbf{y}_{\text{mis}} \mid \mathbf{y}_{\text{obs}}, \delta; \hat{\eta}_p) d\mathbf{y}_{mis}. \end{aligned}$$

The following lemma presents the asymptotic properties of $\hat{\eta}_{\text{I},\infty}^*$. Here, we let η_0 be the true parameter value in the joint density $f(\mathbf{y}, \delta; \eta)$.

Lemma 4.2. *Let $\hat{\eta}_{\text{I},\infty}^*$ be the solution to (4.7). Assume that $\hat{\eta}_p$ converges in probability to η_0. Then, under some regularity conditions,*

$$\hat{\eta}_{\text{I},\infty}^* - \eta_0 \cong \hat{\eta}_{\text{MLE}} - \eta_0 + \mathcal{I}_{\text{com}}^{-1} \mathcal{I}_{\text{mis}} (\hat{\eta}_p - \hat{\eta}_{\text{MLE}}) \tag{4.8}$$

and

$$V(\hat{\eta}_{\text{I},\infty}^*) \doteq \mathcal{I}_{\text{obs}}^{-1} + \mathcal{J}_{\text{mis}} \{V(\hat{\eta}_p) - V(\hat{\eta}_{\text{MLE}})\} \mathcal{J}_{\text{mis}}', \tag{4.9}$$

where $\mathcal{J}_{\text{mis}} = \mathcal{I}_{\text{com}}^{-1} \mathcal{I}_{\text{mis}}$ is the fraction of missing information. The notation $\doteq$ denotes approximation.

Proof. Write

$$\bar{S}(\eta \mid \hat{\eta}_p) = S_{\text{obs}}(\eta) + E\{S_{\text{mis}}(\eta) \mid \mathbf{y}_{\text{obs}}, \delta; \hat{\eta}_p\}$$

where $S_{\text{mis}}(\eta) = S_{\text{com}}(\eta) - S_{\text{obs}}(\eta)$. Taking a Taylor expansion of $E\{S_{\text{mis}}(\eta_0) \mid \mathbf{y}_{\text{obs}}, \delta; \hat{\eta}_p\}$ around $\hat{\eta}_p = \eta_0$ leads to

$$E\{S_{\text{mis}}(\eta_0) \mid \mathbf{y}_{\text{obs}}, \delta; \hat{\eta}_p\} \cong E\{S_{\text{mis}}(\eta_0) \mid \mathbf{y}_{\text{obs}}, \delta; \eta_0\} + E\left\{S_{\text{mis}}(\eta_0)^{\otimes 2} \mid \mathbf{y}_{\text{obs}}, \delta; \eta_0\right\} (\hat{\eta}_p - \eta_0),$$

where we used (3.37) to obtain the partial derivative of $E\{S_{\text{mis}}(\eta_0) \mid \mathbf{y}_{\text{obs}}, \delta; \hat{\eta}_p\}$ with respect to $\hat{\eta}_p$. Thus, since $E\{S_{\text{mis}}(\eta_0) \mid \mathbf{y}_{\text{obs}}, \delta; \eta_0\} = 0$ and $I_{\text{mis}}(\eta_0) = E\{S_{\text{mis}}(\eta_0)^{\otimes 2} \mid \mathbf{y}_{\text{obs}}, \delta; \eta_0\}$ converges to $\mathcal{I}_{\text{mis}}$, we have

$$\begin{aligned}\bar{S}(\eta_0 \mid \hat{\eta}_p) &\cong S_{\text{obs}}(\eta_0) + \mathcal{I}_{\text{mis}}(\hat{\eta}_p - \eta_0) \\ &= S_{\text{obs}}(\eta_0) + \mathcal{I}_{\text{mis}}(\hat{\eta}_{\text{MLE}} - \eta_0) + \mathcal{I}_{\text{mis}}(\hat{\eta}_p - \hat{\eta}_{\text{MLE}})\end{aligned}$$

where $\hat{\eta}_{MLE}$ is the MLE of η_0. Using (4.6), we have

$$\begin{aligned}\bar{S}(\eta_0 \mid \hat{\eta}_p) &\cong S_{\text{obs}}(\eta_0) + \mathcal{I}_{\text{mis}}\mathcal{I}_{\text{obs}}^{-1}S_{\text{obs}}(\eta_0) + \mathcal{I}_{\text{mis}}(\hat{\eta}_p - \hat{\eta}_{MLE}) \\ &\cong \mathcal{I}_{\text{com}}\mathcal{I}_{\text{obs}}^{-1}S_{\text{obs}}(\eta_0) + \mathcal{I}_{\text{mis}}(\hat{\eta}_p - \hat{\eta}_{MLE}) \\ &= \mathcal{I}_{\text{com}}(\hat{\eta}_{MLE} - \eta_0) + \mathcal{I}_{\text{mis}}(\hat{\eta}_p - \hat{\eta}_{MLE}),\end{aligned} \tag{4.10}$$

where $\mathcal{I}_{\text{com}} = \mathcal{I}_{\text{obs}} + \mathcal{I}_{\text{mis}}$. Since $\hat{\eta}^*_{\text{I},\infty}$ is the solution to $\bar{S}(\eta \mid \hat{\eta}_p) = 0$, we can use the Taylor expansion with respect to η to get

$$\begin{aligned}0 &= \bar{S}(\hat{\eta}^*_{\text{I},\infty} \mid \hat{\eta}_p) \\ &\cong \bar{S}(\eta_0 \mid \hat{\eta}_p) + E\left\{\frac{\partial}{\partial \eta'}S_{\text{com}}(\eta_0) \mid \mathbf{y}_{\text{obs}}, \delta\right\}(\hat{\eta}^*_{\text{I},\infty} - \eta_0) \\ &\cong \bar{S}(\eta_0 \mid \hat{\eta}_p) - \mathcal{I}_{\text{com}}(\hat{\eta}^*_{\text{I},\infty} - \eta_0)\end{aligned}$$

and so

$$\hat{\eta}^*_{\text{I},\infty} - \eta_0 \cong \mathcal{I}_{\text{com}}^{-1}\bar{S}(\eta_0 \mid \hat{\eta}_p). \tag{4.11}$$

Thus, combining (4.10) and (4.11), we have

$$\hat{\eta}^*_{\text{I},\infty} - \eta_0 \cong (\hat{\eta}_{MLE} - \eta_0) + \mathcal{I}_{\text{com}}^{-1}\mathcal{I}_{\text{mis}}(\hat{\eta}_p - \hat{\eta}_{MLE}),$$

which proves (4.8). In (4.9), we used the fact that the asymptotic covariance between $\hat{\eta}_{\text{MLE}}$ and $\hat{\eta}_p - \hat{\eta}_{\text{MLE}}$ is zero.

□

Equation (4.8) implies that

$$\hat{\eta}^*_{\text{I},\infty} = (I - \mathcal{J}_{\text{mis}})\hat{\eta}_{MLE} + \mathcal{J}_{\text{mis}}\hat{\eta}_p. \tag{4.12}$$

That is, $\hat{\eta}^*_{\text{I},\infty}$ is a convex combination of $\hat{\eta}_{MLE}$ and $\hat{\eta}_p$. Recall that $\hat{\eta}^*_{\text{I},\infty}$ is the one-step update of $\hat{\eta}_p$ using the mean score equation in (4.7). Writing $\hat{\eta}^{(t)}$ to be the t-th EM update of η that is computed by solving

$$\bar{S}\left(\eta \mid \hat{\eta}^{(t-1)}\right) = 0$$

with $\hat{\eta}^{(0)} = \hat{\eta}_p$. Equation (4.12) implies that

$$\hat{\eta}^{(t)} = (I - \mathcal{J}_{\text{mis}})\hat{\eta}_{MLE} + \mathcal{J}_{\text{mis}}\hat{\eta}^{(t-1)}.$$

Thus, we can obtain

$$\hat{\eta}^{(t)} = \hat{\eta}_{MLE} + (\mathcal{J}_{\text{mis}})^{t-1}\left(\hat{\eta}^{(0)} - \hat{\eta}_{MLE}\right),$$

which justifies $\lim_{t\to\infty}\hat{\eta}^{(t)} = \hat{\eta}_{MLE}$. Therefore, using the same argument for (4.9), we can establish

$$V(\hat{\eta}^{(t)}) \doteq \mathcal{I}_{\text{obs}}^{-1} + (\mathcal{J}_{\text{mis}})^{t-1}\{V(\hat{\eta}_p) - V(\hat{\eta}_{\text{MLE}})\}(\mathcal{J}'_{\text{mis}})^{t-1}.$$

Now, we consider the asymptotic properties of $\hat{\eta}^*_{\text{I,m}}$ obtained from (4.5) for finite m. Writing (4.5) as

$$\bar{S}^*_{\text{I,m}}(\eta \mid \hat{\eta}_p) = \hat{E}_m\{S_{\text{com}}(\eta) \mid \mathbf{y}_{\text{obs}}, \delta; \hat{\eta}_p\}, \tag{4.13}$$

where $\hat{E}_m(\cdot)$ denotes the (Monte Carlo) sample mean, we can apply a similar argument for (4.11) to get

$$\begin{aligned}\hat{\eta}^*_{\mathrm{I,m}} - \eta_0 &\cong \mathcal{I}^{-1}_{\mathrm{com}} \bar{S}^*_{\mathrm{I,m}}(\eta_0 \mid \hat{\eta}_p) \\ &= \mathcal{I}^{-1}_{\mathrm{com}} \bar{S}(\eta_0 \mid \hat{\eta}_p) + \mathcal{I}^{-1}_{\mathrm{com}} \left\{ \bar{S}^*_{\mathrm{I,m}}(\eta_0 \mid \hat{\eta}_p) - \bar{S}(\eta_0 \mid \hat{\eta}_p) \right\}\end{aligned}$$

and the two terms are independent. The first term is asymptotically equal to $\hat{\eta}^*_{\mathrm{I},\infty} - \eta_0$ by (4.11) and so its asymptotic variance is equal to (4.9). The second term has variance

$$\begin{aligned}V_{\mathrm{imp}} &= m^{-1} \mathcal{I}^{-1}_{\mathrm{com}} V\left\{ S_{\mathrm{com}}(\eta_0) - S_{\mathrm{obs}}(\eta_0) \right\} \mathcal{I}^{-1}_{\mathrm{com}} \\ &= m^{-1} \mathcal{I}^{-1}_{\mathrm{com}} \mathcal{I}_{\mathrm{mis}} \mathcal{I}^{-1}_{\mathrm{com}}\end{aligned}$$

if the m imputed values are independently generated. The variance term V_{imp} is called the *imputation variance* because it is the variance due to random generation of the imputed values. Here we used

$$\begin{aligned}V\left\{ \bar{S}^*_m(\eta \mid \hat{\eta}_p) - \bar{S}(\eta \mid \hat{\eta}_p) \right\} &= V\left\{ m^{-1} \sum_{j=1}^{m} S_{\mathrm{com}}(\eta; \mathbf{y}^{*(j)}, \delta) \mid \mathbf{y}_{obs}, \delta; \hat{\eta}_p \right\} \\ &= \frac{1}{m} V\left\{ S_{\mathrm{com}}(\eta) \mid \mathbf{y}_{\mathrm{obs}}, \delta; \hat{\eta}_p \right\} \\ &= \frac{1}{m} V\left\{ S_{\mathrm{mis}}(\eta) \mid \mathbf{y}_{\mathrm{obs}}, \delta; \hat{\eta}_p \right\} \\ &= \frac{1}{m} E\left\{ S_{\mathrm{mis}}(\eta)^{\otimes 2} \mid \mathbf{y}_{\mathrm{obs}}, \delta; \hat{\eta}_p \right\}. \qquad (4.14)\end{aligned}$$

Therefore, we can establish the following theorem, originally proved by Wang and Robins (1998).

Theorem 4.1. *Let $\hat{\eta}_p$ be a preliminary $\sqrt{n}$-consistent estimator of η with variance V_p. Under some regularity conditions, the solution $\hat{\eta}^*_{\mathrm{I,m}}$ to (4.5) has mean η_0 and asymptotic variance*

$$V\left(\hat{\eta}^*_{\mathrm{I,m}}\right) \doteq \mathcal{I}^{-1}_{\mathrm{obs}} + \mathcal{J}_{\mathrm{mis}} \left\{ V_p - \mathcal{I}^{-1}_{\mathrm{obs}} \right\} \mathcal{J}'_{\mathrm{mis}} + m^{-1} \mathcal{I}^{-1}_{\mathrm{com}} \mathcal{I}_{\mathrm{mis}} \mathcal{I}^{-1}_{\mathrm{com}}, \qquad (4.15)$$

where $\mathcal{J}_{\mathrm{mis}} = \mathcal{I}^{-1}_{\mathrm{com}} \mathcal{I}_{\mathrm{mis}}$. In particular, if we use $\hat{\eta}_p = \hat{\eta}_{MLE}$, then the asymptotic variance is

$$V\left(\hat{\eta}^*_{\mathrm{I,m}}\right) \doteq \mathcal{I}^{-1}_{\mathrm{obs}} + m^{-1} \mathcal{I}^{-1}_{\mathrm{com}} \mathcal{I}_{\mathrm{mis}} \mathcal{I}^{-1}_{\mathrm{com}}. \qquad (4.16)$$

We now consider a more general case of estimating ψ_0, which is defined through an estimating equation $E\{U(\psi; \mathbf{y})\} = 0$. Under complete response, a consistent estimator of ψ can be obtained by solving $U(\psi; \mathbf{y}) = 0$. Assume that some part of $\mathbf{y}$, denoted by $\mathbf{y}_{\mathrm{mis}}$, is not observed and m imputed values, say $\mathbf{y}^{*(1)}_{\mathrm{mis}}, \cdots, \mathbf{y}^{*(m)}_{\mathrm{mis}}$, are generated from $h(\mathbf{y}_{\mathrm{mis}} \mid \mathbf{y}_{\mathrm{obs}}, \delta; \hat{\eta}_{MLE})$, where $h(\mathbf{y}_{\mathrm{mis}} \mid \mathbf{y}_{\mathrm{obs}}, \delta; \hat{\eta}_{MLE})$ is defined in (3.56) and $\hat{\eta}_{MLE}$ is the MLE of η_0. The imputed estimating function using m imputed values is computed as

$$\bar{U}^*_{\mathrm{I,m}}(\psi \mid \hat{\eta}_{MLE}) = \frac{1}{m} \sum_{j=1}^{m} U(\psi; \mathbf{y}^{*(j)}), \qquad (4.17)$$

where $\mathbf{y}^{*(j)} = (\mathbf{y}_{\mathrm{obs}}, \mathbf{y}^{*(j)}_{\mathrm{mis}})$. Let $\hat{\psi}^*_{\mathrm{I,m}}$ be the solution to $\bar{U}^*_{\mathrm{I,m}}(\psi \mid \hat{\eta}_{MLE}) = 0$. Let $\bar{U}(\phi \mid \hat{\eta}_{MLE})$ be the probability limit of $\bar{U}^*_{\mathrm{I,m}}(\psi \mid \hat{\eta}_{MLE})$ as $m \to \infty$.

The following theorem, originally proved by Robins and Wang (2000), presents some asymptotic properties of the estimator that is a solution to $\bar{U}(\psi \mid \hat{\eta}_{MLE}) = 0$.

Theorem 4.2. *Suppose that the parameter of interest ψ_0 is estimated by solving $U(\psi) = 0$ under complete response. Then, under some regularity conditions, the solution to*

$$E\{U(\psi) \mid \mathbf{y}_{\mathrm{obs}}, \delta; \hat{\eta}_{MLE}\} = 0$$

has mean ψ_0 *and asymptotic variance* $\tau^{-1}\Omega_0\tau^{-1'}$, *where*

$$\begin{aligned} \tau &= -E\{\partial U(\psi_0)/\partial\psi'\} \\ \Omega_0 &= V\{\bar{U}(\psi_0 \mid \eta_0) + \kappa S_{\text{obs}}(\eta_0)\} \end{aligned} \tag{4.18}$$

and

$$\kappa = E\{U(\psi_0)S_{\text{mis}}(\eta_0)\}\mathcal{I}_{\text{obs}}^{-1}. \tag{4.19}$$

Proof. Define

$$\bar{U}(\psi \mid \hat{\eta}_{MLE}) = E\{U(\psi) \mid \mathbf{y}_{\text{obs}}, \delta; \hat{\eta}_{MLE}\}.$$

By a Taylor expansion of $\bar{U}(\psi \mid \hat{\eta}_{MLE})$ around $\hat{\eta}_{MLE} = \eta_0$, we have

$$\bar{U}(\psi \mid \hat{\eta}) \cong \bar{U}(\psi \mid \eta_0) + \{\partial\bar{U}(\psi \mid \eta_0)/\partial\eta'\}(\hat{\eta}_{MLE} - \eta_0). \tag{4.20}$$

Since

$$\bar{U}(\psi \mid \eta) = \int U(\psi;\mathbf{y}) f(\mathbf{y}_{\text{mis}} \mid \mathbf{y}_{\text{obs}}, \delta; \eta)\, d\mu(\mathbf{y}_{\text{mis}})$$

we have, similar to (3.37),

$$\begin{aligned} \frac{\partial}{\partial\eta'}\bar{U}(\psi \mid \eta) &= \int U(\psi;\mathbf{y}) \frac{\partial}{\partial\eta'} f(\mathbf{y}_{\text{mis}} \mid \mathbf{y}_{\text{obs}}, \delta; \eta)\, d\mu(\mathbf{y}_{\text{mis}}) \\ &= E\{U(\psi)S_{\text{mis}}(\eta)' \mid \mathbf{y}_{\text{obs}}, \delta; \eta\}. \end{aligned}$$

Thus, using (4.6), the expansion in (4.20) reduces to

$$\bar{U}(\psi \mid \hat{\eta}) \cong \bar{U}(\psi \mid \eta_0) + E\{U(\psi)S_{\text{mis}}(\eta_0)'\}\mathcal{I}_{\text{obs}}^{-1}S_{\text{obs}}(\eta_0) \tag{4.21}$$

Writing

$$\bar{U}_l(\psi \mid \eta_0) \equiv \bar{U}(\psi \mid \eta_0) + E\{U(\psi)S_{\text{mis}}(\eta_0)'\}\mathcal{I}_{\text{obs}}^{-1}S_{\text{obs}}(\eta_0),$$

with subscript l denoting linearization, we can express, by Lemma 4.1

$$\hat{\psi}^*_{\text{I},\infty} - \psi_0 \cong -\left[E\left\{\frac{\partial}{\partial\psi'}\bar{U}_l(\psi_0 \mid \eta_0)\right\}\right]^{-1}\bar{U}_l(\psi_0 \mid \eta_0).$$

Since $E\{S_{\text{obs}}(\eta_0)\} = 0$, we have

$$E\left\{\frac{\partial}{\partial\psi'}\bar{U}_l(\psi_0 \mid \eta_0)\right\} = E\left\{\frac{\partial}{\partial\psi'}\bar{U}(\psi_0 \mid \eta_0)\right\} = E\left\{\frac{\partial}{\partial\psi'}U(\psi_0)\right\}$$

and we can write

$$\hat{\psi}^*_{\text{I},\infty} - \psi_0 \cong \tau^{-1}\{\bar{U}(\psi_0 \mid \eta_0) + \kappa S_{\text{obs}}(\eta_0)\} \tag{4.22}$$

where κ is as defined in (4.19), and the result follows. □

Because we can write

$$\hat{\psi}^*_{\text{I},m} = \hat{\psi}^*_{\text{I},\infty} + \left(\hat{\psi}^*_{\text{I},m} - \hat{\psi}^*_{\text{I},\infty}\right)$$

and the two terms are independent, we have

$$V\left(\hat{\psi}^*_{\text{I},m}\right) = V\left(\hat{\psi}^*_{\text{I},\infty}\right) + V\left(\hat{\psi}^*_{\text{I},m} - \hat{\psi}^*_{\text{I},\infty}\right).$$

By the same argument for (4.22), we have

$$\hat{\psi}^*_{\text{I},m} - \psi_0 \cong \tau^{-1}\left\{\bar{U}^*_{I,m}(\psi_0 \mid \eta_0) + \kappa S_{\text{obs}}(\eta_0)\right\}$$

and so

$$\hat{\psi}^*_{\mathrm{I},m} - \hat{\psi}^*_{\mathrm{I},\infty} \cong \tau^{-1} \left\{ \bar{U}^*_{I,m}(\psi_0 \mid \eta_0) - \bar{U}^*_{I,\infty}(\psi_0 \mid \eta_0) \right\}.$$

Thus, writing $V_{\mathrm{imp}}(U) = E\{V(U \mid \mathbf{y}_{\mathrm{obs}}, \delta)\}$, we can obtain $V\left(\hat{\psi}^*_{\mathrm{I},m} - \hat{\psi}^*_{\mathrm{I},\infty}\right) \cong m^{-1}\tau^{-1}V_{\mathrm{imp}}(U)\tau^{-1'}$ and

$$V\left(\hat{\psi}^*_{\mathrm{I},m}\right) \cong \tau^{-1}\left\{\Omega_0 + m^{-1}V_{\mathrm{imp}}(U)\right\}\tau^{-1'},$$

where Ω_0 is defined in (4.18).

Example 4.2. *We now go back to Example 4.1. In this case, we can write*

$$U(\theta) = \sum_{i=1}^{n} (y_i - \theta)/\sigma_e^2$$

and the score function for β under complete response is, under joint normality,

$$S(\beta) = \frac{1}{\sigma_e^2}\sum_{i=1}^{n}(y_i - \beta_0 - \beta_1 x_i)(1, x_i).$$

Assuming that the response mechanism is ignorable, we have

$$S_{\mathrm{obs}}(\beta) = \frac{1}{\sigma_e^2}\sum_{i=1}^{n}\delta_i(y_i - \beta_0 - \beta_1 x_i)(1, x_i)'.$$

Let $\hat{\beta} = (\hat{\beta}_0, \hat{\beta}_1)$ be the solution to $S_{\mathrm{obs}}(\beta) = 0$. Note that the imputed estimator (4.2) with $y_i^* = \hat{\beta}_0 + \hat{\beta}_1 x_i$ can be written as the solution to

$$E\left\{U(\theta) \mid \mathbf{y}_{\mathrm{obs}}, \delta; \hat{\beta}\right\} = 0$$

since

$$E\{U(\theta) \mid \mathbf{y}_{\mathrm{obs}}, \delta; \beta\} \equiv \bar{U}(\theta \mid \beta) = \sum_{i=1}^{n}\left\{\delta_i y_i + (1-\delta_i)(\beta_0 + \beta_1 x_i) - \theta\right\}/\sigma_e^2.$$

Thus, using the linearization formula (4.21), we have

$$\bar{U}_l(\theta \mid \beta) = \bar{U}(\theta \mid \beta) + (\kappa_1, \kappa_2)S_{\mathrm{obs}}(\beta), \tag{4.23}$$

where

$$(\kappa_1, \kappa_2)' = \mathcal{I}_{\mathrm{obs}}^{-1}E\left\{S_{\mathrm{mis}}(\beta)U(\theta)\right\}. \tag{4.24}$$

In this example, we have

$$\begin{aligned}
\begin{pmatrix} \kappa_0 \\ \kappa_1 \end{pmatrix} &= \left[E\left\{\sum_{i=1}^{n}\delta_i(1, x_i)'(1, x_i)\right\}\right]^{-1} E\left\{\sum_{i=1}^{n}(1-\delta_i)(1, x_i)'\right\} \\
&\cong E\left\{(-1 + (n/r)(1 - g\bar{x}_r), (n/r)g)'\right\},
\end{aligned}$$

where $g = (\bar{x}_n - \bar{x}_r)/\{\sum_{i=1}^{n}\delta_i(x_i - \bar{x}_r)^2/r\}$. Thus,

$$\begin{aligned}
\bar{U}_l(\theta \mid \beta)\sigma_e^2 &= \sum_{i=1}^{n}\delta_i(y_i - \theta) + \sum_{i=1}^{n}(1-\delta_i)(\beta_0 + \beta_1 x_i - \theta) \\
&\quad + \sum_{i=1}^{n}\delta_i(y_i - \beta_0 - \beta_1 x_i)(\kappa_0 + \kappa_1 x_i).
\end{aligned}$$

Note that the solution to $\bar{U}_I(\theta \mid \beta) = 0$ *leads to*

$$\hat{\theta}_I = \frac{1}{n}\sum_{i=1}^{n}\{\beta_0 + \beta_1 x_i + \delta_i(1+\kappa_0+\kappa_1 x_i)(y_i - \beta_0 - \beta_1 x_i)\} = \frac{1}{n}\sum_{i=1}^{n} d_i, \tag{4.25}$$

where $1+\kappa_0+\kappa_1 x_i = (n/r)\{1+g(x_i-\bar{x}_r)\}$. *Kim and Rao (2009) called* d_i *the* linearized pseudo values. *For variance estimation, one can just apply the standard variance estimation formula to the pseudo values. Under a uniform response mechanism,* $1+\kappa_0+\kappa_1 x_i \cong n/r$ *and the asymptotic variance of* $\hat{\theta}_I$ *is equal to*

$$\frac{1}{n}\beta_1^2\sigma_x^2 + \frac{1}{r}\sigma_e^2 = \frac{1}{n}\sigma_y^2 + \left(\frac{1}{r}-\frac{1}{n}\right)\sigma_e^2$$

which is consistent with the result in (4.4).

4.3 Variance estimation after imputation

We now discuss variance estimation of the imputed point estimators. Variance estimation can be implemented using either a linearization method or a replication method. We first discuss the linearization method. The replication method will be discussed in Section 4.4.

In the linearization method, we first decompose the imputed estimator into a sum of two components, a deterministic part and a stochastic part. The linearization method is applied to the deterministic component. For example, in Example 4.1, the imputed estimator based on the regression imputation can be written as a function of $\hat{\beta} = (\hat{\beta}_0, \hat{\beta}_1)$. That is, we can write

$$\hat{\theta}_I(\hat{\beta}) = \hat{\theta}_{Id} = n^{-1}\sum_{i=1}^{n}\left\{\delta_i y_i + (1-\delta_i)\left(\hat{\beta}_0 + \hat{\beta}_1 x_i\right)\right\}. \tag{4.26}$$

By the linearization result in (4.25), we can find $d_i = d_i(\beta)$ such that

$$\hat{\theta}_I(\hat{\beta}) \cong n^{-1}\sum_{i=1}^{n} d_i(\beta)$$

where

$$d_i(\beta) = \beta_0 + \beta_1 x_i + \delta_i\left(\frac{n}{r}\right)\{1+g(x_i-\bar{x}_r)\}(y_i - \beta_0 - \beta_1 x_i).$$

Note that, if (x_i, y_i, δ_i) are IID, then $d_i = d(x_i, y_i, \delta_i)$ are also IID. Thus, the variance of $\bar{d}_n = n^{-1}\sum_{i=1}^{n} d_i$ is unbiasedly estimated by

$$\hat{V}(\bar{d}_n) = \frac{1}{n}\frac{1}{n-1}\sum_{i=1}^{n}\left(d_i - \bar{d}_n\right)^2. \tag{4.27}$$

Unfortunately, we cannot compute $\hat{V}(\bar{d}_n)$ in (4.27) since $d_i = d_i(\beta)$ is a function of an unknown parameter. Thus, we use $\hat{d}_i = d_i(\hat{\beta})$ in (4.27) to get a consistent variance estimator of the imputed estimator.

Instead of the deterministic imputation, suppose that a stochastic imputation is used such that

$$\hat{\theta}_I = n^{-1}\sum_{i=1}^{n}\left\{\delta_i y_i + (1-\delta_i)\left(\hat{\beta}_0 + \hat{\beta}_1 x_i + \hat{e}_i^*\right)\right\},$$

where $\hat{e}_i^*$ are the additional noise terms in the stochastic imputation. Often $\hat{e}_i^*$ are randomly selected from the empirical distribution of the sample residuals in the respondents. That is, $\hat{e}_i^*$ is randomly selected among the set $\hat{\mathbf{e}}_R = \{\hat{e}_i; \delta_i = 1\}$ where $\hat{e}_i = y_i - \hat{\beta}_0 - \hat{\beta}_1 x_i$ and $\sum_{i=1}^{n}\delta_i\hat{e}_i = 0$. The variance of the imputed estimator can be decomposed into two parts:

$$V\left(\hat{\theta}_I\right) = V\left(\hat{\theta}_{Id}\right) + V\left(\hat{\theta}_I - \hat{\theta}_{Id}\right) \tag{4.28}$$

where the first part is the deterministic part and the second part is the additional increase in the variance due to stochastic imputation. The first part can be estimated by the linearization method discussed above. The second part is called the *imputation variance*. If we require the imputation mechanism to satisfy

$$\sum_{i=1}^{n} (1-\delta_i)\hat{e}_i^* = 0$$

then the imputation variance is equal to zero. Also, if we impute more than one imputed values so that

$$\hat{\theta}_I = n^{-1}m^{-1}\sum_{i=1}^{n}\sum_{j=1}^{m}\left\{\delta_i y_i + (1-\delta_i)\left(\hat{\beta}_0 + \hat{\beta}_1 x_i + \hat{e}_i^{*(j)}\right)\right\},$$

then the imputation variance is reduced by increasing the imputation size m. Often the variance of $\hat{\theta}_I - \hat{\theta}_{Id} = n^{-1}\sum_{i=1}^{n}(1-\delta_i)\hat{e}_i^*$ can be computed under the known imputation mechanism. For example, if simple random sampling without replacement is used then

$$V\left(\hat{\theta}_I - \hat{\theta}_{Id}\right) = E\left\{V\left(\hat{\theta}_I \mid \mathbf{y}_{obs}, \delta\right)\right\} = \frac{n-r}{n^2}\left(2-\frac{n}{r}\right)\frac{1}{r-1}\sum_{i=1}^{n}\delta_i \hat{e}_i^2,$$

if $n-r<r$. If we can write $\hat{e}_i^* = \sum_{k=1}^{n} d_{ki}\delta_k \hat{e}_k$ for some d_{ji}, where d_{ji} takes the value one if $\hat{e}_j$ is used for $\hat{e}_i^*$ and takes the value zero otherwise, then $\hat{\theta}_I - \hat{\theta}_{Id} = \sum_{i=1}^{n}\delta_i d_i \hat{e}_i$ where $d_i = \sum_{j=1}^{n}(1-\delta_j)d_{ij}$ is the number of times that $\hat{e}_i$ is used for imputation and

$$\hat{V}_{imp} = n^{-2}\sum_{i=1}^{n}\delta_i d_i^2 \hat{e}_i^2$$

can be used to estimate the imputation variance.

Instead of the above tailor-made estimation method for imputation variance, we may consider an alternative approach when the imputed values are independently generated and m is greater than one. Write

$$\hat{\theta}_I = \frac{1}{m}\sum_{j=1}^{m}\hat{\theta}_I^{(j)} = \bar{\theta}_I^{(\cdot)}$$

where

$$\hat{\theta}_I^{(j)} = n^{-1}\sum_{i=1}^{n}\left\{\delta_i y_i + (1-\delta_i)\left(\hat{\beta}_0 + \hat{\beta}_1 x_i + \hat{e}_i^{*(j)}\right)\right\},$$

the imputation variance is unbiasedly estimated by $m^{-1}B_m$, where

$$B_m = \frac{1}{m-1}\sum_{j=1}^{m}\left(\hat{\theta}_I^{(j)} - \bar{\theta}_I^{(\cdot)}\right)^2. \tag{4.29}$$

The following lemma presents the properties of B_m in (4.29).

Lemma 4.3. *Let $X_1,\cdots,X_m$ be identically distributed (not necessarily independent) with mean θ and covariance*

$$Cov(X_i,X_j) = \begin{cases} c_{11} & \text{if } i=j \\ c_{12} & \text{otherwise.}\end{cases}$$

Let $\bar{X}_m = m^{-1}\sum_{i=1}^{m}X_i$ and $B_m = (m-1)^{-1}\sum_{i=1}^{m}(X_i-\bar{X}_m)^2$. Then, we have

$$V(\bar{X}_m) = c_{12} + m^{-1}(c_{11}-c_{12}), \tag{4.30}$$

and

$$E(B_m) = c_{11} - c_{12}. \tag{4.31}$$

Proof. Result (4.30) can be easily proved by

$$V(\bar{X}_m) = m^{-2}\left\{\sum_{i=1}^{n} V(X_i) + \sum_{i=1}^{n}\sum_{j\neq i} Cov(X_i, X_j)\right\}.$$

Result (4.31) can be proved using the equality

$$\sum_{i=1}^{m} (X_i - \bar{X}_m)^2 = \sum_{i=1}^{m} (X_i - c) - m(\bar{X}_m - c)^2$$

for any constant c. Choosing c equal to the mean of X_i and taking the expectation on both sides of the above equality, we have

$$(m-1)E(B_m) = mc_{11} - m\left\{c_{12} + m^{-1}(c_{11} - c_{12})\right\},$$

and (4.31) follows. □

If we apply Lemma 4.3 to the imputed estimator with $X_j = \hat{\theta}_I^{*(j)} - \hat{\theta}_{Id}$, then

$$c_{12} = Cov\left\{n^{-1}\sum_{i=1}^{n}(1-\delta_i)\hat{e}_i^{*(1)}, n^{-1}\sum_{i=1}^{n}(1-\delta_i)\hat{e}_i^{*(2)}\right\} = 0,$$

if $\hat{e}_i^*$ are randomly selected among $\hat{\mathbf{e}}_R$. Thus, we have

$$E\left(m^{-1}B_m\right) = m^{-1}V\left(\bar{\theta}_I^{(1)} - \hat{\theta}_{Id}\right) = V\left(\bar{\theta}_I^{(\cdot)} - \hat{\theta}_{Id}\right) \quad (4.32)$$

and the imputation variance is unbiasedly estimated by $m^{-1}B_m$.

We now discuss a general case of parameter estimation when the parameter of interest ψ is estimated by the solution $\hat{\psi}_n$ to

$$\sum_{i=1}^{n} U(\psi; \mathbf{y}_i) = 0 \quad (4.33)$$

under complete response of $\mathbf{y}_1, \cdots, \mathbf{y}_n$. The complete-sample variance estimator of $\hat{\psi}_n$ is

$$\hat{V}(\hat{\psi}_n) = \hat{\tau}_u^{-1}\hat{\Omega}_u\hat{\tau}_u^{-1'}, \quad (4.34)$$

where

$$\begin{aligned}\hat{\tau}_u &= n^{-1}\sum_{i=1}^{n}\dot{U}(\hat{\psi}_n; \mathbf{y}_i)\\ \hat{\Omega}_u &= n^{-1}(n-1)^{-1}\sum_{i=1}^{n}(\hat{u}_i - \bar{u}_n)^{\otimes 2},\end{aligned}$$

$\dot{U}(\psi; \mathbf{y}) = \partial U(\psi; \mathbf{y})/\partial\psi'$, $\bar{u}_n = n^{-1}\sum_{i=1}^{n}\hat{u}_i$, and $\hat{u}_i = U(\hat{\psi}_n; \mathbf{y}_i)$. The variance estimator in (4.34) is often called the *sandwich variance estimator*.

Under the existence of missing data, let $\mathbf{y}_{i,\text{obs}}$ be the observed part of $\mathbf{y}_i$ and $\mathbf{y}_{i,\text{mis}}$ the missing part of $\mathbf{y}_i$. Assume that m imputed values of $\mathbf{y}_{i,\text{mis}}$, denoted by $\mathbf{y}_{i,\text{mis}}^{*(1)}, \cdots, \mathbf{y}_{i,\text{mis}}^{*(m)}$, are randomly generated from the conditional distribution $h(\mathbf{y}_{i,\text{mis}} \mid \mathbf{y}_{i,\text{obs}}, \delta_i; \hat{\eta}_p)$ where $\hat{\eta}_p$ is estimated by solving

$$\hat{U}_p(\eta) \equiv \sum_{i=1}^{n} U_p(\eta; \mathbf{y}_{i,\text{obs}}) = 0. \quad (4.35)$$

If we apply the m imputed values to (4.33), we can get the imputed estimating equation

$$\bar{U}_m^*(\psi) \equiv m^{-1}\sum_{i=1}^{n}\sum_{j=1}^{m} U(\psi;\mathbf{y}_i^{*(j)}) = 0, \tag{4.36}$$

where $\mathbf{y}_i^{*(j)} = \left(\mathbf{y}_{i,\text{obs}}, \mathbf{y}_{i,mis}^{*(j)}\right)$. To apply the linearization method, we first compute the conditional expectation of $U(\psi;\mathbf{y}_i)$ given $(\mathbf{y}_{i,\text{obs}},\delta_i)$ evaluated at $\hat{\eta}_p$. That is, compute

$$\bar{U}(\psi \mid \hat{\eta}_p) = \sum_{i=1}^{n} \bar{U}_i(\psi \mid \hat{\eta}_p) = \sum_{i=1}^{n} E\left\{U(\psi;\mathbf{y}_i) \mid \mathbf{y}_{i,\text{obs}},\delta_i;\hat{\eta}_p\right\}. \tag{4.37}$$

Let $\hat{\psi}_R$ be the solution to $\bar{U}(\psi \mid \hat{\eta}_p) = 0$. Using the linearization technique, we have

$$\bar{U}(\psi \mid \hat{\eta}_p) \cong \bar{U}(\psi \mid \eta_0) + E\left\{\frac{\partial}{\partial \eta'}\bar{U}(\psi \mid \eta_0)\right\}(\hat{\eta}_p - \eta_0) \tag{4.38}$$

and

$$0 = \hat{U}_p(\hat{\eta}_p) = \hat{U}_p(\eta_0) + E\left\{\frac{\partial}{\partial \eta'}\hat{U}_p(\eta_0)\right\}(\hat{\eta}_p - \eta_0). \tag{4.39}$$

Thus, combining (4.38) and (4.39), we have

$$\bar{U}(\psi \mid \hat{\eta}_p) \cong \bar{U}(\psi \mid \eta_0) + \kappa(\psi)\hat{U}_p(\eta_0) = \sum_{i=1}^{n}\left\{\bar{U}_i(\psi \mid \eta_0) + \kappa(\psi) U_p(\eta_0;\mathbf{y}_{i,\text{obs}})\right\}, \tag{4.40}$$

where

$$\kappa(\psi) = -E\left\{\frac{\partial}{\partial \eta'}\bar{U}(\psi \mid \eta_0)\right\}\left[E\left\{\frac{\partial}{\partial \eta'}\hat{U}_p(\eta_0)\right\}\right]^{-1}.$$

Write

$$\bar{U}_l(\psi \mid \eta_0) = \sum_{i=1}^{n}\left\{\bar{U}_i(\psi \mid \eta_0) + \kappa(\psi)\hat{U}_p(\eta_0;\mathbf{y}_{i,\text{obs}})\right\} = \sum_{i=1}^{n} q_i(\psi \mid \eta_0),$$

and $q_i(\psi \mid \eta_0) = \bar{U}_i(\psi \mid \eta_0) + \kappa(\psi)\hat{U}_p(\eta_0;\mathbf{y}_{i,\text{obs}})$, and the variance of $\bar{U}(\psi \mid \hat{\eta}_p)$ is asymptotically equal to the variance of $\bar{U}_l(\psi \mid \eta_0)$. Thus, the sandwich-type variance estimator for $\hat{\psi}_R$ is

$$\hat{V}(\hat{\psi}_R) = \hat{\tau}_q^{-1}\hat{\Omega}_q\hat{\tau}_q^{-1'}, \tag{4.41}$$

where

$$\begin{aligned}
\hat{\tau}_q &= n^{-1}\sum_{i=1}^{n}\dot{q}_i(\hat{\psi}_R \mid \hat{\eta}_p) \\
\hat{\Omega}_q &= n^{-1}(n-1)^{-1}\sum_{i=1}^{n}(\hat{q}_i - \bar{q}_n)^{\otimes 2},
\end{aligned}$$

$\dot{q}_i(\psi \mid \eta) = \partial q_i(\psi \mid \eta)/\partial \psi'$, $\bar{q}_n = n^{-1}\sum_{i=1}^{n}\hat{q}_i$, and $\hat{q}_i = q_i(\hat{\psi}_R \mid \hat{\eta}_p)$. Note that

$$\begin{aligned}
\hat{\tau}_q &= n^{-1}\sum_{i=1}^{n}\dot{q}_i(\hat{\psi}_R \mid \hat{\eta}_p) \\
&= n^{-1}\sum_{i=1}^{n} E\left\{\dot{U}(\hat{\psi}_R;\mathbf{y}_i) \mid \mathbf{y}_{i,\text{obs}},\delta_i;\hat{\eta}_p\right\}
\end{aligned}$$

because $\hat{\eta}_p$ is the solution to (4.35).

Remark 4.1. *The variance estimator (4.41) can be understood as a sandwich formula based on the joint estimating equations (4.35) and (4.37). Because* $(\hat{\psi}_R, \hat{\eta}_p)$ *is the solution to*

$$\mathbf{U}(\psi,\eta) \equiv \begin{bmatrix} U_1(\psi,\eta) \\ U_2(\eta) \end{bmatrix} = \mathbf{0},$$

where $U_1(\psi,\eta) = \bar{U}(\psi \mid \eta)$ *and* $U_2(\eta) = \hat{U}_p(\eta)$*, we can apply the Taylor expansion to get*

$$\begin{pmatrix} \hat{\psi}_R \\ \hat{\eta}_p \end{pmatrix} \cong \begin{pmatrix} \psi_0 \\ \eta_0 \end{pmatrix} - \begin{pmatrix} B_{11} & B_{12} \\ B_{21} & B_{22} \end{pmatrix}^{-1} \begin{bmatrix} U_1(\psi_0,\eta_0) \\ U_2(\eta_0) \end{bmatrix}$$

where

$$\begin{pmatrix} B_{11} & B_{12} \\ B_{21} & B_{22} \end{pmatrix} = \begin{bmatrix} E(\partial U_1/\partial \psi') & E(\partial U_1/\partial \eta') \\ E(\partial U_2/\partial \psi') & E(\partial U_2/\partial \eta') \end{bmatrix}.$$

Since $B_{21} = 0$*, we have*

$$\begin{pmatrix} B_{11} & B_{12} \\ 0 & B_{22} \end{pmatrix}^{-1} = \begin{pmatrix} B_{11}^{-1} & -B_{11}^{-1}B_{12}B_{22}^{-1} \\ 0 & B_{22}^{-1} \end{pmatrix}$$

and

$$\hat{\psi}_R \cong \psi_0 - B_{11}^{-1}\left\{U_1(\psi_0,\eta_0) - B_{12}B_{22}^{-1}U_2(\eta_0)\right\}.$$

Thus, the result in (4.41) follows directly.

Finally, we consider the variance estimation of the imputed estimator $\hat{\psi}_m^*$, which is the solution to the imputed estimating equation (4.36). Writing

$$\bar{U}_m^*(\psi \mid \hat{\eta}_p) = \bar{U}(\psi \mid \hat{\eta}_p) + \{\bar{U}_m^*(\psi \mid \hat{\eta}_p) - \bar{U}(\psi \mid \hat{\eta}_p)\},$$

we can express

$$V\{\bar{U}_m^*(\psi \mid \hat{\eta}_p)\} = V\{\bar{U}(\psi \mid \hat{\eta}_p)\} + V_{imp}(\bar{U}_m^*),$$

where $V_{imp}(\bar{U}_m^*)$ is the imputation variance of $\bar{U}_m^* = \bar{U}_m^*(\psi \mid \hat{\eta}_p)$. The imputation variance can be estimated, for example, by

$$\hat{V}_{imp} = m^{-1}B_m(\bar{U}_m^*)$$

where

$$B_m(\bar{U}_m^*) = \frac{1}{m-1}\sum_{j=1}^{m}\left(U^{*(j)} - \bar{U}_m^*\right)^{\otimes 2},$$

$U^{*(j)} = n^{-1}\sum_{i=1}^{n} U(\hat{\psi}^*; \mathbf{y}_i^{*(j)})$ and $\bar{U}_m^* = m^{-1}\sum_{j=1}^{m} U^{*(j)}$. Thus, the sandwich-type variance estimator of $\hat{\psi}_m^*$ is

$$\hat{V}(\hat{\psi}_m^*) = \left(\hat{\tau}_q^*\right)^{-1}\{\hat{\Omega}_q^* + \hat{V}_{imp}\}\left(\hat{\tau}_q^{*\prime}\right)^{-1}, \tag{4.42}$$

where

$$\hat{\tau}_q^* = n^{-1}\sum_{i=1}^{n}\dot{U}(\hat{\psi}_m^*; \mathbf{y}_i^*)$$

$$\hat{\Omega}_q = n^{-1}(n-1)^{-1}\sum_{i=1}^{n}(\hat{q}_i^* - \bar{q}_n^*)^{\otimes 2}$$

and $\hat{q}_i^* = q_i^*(\hat{\psi}_m^* \mid \hat{\eta}_p)$.

Example 4.3. *Assume that the original sample is decomposed into G disjoint groups (often called* imputation cells*) and the sample observations are independently and identically distributed within the same cell. That is,*

$$y_i \mid i \in A_g \overset{i.i.d.}{\sim} (\mu_g, \sigma_g^2) \tag{4.43}$$

where A_g *is the set of sample indices in cell g. Assume there are* n_g *sample elements in cell g and* r_g *elements are observed. Assume that the response mechanism is MAR. The parameter of interest is* $\theta = E(Y)$. *Model (4.43) is often called the* cell mean model.

In this case, a deterministic imputation can be used with $\hat{\eta} = (\hat{\mu}_1, \cdots, \hat{\mu}_G)$. *Let* $\hat{\mu}_g = r_g^{-1} \sum_{i \in A_g} \delta_i y_i$ *be the g-th cell mean of y among respondents. The imputed estimator of* θ *is*

$$\hat{\theta}_{Id} = n^{-1} \sum_{g=1}^{G} \sum_{i \in A_g} \{\delta_i y_i + (1-\delta_i)\hat{\mu}_g\} = n^{-1} \sum_{g=1}^{G} n_g \hat{\mu}_g. \tag{4.44}$$

By the linearization technique in (4.40), the imputed estimator can be expressed as

$$\hat{\theta}_{Id} \cong n^{-1} \sum_{g=1}^{G} \sum_{i \in A_g} \left\{ \mu_g + \frac{n_g}{r_g} \delta_i (y_i - \mu_g) \right\} \tag{4.45}$$

and the plug-in variance estimator can be expressed as

$$\hat{V}(\hat{\theta}_{Id}) = \frac{1}{n} \frac{1}{n-1} \sum_{i=1}^{n} \left(\hat{d}_i - \bar{d}_n\right)^2, \tag{4.46}$$

where $\hat{d}_i = \hat{\mu}_g + (n_g/r_g)\delta_i (y_i - \hat{\mu}_g)$ *and* $\bar{d}_n = n^{-1} \sum_{i=1}^{n} \hat{d}_i$.

If a stochastic imputation is used where an imputed value is randomly selected from the set of respondents in the same cell, then we can write

$$\hat{\theta}_{Is} = n^{-1} \sum_{g=1}^{G} \sum_{i \in A_g} \{\delta_i y_i + (1-\delta_i) y_i^*\}. \tag{4.47}$$

Such an imputation method is often called hot deck imputation *(within cells). Write*

$$\hat{\theta}_{Is} = \hat{\theta}_{Id} + n^{-1} \sum_{g=1}^{G} \sum_{i \in A_g} (1-\delta_i)\left(y_i^* - \hat{\mu}_g\right),$$

the variance of the first term can be estimated by (4.46) and the variance of the second term in (4.47) can be estimated by

$$n^{-2} \sum_{g=1}^{G} \sum_{i \in A_g} (1-\delta_i)\left(y_i^* - \hat{\mu}_g\right)^2,$$

if the imputed values are generated independently, conditional on the respondents.

An extension of Example 4.3 can be made by considering a general model for imputation

$$y_i \mid \mathbf{x}_i \overset{i.i.d.}{\sim} \{E(y_i \mid \mathbf{x}_i), V(y_i \mid \mathbf{x}_i)\}. \tag{4.48}$$

If $E(y_i \mid \mathbf{x}_i)$ is a known function of unknown parameters, such as $E(y_i \mid \mathbf{x}_i) = m(\mathbf{x}_i; \eta)$, then we can use the linearization technique as discussed in Kim and Rao (2009). If $E(y_i \mid \mathbf{x}_i)$ is unknown, then we can use a nonparametric regression technique as in Wang and Chen (2009). The nearest neighbor imputation is a special case of the nonparametric regression imputation. See Chen and Shao (2001) and Beaumont and Bocci (2009) for variance estimation after nearest neighbor imputation.

4.4 Replication variance estimation

Replication variance estimation is a simulation-based method using replicates of the given point estimator. Let $\hat{\theta}_n$ be the complete-sample estimator of θ. The replication variance estimator of $\hat{\theta}_n$ takes the form of

$$\hat{V}_{rep}(\hat{\theta}_n) = \sum_{k=1}^{L} c_k \left(\hat{\theta}_n^{(k)} - \hat{\theta}_n \right)^2 \tag{4.49}$$

where L is the number of replicates, c_k is the replication factor associated with replication k, and $\hat{\theta}_n^{(k)}$ is the k-th replicate of $\hat{\theta}_n$. If $\hat{\theta}_n = \sum_{i=1}^n y_i/n$, then we can write $\hat{\theta}_n^{(k)} = \sum_{i=1}^n w_i^{(k)} y_i$ for some replication weights $w_1^{(k)}, w_2^{(k)}, \cdots, w_n^{(k)}$. For example, in the jackknife method, we have $L = n$, $c_k = (n-1)/n$, and

$$w_i^{(k)} = \begin{cases} (n-1)^{-1} & \text{if } i \neq k \\ 0 & \text{if } i = k. \end{cases}$$

If we use the above jackknife method to $\hat{\theta}_n = \sum_{i=1}^n y_i/n$, the resulting jackknife estimator in (4.49) is algebraically equivalent to $n^{-1}(n-1)^{-1} \sum_{i=1}^n (y_i - \bar{y}_n)^2$. Furthermore, if we apply the jackknife to $\hat{\theta}_n = \sum_{i=1}^n y_i / (\sum_{i=1}^n x_i)$, then

$$\hat{V}_{rep}(\hat{\theta}_n) = \frac{1}{n} \frac{1}{n-1} \sum_{k=1}^{n} \left(\frac{1}{\bar{x}_n^{(k)}} \right)^2 \left(y_k - \hat{\theta}_n x_k \right)^2$$

which is close to the linearized variance estimator

$$\hat{V}_l(\hat{\theta}_n) = \frac{1}{(\bar{x}_n)^2} \frac{1}{n} \frac{1}{n-1} \sum_{k=1}^{n} \left(y_k - \hat{\theta}_n x_k \right)^2.$$

In general, under some regularity conditions, for $\hat{\theta}_n = g(\bar{y}_n)$ which is a smooth function of $\bar{y}_n$, the replication variance estimator of $\hat{\theta}_n$, defined by

$$\hat{V}_{rep}\left(\hat{\theta}_n\right) = \sum_{k=1}^{L} c_k \left(\hat{\theta}_n^{(k)} - \hat{\theta}_n \right)^2, \tag{4.50}$$

where $\hat{\theta}_n^{(k)} = g(\bar{y}_n^{(k)})$, satisfies

$$\hat{V}_{rep}\left(\hat{\theta}_n\right) \cong \{g'(\bar{y}_n)\}^2 \hat{V}_{rep}(\bar{y}_n).$$

That is, the replication variance estimator is asymptotically equivalent to the linearized variance estimator.

We now look at parameters other than regression parameters such as β_0, β_1 and σ_e^2, often denoted by θ. Denote one such nonregression parameter, an example of which could be a proportion parameter, as ψ, and estimate it by $\hat{\psi}_n$ obtained by solving an estimating equation $\sum_{i=1}^n U(\psi; y_i) = 0$. A consistent variance estimator can be obtained by the sandwich formula in (4.34). If we want to use the replication method of the form (4.49), we can construct the replication variance estimator of $\hat{\psi}_n$ by

$$\hat{V}_{rep}(\hat{\psi}_n) = \sum_{k=1}^{L} c_k \left(\hat{\psi}_n^{(k)} - \hat{\psi}_n \right)^2, \tag{4.51}$$

where $\hat{\psi}_n^{(k)}$ is computed by

$$\hat{U}^{(k)}(\psi) \equiv \sum_{i=1}^{n} w_i^{(k)} U(\psi; y_i) = 0. \tag{4.52}$$

The replication variance estimator (4.51) is asymptotically equivalent to the sandwich-type variance estimator. Note that the replication variance estimator does require computing partial derivatives in variance estimation. In some cases, finding the solution to (4.52) can be computationally challenging. In this case, the one-step approximation method can be used. The one-step approximation method is based on Taylor expansion, as described below.

$$\begin{aligned} 0 &= \hat{U}^{(k)}(\hat{\psi}^{(k)}) \\ &\cong \hat{U}^{(k)}(\hat{\psi}) + \dot{U}^{(k)}(\hat{\psi})\left(\hat{\psi}^{(k)} - \hat{\psi}\right), \end{aligned}$$

where $\dot{U}^{(k)}(\psi) = \partial \hat{U}^{(k)}(\psi)/\partial \psi'$. Thus, the one-step approximation of $\hat{\psi}^{(k)}$ is to use

$$\hat{\psi}_1^{(k)} = \hat{\psi} - \left\{\dot{U}^{(k)}(\hat{\psi})\right\}^{-1} \hat{U}^{(k)}(\hat{\psi}) \tag{4.53}$$

or, even more simply, use

$$\hat{\psi}_1^{(k)} = \hat{\psi} - \{\dot{U}(\hat{\psi})\}^{-1} \hat{U}^{(k)}(\hat{\psi}). \tag{4.54}$$

The replication variance estimator of (4.54) is algebraically equivalent to

$$\{\dot{U}(\hat{\psi})\}^{-1} \left[\sum_{k=1}^{n} c_k \left\{\hat{U}^{(k)}(\hat{\psi}) - \hat{U}(\hat{\psi})\right\}^{\otimes 2}\right] \{\dot{U}(\hat{\psi})\}^{-1},$$

which is very close to the sandwich variance formula in (4.34).

We now discuss replication variance estimation after a deterministic imputation. The replication method can be applied to the deterministic part naturally. For example, in the regression imputation estimator of the form (4.26), the replication variance estimator can be computed by

$$\hat{V}_{rep}\left(\hat{\theta}_{Id}\right) = \sum_{k=1}^{L} c_k \left(\hat{\theta}_{Id}^{(k)} - \hat{\theta}_{Id}\right)^2, \tag{4.55}$$

where the k-th replicate of the imputed estimator is

$$\hat{\theta}_{Id}^{(k)} = \sum_{i=1}^{n} w_i^{(k)} \left\{\delta_i y_i + (1-\delta_i)\left(\hat{\beta}_0^{(k)} + \hat{\beta}_1^{(k)} x_i\right)\right\}$$

and $(\hat{\beta}_0^{(k)}, \hat{\beta}_1^{(k)})$ is the solution to

$$\sum_{i=1}^{n} w_i^{(k)} \delta_i (y_i - \beta_0 - \beta_1 x_i)(1, x_i) = (0,0).$$

To explain the validity of the above variance estimator, note that we can write $\hat{\theta}_{Id} = \hat{\theta}_{Id}(\hat{\beta})$ and

$$\hat{\theta}_{Id}(\hat{\beta}) \cong \hat{\theta}_{Id}(\beta) + d\left(\hat{\beta} - \beta\right) \tag{4.56}$$

for some $d = E\left\{\partial \hat{\theta}_{Id}(\beta)/\partial \beta\right\}$, where

$$\hat{\theta}_{Id}(\beta) = n^{-1} \sum_{i=1}^{n} \{\delta_i y_i + (1-\delta_i)(\beta_0 + \beta_1 x_i)\}.$$

By (4.56), we can write

$$V\left\{\hat{\theta}_{Id}(\hat{\beta})\right\} \cong V\left\{\hat{\theta}_{Id}(\beta)\right\} + 2Cov\left\{\hat{\theta}_{Id}(\beta), d\left(\hat{\beta} - \beta\right)\right\} + V\left\{d\left(\hat{\beta} - \beta\right)\right\}. \tag{4.57}$$

Now, writing $\hat{\theta}_{Id}^{(k)} = \hat{\theta}_{Id}^{(k)}\left(\hat{\beta}^{(k)}\right)$, where

$$\hat{\theta}_{Id}^{(k)}(\beta) = \sum_{i=1}^{n} w_i^{(k)} \{\delta_i y_i + (1-\delta_i)(\beta_0 + \beta_1 x_i)\},$$

we can apply the Taylor linearization to get

$$\hat{\theta}_{Id}^{(k)}(\hat{\beta}^{(k)}) \cong \hat{\theta}_{Id}^{(k)}(\beta) + d^{(k)}\left(\hat{\beta}^{(k)} - \beta\right), \tag{4.58}$$

where $d^{(k)} = \partial\hat{\theta}_{Id}^{(k)}(\beta)/\partial\beta$ evaluated at $\beta = \hat{\beta}$. Since $\sum_{k=1}^{L} c_k \left(d^{(k)} - d\right)^2$ converges to the variance of the partial derivatives of $\hat{\theta}_{Id}(\beta)$, which is $O(n^{-1})$, we have $d^{(k)} - d = o(1)$ and (4.58) becomes

$$\hat{\theta}_{Id}^{(k)}(\hat{\beta}^{(k)}) \cong \hat{\theta}_{Id}^{(k)}(\beta) + d\left(\hat{\beta}^{(k)} - \beta\right). \tag{4.59}$$

Combining (4.56) with (4.59), we have

$$\hat{\theta}_{Id}^{(k)}(\hat{\beta}^{(k)}) - \hat{\theta}_{Id}(\hat{\beta}) \cong \hat{\theta}_{Id}^{(k)}(\beta) - \hat{\theta}_{Id}(\beta) + d\left(\hat{\beta}^{(k)} - \hat{\beta}\right). \tag{4.60}$$

Therefore, we can write

$$\begin{aligned}
\sum_{k=1}^{L} c_k \left\{\hat{\theta}_{Id}^{(k)}(\hat{\beta}^{(k)}) - \hat{\theta}_{Id}(\hat{\beta})\right\}^2 &\cong \sum_{k=1}^{L} c_k \left\{\hat{\theta}_{Id}^{(k)}(\beta) - \hat{\theta}_{Id}(\beta)\right\}^2 \\
&+ 2d\sum_{k=1}^{L} c_k \left\{\hat{\theta}_{Id}^{(k)}(\beta) - \hat{\theta}_{Id}(\beta)\right\}\left(\hat{\beta}^{(k)} - \hat{\beta}\right) \\
&+ d\sum_{k=1}^{L} c_k \left(\hat{\beta}^{(k)} - \hat{\beta}\right)^{\otimes 2} d'
\end{aligned}$$

which estimates the variance term in (4.57).

If a stochastic imputation is used such that the imputed estimator can be written as

$$\hat{\theta}_{I,s} = n^{-1}\sum_{i=1}^{n}\left\{\delta_i y_i + (1-\delta_i)\left(\hat{\beta}_0 + \hat{\beta}_1 x_i + \hat{e}_i^*\right)\right\}$$

then the k-th replicate of $\hat{\theta}_{Is}$ can be computed by

$$\hat{\theta}_{Is}^{(k)} = \sum_{i=1}^{n} w_i^{(k)}\left\{\delta_i y_i + (1-\delta_i)\left(\hat{\beta}_0^{(k)} + \hat{\beta}_1^{(k)} x_i + \hat{e}_i^*\right)\right\}.$$

The replication variance estimator defined by

$$\sum_{i=1}^{L} c_k \left(\hat{\theta}_{Is}^{(k)} - \hat{\theta}_{Is}\right)^2$$

can be shown to be consistent for the variance of the imputed estimator $\hat{\theta}_{Is}$. For details, see Rao and Shao (1992) and Rao and Sitter (1995).

Example 4.4. *We now return to the setup of Example 3.11. In this case, the deterministically imputed estimator of $\theta = E(Y)$ is constructed by*

$$\hat{\theta}_{Id} = n^{-1}\sum_{i=1}^{n}\{\delta_i y_i + (1-\delta_i)\hat{p}_{0i}\} \tag{4.61}$$

where $\hat{p}_{0i}$ is the predictor of y_i given $\mathbf{x}_i$ and $\delta_i = 0$. That is,

$$\hat{p}_{0i} = \frac{p(\mathbf{x}_i;\hat{\beta})\{1-\pi(\mathbf{x}_i,1;\hat{\phi})\}}{\{1-p(\mathbf{x}_i;\hat{\beta})\}\{1-\pi(\mathbf{x}_i,0;\hat{\phi})\}+p(\mathbf{x}_i;\hat{\beta})\{1-\pi(\mathbf{x}_i,1;\hat{\phi})\}},$$

where $\hat{\beta}$ and $\hat{\phi}$ are jointly estimated by the EM algorithm described in Example 3.11. For replication variance estimation, we can use (4.55) with

$$\hat{\theta}_{Id}^{(k)} = \sum_{i=1}^{n} w_i^{(k)} \left\{ \delta_i y_i + (1-\delta_i)\hat{p}_{0i}^{(k)} \right\}. \tag{4.62}$$

In the above formula,

$$\hat{p}_{0i}^{(k)} = \frac{p(\mathbf{x}_i;\hat{\beta}^{(k)})\{1-\pi(\mathbf{x}_i,1;\hat{\phi}^{(k)})\}}{\{1-p(\mathbf{x}_i;\hat{\beta}^{(k)})\}\{1-\pi(\mathbf{x}_i,0;\hat{\phi}^{(k)})\}+p(\mathbf{x}_i;\hat{\beta}^{(k)})\{1-\pi(\mathbf{x}_i,1;\hat{\phi}^{(k)})\}},$$

and $(\hat{\beta}^{(k)},\hat{\phi}^{(k)})$ is obtained by solving the mean score equations with original weights replaced by replication weights $w_i^{(k)}$. That is, $(\hat{\beta}^{(k)},\hat{\phi}^{(k)})$ is the solution to

$$\begin{aligned}
\bar{S}_1^{(k)}(\beta,\phi) &\equiv \sum_{\delta_i=1} w_i^{(k)}\{y_i - p(\mathbf{x}_i;\beta)\}\mathbf{x}_i + \sum_{\delta_i=0} w_i^{(k)} \sum_{y=0}^{1} w_{iy}^{*}(\beta,\phi)\{y-p(\mathbf{x}_i;\beta)\}\mathbf{x}_i = 0 \\
\bar{S}_2^{(k)}(\beta,\phi) &\equiv \sum_{\delta_i=1} w_i^{(k)}\{\delta_i - \pi(\mathbf{x}_i,y_i;\phi)\}(\mathbf{x}_i',y_i)' + \sum_{\delta_i=0} w_i^{(k)} \sum_{y=0}^{1} w_{iy}^{*}(\beta,\phi)\{\delta_i-\pi(\mathbf{x}_i,y;\beta)\}(\mathbf{x}_i',y)' = 0
\end{aligned}$$

and

$$w_{iy}^{*}(\beta,\phi) = \frac{p(\mathbf{x}_i;\beta)\{1-\pi(\mathbf{x}_i,1;\phi)\}}{\{1-p(\mathbf{x}_i;\beta)\}\{1-\pi(\mathbf{x}_i,0;\phi)\}+p(\mathbf{x}_i;\beta)\{1-\pi(\mathbf{x}_i,1;\phi)\}}.$$

Thus, we may apply the same EM algorithm to compute $(\hat{\beta}^{(k)},\hat{\phi}^{(k)})$ iteratively.

Under MAR, $\hat{p}_{0i}^{(k)} = p(\mathbf{x}_i;\hat{\beta}^{(k)})$ and $\hat{\beta}^{(k)}$ is computed by

$$\sum_{i=1}^{n} w_i^{(k)} \delta_i \{y_i - p_i(\hat{\beta}^{(k)})\}\mathbf{x}_i = \mathbf{0}. \tag{4.63}$$

Instead of solving (4.63), one can use a one-step approximation

$$\hat{\beta}^{(k)} = \hat{\beta} + \left\{ \sum_{i=1}^{n} w_i^{(k)} \delta_i \hat{p}_i (1-\hat{p}_i)\mathbf{x}_i\mathbf{x}_i' \right\}^{-1} \sum_{i=1}^{n} w_i^{(k)} \delta_i (y_i - \hat{p}_i)\mathbf{x}_i.$$

4.5 Multiple imputation

Multiple imputation, proposed by Rubin (1978) and further developed by Rubin (1987), is an approach of generating imputed values with simplified variance estimation. In this procedure, Bayesian methods of generating imputed values, discussed in Section 3.6, are considered, where $m > 1$ imputed values are generated from the posterior predictive distribution as in (3.63). When the observed posterior distribution $p_{\text{obs}}(\eta \mid \mathbf{y}_{\text{obs}}, \delta)$ is available, we have the following two steps in multiple imputation:

[Step 1] Generate $\eta_p^{*(1)}, \cdots, \eta_p^{*(m)}$ independently from $p_{\text{obs}}(\eta \mid \mathbf{y}_{\text{obs}}, \delta)$.

[Step 2] Given the j-th parameter value $\eta_p^{*(j)} = (\theta_p^{*(j)}, \phi_p^{*(j)})$ generated from [Step 1], generate $\mathbf{y}_{\text{mis}}^{*(j)}$ from the conditional distribution $h(\mathbf{y}_{\text{mis}} \mid \mathbf{y}_{\text{obs}}, \delta; \eta_p^{*(j)})$, where

$$h(\mathbf{y}_{mis} \mid \mathbf{y}_{\text{obs}}, \delta; \eta_p^{*(j)}) = \frac{f(\mathbf{y}; \theta_p^{*(j)}) P(\delta \mid \mathbf{y}; \phi_p^{*(j)})}{\int f(\mathbf{y}; \theta_p^{*(j)}) P(\delta \mid \mathbf{y}; \phi_p^{*(j)}) d\mathbf{y}_{mis}}. \tag{4.64}$$

Use the imputed values, $\mathbf{y}^{*(1)}, \cdots, \mathbf{y}^{*(m)}$, and the multiple imputation (MI) estimator of η, denoted by $\hat{\eta}_{MI}$, can be obtained by

$$\hat{\eta}_{MI} = \frac{1}{m} \sum_{j=1}^{m} \hat{\eta}^{(j)}, \tag{4.65}$$

where $\hat{\eta}^{(j)}$ is obtained by solving $S(\eta; \mathbf{y}^{*(j)}) = 0$ for η. Note that $\hat{\eta}^{(j)}$ is an one-step update of $\hat{\eta}_p^{*(j)}$. If another parameter of interest, denoted by ψ, is defined by $E\{U(\psi)\} = 0$. In this case, the MI estimator of ψ, denoted by $\hat{\psi}_{MI}$, can be obtained by $\hat{\psi}_{MI} = m^{-1} \sum_{j=1}^{m} \hat{\psi}_I^{(j)}$ where $\hat{\psi}_I^{(j)}$ is obtained by solving $U(\psi; \mathbf{y}^{*(j)}) = 0$ for ψ.

In multiple imputation, a simple variance estimation formula was proposed by Rubin (1987), which is given by

$$\hat{V}_{MI}(\hat{\psi}_{MI}) = W_m + \left(1 + \frac{1}{m}\right) B_m, \tag{4.66}$$

where

$$W_m = \frac{1}{m} \sum_{j=1}^{m} \hat{V}_I^{(j)}(\hat{\psi}),$$

with $\hat{V}_I^{(j)}(\hat{\psi})$ being the imputed version of the complete-sample variance estimator of $\hat{\psi}$ based on the j-th imputed data, and

$$B_m = \frac{1}{m-1} \sum_{j=1}^{m} \left(\hat{\psi}_I^{(j)} - \hat{\psi}_{MI}\right)^{\otimes 2}.$$

For example, if the parameter of interest is the population mean of y, then the MI estimator of $\psi = E(Y)$ is

$$\hat{\psi}_{MI} = \frac{1}{m} \sum_{j=1}^{m} \hat{\psi}_I^{(j)} = \frac{1}{m} \sum_{j=1}^{m} \left[\frac{1}{n} \sum_{i=1}^{n} \left\{ \delta_i y_i + (1 - \delta_i) y_i^{*(j)} \right\} \right].$$

Write $\hat{\psi}_I^{*(j)}$ as the sample mean of $\tilde{y}_i^{(j)} = \delta_i y_i + (1 - \delta_i) y_i^{*(j)}$, and Rubin's variance estimator of $\hat{\psi}_{MI}$ is given by (4.66) with

$$\hat{V}_I^{(j)}(\hat{\psi}) = \frac{1}{n} \frac{1}{n-1} \sum_{i=1}^{n} \left(\tilde{y}_i^{(j)} - \bar{y}^{(j)}\right)^2,$$

and $B_m = (m-1)^{-1} \sum_{j=1}^{m} \left(\hat{\psi}_I^{(j)} - \hat{\psi}_{MI}\right)^2$, where $\bar{y}^{(j)} = n^{-1} \sum_{i=1}^{n} \tilde{y}_i^{(j)}$. The variance formula (4.66) is easy to compute since we only need to apply the complete-sample point estimators and the complete-sample variance estimators to the imputed data set, treating imputed values as if they were real observations.

In multiple imputation, m independent realizations of η, denoted by $\eta_p^{*(1)}, \cdots, \eta_p^{*(m)}$, are first generated from $p_{\text{obs}}(\eta \mid \mathbf{y}_{\text{obs}}, \delta)$ and the solution $\hat{\eta}^{(j)}$ is computed by solving $S(\eta; \mathbf{y}_{\text{obs}}, \mathbf{y}_{\text{mis}}^{*(j)}) = 0$ for η when $\mathbf{y}_{\text{mis}}^{*(j)}$ is generated from $h(\mathbf{y}_{mis} \mid \mathbf{y}_{\text{obs}}, \delta; \eta_p^{*(j)})$ in (4.64). The MI estimator $\hat{\eta}_{MI}$ is computed by (4.65). To discuss the asymptotic properties of $\hat{\eta}_{MI}$, we first establish the following lemma.

Lemma 4.4. *Let $S_{\mathrm{I}}^{*}(\eta \mid \hat{\eta}_p) = S_{\mathrm{com}}(\eta; \mathbf{y}^*)$ be the imputed score function evaluated with $\mathbf{y}^* = (\mathbf{y}_{\mathrm{obs}}, \mathbf{y}_{\mathrm{mis}}^*)$ where $\mathbf{y}_{\mathrm{mis}}^*$ is generated from $h(\mathbf{y}_{\mathrm{mis}} \mid \mathbf{y}_{\mathrm{obs}}, \delta; \hat{\eta}_p)$ in (4.64). Assume that $\hat{\eta}_p$ converges in probability to η_0. Then, under some regularity conditions,*

$$S_I^*(\eta_0 \mid \hat{\eta}_p) \cong \mathcal{I}_{\mathrm{com}}(\hat{\eta}_{MLE} - \eta_0) + \mathcal{I}_{\mathrm{mis}}(\hat{\eta}_p - \hat{\eta}_{MLE}) + S_{\mathrm{mis}}^*(\eta_0 \mid \hat{\eta}_p) \tag{4.67}$$

where $S_{\mathrm{mis}}^*(\eta_0 \mid \hat{\eta}_p) = S_{\mathrm{I}}^*(\eta_0 \mid \hat{\eta}_p) - \bar{S}(\eta_0 \mid \hat{\eta}_p)$, with $\bar{S}(\eta_0 \mid \hat{\eta}_p)$ defined in (4.7). Also, the solution $\hat{\eta}^*$ to $S_{\mathrm{I}}^*(\eta \mid \hat{\eta}_p) = 0$ satisfies

$$\hat{\eta}^* - \eta_0 \cong (\hat{\eta}_{MLE} - \eta_0) + \mathcal{J}_{\mathrm{mis}}(\hat{\eta}_p - \hat{\eta}_{MLE}) + \mathcal{I}_{\mathrm{com}}^{-1} S_{\mathrm{mis}}^*(\eta_0 \mid \hat{\eta}_p), \tag{4.68}$$

where $\mathcal{J}_{\mathrm{mis}} = \mathcal{I}_{\mathrm{com}}^{-1}\mathcal{I}_{\mathrm{mis}}$ is the fraction of missing information.

Proof. Writing

$$S_{\mathrm{I}}^*(\eta_0 \mid \hat{\eta}_p) = \bar{S}(\eta_0 \mid \hat{\eta}_p) + S_{\mathrm{mis}}^*(\eta_0 \mid \hat{\eta}_p)$$

and using (4.10), we have (4.67). Now, use the same argument for (4.11), and the solution to $S^*(\eta \mid \hat{\eta}_p) = 0$ satisfies

$$\hat{\eta}^* - \eta_0 \cong \mathcal{I}_{\mathrm{com}}^{-1}\{S_{\mathrm{I}}^*(\eta_0 \mid \hat{\eta}_p)\},$$

which proves (4.68). □

Note that the three terms in (4.68) are mutually independent. In multiple imputation, the solution $\hat{\eta}^{(j)}$ to $S_I^{*(j)}(\eta \mid \hat{\eta}_p) = S_{\mathrm{com}}(\eta; \mathbf{y}^{*(j)})$, where $\mathbf{y}^{*(j)} = (\mathbf{y}_{\mathrm{obs}}, \mathbf{y}_{\mathrm{mis}}^{*(j)})$ and $\mathbf{y}_{\mathrm{mis}}^{*(j)}$ are generated from $h(\mathbf{y}_{\mathrm{mis}} \mid \mathbf{y}_{\mathrm{obs}}; \eta_p^{*(j)})$, satisfies

$$\hat{\eta}^{(j)} - \eta_0 \cong (\hat{\eta}_{MLE} - \eta_0) + \mathcal{J}_{\mathrm{mis}}\left(\eta_p^{*(j)} - \hat{\eta}_{MLE}\right) + \mathcal{I}_{\mathrm{com}}^{-1} S_{\mathrm{mis}}^{*(j)}(\eta_0 \mid \eta_p^{*(j)}), \tag{4.69}$$

where $S_{\mathrm{mis}}^{*(j)}(\eta_0 \mid \eta_p^{*(j)}) = S_{\mathrm{I}}^{*(j)}(\eta_0 \mid \eta_p^{*(j)}) - \bar{S}(\eta_0 \mid \eta_p^{*(j)})$. Taking the sample mean of the m solutions, we have

$$\hat{\eta}_{MI} \cong \hat{\eta}_{MLE} + m^{-1}\sum_{j=1}^{m} \mathcal{J}_{\mathrm{mis}}\left(\eta_p^{*(j)} - \hat{\eta}_{MLE}\right) + m^{-1}\sum_{j=1}^{m} \mathcal{I}_{\mathrm{com}}^{-1} S_{\mathrm{mis}}^{*(j)}(\eta_0 \mid \eta_p^{*(j)}). \tag{4.70}$$

If the posterior distribution of η given the observed data $(\mathbf{y}_{\mathrm{obs}}, \delta)$ is asymptotically normal with mean $\hat{\eta}_{MLE}$ and variance matrix $\{I_{\mathrm{obs}}(\hat{\eta}_{MLE})\}^{-1}$ almost surely on $(\mathbf{y}_{\mathrm{obs}}, \delta)$, then the second term in (4.70) is asymptotically distributed as

$$m^{-1}\sum_{j=1}^{m} \mathcal{J}_{\mathrm{mis}}\left(\eta_p^{*(j)} - \hat{\eta}_{MLE}\right) \sim N\left(0, m^{-1}\mathcal{J}_{\mathrm{mis}}\mathcal{I}_{\mathrm{obs}}^{-1}\mathcal{J}_{\mathrm{mis}}'\right).$$

Also, by (4.14),

$$m^{-1}\sum_{j=1}^{m} S_{\mathrm{mis}}^{*(j)}(\eta_0 \mid \eta_p^{*(j)}) \mid (\mathbf{y}_{\mathrm{obs}}, \delta, \eta_p^*) \sim (0, m^{-1}\mathcal{I}_{\mathrm{mis}}),$$

where $\eta_p^* = (\eta_p^{*(1)}, \cdots, \eta_p^{*(m)})$. Thus, the MI estimator $\hat{\eta}_{MI}$ is approximately unbiased for η_0 and has the asymptotic variance

$$V(\hat{\eta}_{MI}) \cong \mathcal{I}_{\mathrm{obs}}^{-1} + m^{-1}\mathcal{J}_{\mathrm{mis}}\mathcal{I}_{\mathrm{obs}}^{-1}\mathcal{J}_{\mathrm{mis}}' + m^{-1}\mathcal{I}_{\mathrm{com}}^{-1}\mathcal{I}_{\mathrm{mis}}\mathcal{I}_{\mathrm{com}}^{-1}. \tag{4.71}$$

Comparing (4.16) with (4.71), the MI estimator has greater variance than the imputation estimator using the MLE. The first additional variance term, $m^{-1}\mathcal{J}_{\mathrm{mis}}\mathcal{I}_{\mathrm{obs}}^{-1}\mathcal{J}_{\mathrm{mis}}'$, comes from the posterior step, as the parameters $\eta_p^{*(1)}, \cdots, \eta_p^{*(m)}$ are generated from the observed posterior distribution. The second additional variance term, $m^{-1}\mathcal{I}_{\mathrm{com}}^{-1}\mathcal{I}_{\mathrm{mis}}\mathcal{I}_{\mathrm{com}}^{-1}$, represents the variance due to the imputation step given the realized parameter values.

The following theorem presents the conditions for the asymptotic unbiasedness of Rubin's variance estimator.

Theorem 4.3. *Assume that m preliminary values of η, denoted by $\eta_p^{*(1)}, \cdots, \eta_p^{*(m)}$, are independently generated from a normal distribution with mean $\hat{\eta}_{MLE}$ and variance matrix $\{I_{\text{obs}}(\hat{\eta}_{MLE})\}^{-1}$. Assume that the complete sample variance estimator $\hat{V}$ satisfies*

$$E\left\{\hat{V}_I^{(j)}\right\} \cong \mathcal{I}_{\text{com}}^{-1}, \tag{4.72}$$

where $\hat{V}_I^{(j)}$ is the naive variance estimator computed by applying $\hat{V}$ to the j-th imputed data $\mathbf{y}^{*(j)}$. Then, Rubin's variance estimator (4.66) is asymptotically unbiased for the variance of the MI estimator $\hat{\eta}_{MI}$.

Proof. By (4.31) and (4.69), we have

$$E(B_m) = V(\hat{\eta}^{(1)}) - Cov(\hat{\eta}^{(1)}, \hat{\eta}^{(2)}) = \mathcal{J}_{\text{mis}}\mathcal{I}_{\text{obs}}^{-1}\mathcal{J}_{\text{mis}}' + \mathcal{I}_{\text{com}}^{-1}\mathcal{I}_{\text{mis}}\mathcal{I}_{\text{com}}^{-1}. \tag{4.73}$$

Thus, by assumption (4.72), we have

$$E\left\{\hat{V}_{MI}(\hat{\eta}_{MI})\right\} \cong \mathcal{I}_{\text{com}}^{-1} + \left(1+m^{-1}\right)\left(\mathcal{J}_{\text{mis}}\mathcal{I}_{\text{obs}}^{-1}\mathcal{J}_{\text{mis}}' + \mathcal{I}_{\text{com}}^{-1}\mathcal{I}_{\text{mis}}\mathcal{I}_{\text{com}}^{-1}\right). \tag{4.74}$$

Using matrix algebra, we have

$$(A+BCB')^{-1} = A^{-1} - A^{-1}B\left(C^{-1}+B'A^{-1}B\right)^{-1}B'A^{-1}$$

and

$$\left(C^{-1}+B'A^{-1}B\right)^{-1} = C - CB'\left(A+BCB'\right)^{-1}BC,$$

which leads to

$$(A+BCB')^{-1} = A^{-1} - A^{-1}BCB'A^{-1} + A^{-1}BCB'\left(A+BCB'\right)^{-1}BCB'A^{-1}.$$

Applying the above equality to $A = \mathcal{I}_{\text{com}}$, $B = I$, and $C = -\mathcal{I}_{\text{mis}}$, we have

$$I_{\text{obs}}^{-1} = \mathcal{I}_{\text{com}}^{-1} + \mathcal{J}_{\text{mis}}\mathcal{I}_{\text{obs}}^{-1}\mathcal{J}_{\text{mis}}' + \mathcal{I}_{\text{com}}^{-1}\mathcal{I}_{\text{mis}}\mathcal{I}_{\text{com}}^{-1} \tag{4.75}$$

and (4.74) reduces to

$$E\left\{\hat{V}_{MI}(\hat{\eta}_{MI})\right\} \cong \mathcal{I}_{\text{obs}}^{-1} + m^{-1}\mathcal{J}_{\text{mis}}\mathcal{I}_{\text{obs}}^{-1}\mathcal{J}_{\text{mis}}' + m^{-1}\mathcal{I}_{\text{com}}^{-1}\mathcal{I}_{\text{mis}}\mathcal{I}_{\text{com}}^{-1}, \tag{4.76}$$

which shows the asymptotic unbiasedness of Rubin's variance estimator by (4.71). □

Example 4.5. *(Univariate Normal distribution)*

Let $y_1, \cdots, y_n$ be IID observations from $N(\mu, \sigma^2)$ and only the first r elements are observed and the remaining $n-r$ elements are missing. Assume that the response mechanism is ignorable. To summarize, we have

$$y_1, \cdots, y_r \overset{i.i.d.}{\sim} N(\mu, \sigma^2). \tag{4.77}$$

In this case, the j-th posterior values of (μ, σ^2) are generated from

$$\sigma^{*(j)2} \mid \mathbf{y}_r \sim r\hat{\sigma}_r^2/\chi_{r-1}^2 \tag{4.78}$$

and

$$\mu^{*(j)} \mid (\mathbf{y}_r, \sigma^{*(j)2}) \sim N\left(\bar{y}_r, r^{-1}\sigma^{*(j)2}\right) \tag{4.79}$$

where $\mathbf{y}_r = (y_1, \cdots, y_r)$, $\bar{y}_r = r^{-1}\sum_{i=1}^r y_i$, and $\hat{\sigma}_r^2 = r^{-1}\sum_{i=1}^r (y_i - \bar{y}_r)^2$. Given the posterior sample $(\mu^{*(j)}, \sigma^{*(j)2})$, the imputed values are generated from

$$y_i^{*(j)} \mid \left(\mathbf{y}_r, \mu^{*(j)}, \sigma^{*(j)2}\right) \sim N\left(\mu^{*(j)}, \sigma^{*(j)2}\right) \tag{4.80}$$

independently for $i = r+1,\cdots,n$. The m imputed values are generated by independently repeating (4.78)-(4.80) m times.

Let $\theta = E(Y)$ be the parameter of interest and the MI estimator of θ can be expressed as

$$\hat{\theta}_{MI} = \frac{1}{m}\sum_{j=1}^{m}\hat{\theta}_I^{(j)}$$

where

$$\hat{\theta}_I^{(j)} = \frac{1}{n}\left\{\sum_{i=1}^{r} y_i + \sum_{i=r+1}^{n} y_i^{*(j)}\right\}.$$

Then,

$$\hat{\theta}_{MI} = \bar{y}_r + \frac{n-r}{nm}\sum_{j=1}^{m}\left(\mu^{*(j)} - \bar{y}_r\right) + \frac{1}{nm}\sum_{i=r+1}^{n}\sum_{j=1}^{m}\left(y_i^{*(j)} - \mu^{*(j)}\right). \tag{4.81}$$

Asymptotically, the first term has mean μ and variance $r^{-1}\sigma^2$, the second term has mean zero and variance $(1-r/n)^2\sigma^2/(mr)$, the third term has mean zero and variance $\sigma^2(n-r)/(n^2m)$, and the three terms are mutually independent. Thus, the variance of $\hat{\theta}_{MI}$ is

$$V\left(\hat{\theta}_{MI}\right) = \frac{1}{r}\sigma^2 + \frac{1}{m}\left(\frac{n-r}{n}\right)^2\left(\frac{1}{r}\sigma^2 + \frac{1}{n-r}\sigma^2\right), \tag{4.82}$$

which is consistent with the general result in (4.71).

For variance estimation, note that

$$\begin{aligned} V(y_i^{*(j)}) &= V(\bar{y}_r) + V(\mu^{*(j)} - \bar{y}_r) + V(y_i^{*(j)} - \mu^{*(j)}) \\ &= \frac{1}{r}\sigma^2 + \frac{1}{r}\sigma^2\left(\frac{r+1}{r-1}\right) + \sigma^2\left(\frac{r+1}{r-1}\right) \\ &\cong \sigma^2. \end{aligned}$$

Writing

$$\begin{aligned} \hat{V}_I^{(j)}(\hat{\theta}) &= n^{-1}(n-1)^{-1}\sum_{i=1}^{n}\left\{\tilde{y}_i^{*(j)} - \frac{1}{n}\sum_{k=1}^{n}\tilde{y}_k^{*(j)}\right\}^2 \\ &= n^{-1}(n-1)^{-1}\left\{\sum_{i=1}^{n}\left(\tilde{y}_i^{*(j)} - \mu\right)^2 - n\left(\frac{1}{n}\sum_{k=1}^{n}\tilde{y}_k^{*(j)} - \mu\right)^2\right\} \end{aligned}$$

where $\tilde{y}_i^* = \delta_i y_i + (1-\delta_i)y_i^{*(j)}$, we have

$$\begin{aligned} E\left\{\hat{V}_I^{(j)}(\hat{\theta})\right\} &= n^{-1}(n-1)^{-1}\left\{\sum_{i=1}^{n}E\left(\tilde{y}_i^{*(j)} - \mu\right)^2 - nV\left(\frac{1}{n}\sum_{k=1}^{n}\tilde{y}_k^{*(j)}\right)\right\} \\ &\cong n^{-1}(n-1)^{-1}\left[n\sigma^2 - n\left\{\frac{1}{r}\sigma^2 + \left(\frac{n-r}{n}\right)^2\left(\frac{1}{r}\sigma^2 + \frac{1}{n-r}\sigma^2\right)\right\}\right] \\ &\cong n^{-1}\sigma^2, \end{aligned}$$

which satisfies (4.72). By (4.31) and (4.82), we have

$$
\begin{aligned}
E(B_m) &= V\left(\hat{\theta}_I^{*(1)}\right) - Cov\left(\hat{\theta}_I^{*(1)}, \hat{\theta}_I^{*(2)}\right) \\
&= V\left\{\frac{n-r}{n}\left(\mu^{*(1)} - \bar{y}_r\right) + \frac{1}{n}\sum_{i=r+1}^{n}\left(y_i^{*(1)} - \mu^{*(1)}\right)\right\} \\
&\cong \left(\frac{n-r}{n}\right)^2\left(\frac{1}{r} + \frac{1}{n-r}\right)\sigma^2 \\
&= \left(\frac{1}{r} - \frac{1}{n}\right)\sigma^2.
\end{aligned}
$$

Thus, Rubin's variance estimator satisfies

$$
E\left\{\hat{V}_{MI}(\hat{\theta}_{MI})\right\} \cong \frac{1}{r}\sigma^2 + \frac{1}{m}\left(\frac{n-r}{n}\right)^2\left(\frac{1}{r}\sigma^2 + \frac{1}{n-r}\sigma^2\right) \cong V\left(\hat{\theta}_{MI}\right),
$$

which is consistent with the general result in (4.76).

Example 4.6. *Multiple imputation can be implemented nonparametrically using the Bayesian bootstrap of Rubin (1981), in which we first assume that an element of the population takes one of the values $d_1, \cdots, d_K$ with probability $p_1, \cdots, p_K$, respectively. That is, we assume*

$$
P(Y = d_k) = p_k, \quad \sum_{k=1}^{K} p_k = 1. \tag{4.83}
$$

Let $y_1, \cdots, y_n$ be an IID sample from (4.83) and let n_k be the number of y_i equal to d_k. The parameter is a vector of probabilities $\mathbf{p} = (p_1, \cdots, p_K)$, such that $\sum_{i=1}^{K} p_i = 1$. In this case, the population mean $\theta = E(Y)$ can be expressed as $\theta = \sum_{i=1}^{K} p_i d_i$ and we only need to estimate $\mathbf{p}$. If the improper Dirichlet prior with density proportional to $\prod_{k=1}^{K} p_k^{-1}$ is placed on the vector $\mathbf{p}$, then the posterior distribution of $\mathbf{p}$ is proportional to

$$
\prod_{k=1}^{K} p_k^{n_k - 1}
$$

which is a Dirichlet distribution with parameter $(n_1, \cdots, n_K)$. This posterior distribution can be simulated using $n-1$ independent uniform random numbers. Let $u_1, \cdots, u_{n-1}$ be IID $U(0,1)$, and let $g_i = u_{(i)} - u_{(i-1)}, i = 1, 2, \cdots, n-1$ where $u_{(k)}$ is the k-th order statistic of $u_1, \cdots, u_{n-1}$ with $u_{(0)} = 0$ and $u_{(n)} = 1$. Partition the $g_1, \cdots, g_n$ into K collections, with the k-th one having n_k elements, and let p_k be the sum of the g_i in the k-th collection. Then, the realized value of $p_1, \cdots, p_k$ follows a $(K-1)$-variate Dirichlet distribution with parameter $(n_1, \cdots, n_K)$. In particular, if $K = n$, then $(g_1, \cdots, g_n)$ is the vector of probabilities to attach to the data values $y_1, \cdots, y_n$ in that Bayesian bootstrap replication.

To implement Rubin's Bayesian bootstrap to multiple imputation, assume that the first r elements are observed and the remaining $n-r$ elements are missing. The imputed values can be generated with the following steps:

[Step 1] From $\mathbf{y}_r = (y_1, \cdots, y_r)$, generate $\mathbf{p}_r^* = (p_1^*, \cdots, p_r^*)$ from the posterior distribution using the Bayesian bootstrap as follows.

1. *Generate $u_1, \cdots, u_{r-1}$ independently from $U(0,1)$ and sort them to get $0 = u_{(0)} < u_{(1)} < \cdots < u_{(r-1)} < u_{(r)} = 1$.*
2. *Compute $p_i^* = u_{(i)} - u_{(i-1)}$, $i = 1, 2, \cdots, r-1$ and $p_r^* = 1 - \sum_{i=1}^{r-1} p_i^*$.*

[Step 2] Select the imputed value of y_i by

$$
y_i^* = \begin{cases} y_1 & \text{with probability } p_1^* \\ \cdots & \cdots \\ y_r & \text{with probability } p_r^* \end{cases}
$$

independently for each $i = r+1, \cdots, n$.

Using the above Bayesian bootstrap imputation m times independently, we can compute the MI point estimator and the MI variance estimator. For estimating $\theta = E(Y)$, *we can also establish (4.81) with* $\mu^{*(j)} = \sum_{i=1}^{r} p_i^{*(j)} y_i$, *where* $p_i^{*(j)}$ *is the realized value of selection probability* p_i^* *in [Step1] obtained from the j-th application of Rubin's Bayesian bootstrap method. Using a property of the Dirichlet distribution and the multinomial distribution, we can establish the same variance formula as in (4.82). Thus, the above Bayesian bootstrap imputation is asymptotically equivalent to the normal imputation. Also, it can be shown that the MI variance estimator is asymptotically unbiased.*

Rubin and Schenker (1986) proposed an approximation of this Bayesian bootstrap method, called the approximate Bayesian boostrap (ABB) method, which provides an alternative approach of generating imputed values from the empirical distribution. The ABB method can be described as follows:

[Step 1] From $\mathbf{y}_r = (y_1, \cdots, y_r)$, *generate a donor set* $\mathbf{y}_r^* = (y_1^*, \cdots, y_r^*)$ *by bootstrapping. That is, we select*

$$y_i^* = \begin{cases} y_1 & \text{with probability } 1/r \\ \cdots & \cdots \\ y_r & \text{with probability } 1/r \end{cases}$$

independently for each $i = 1, \cdots, r$.

[Step 2] From the donor set $\mathbf{y}_r^* = (y_1^*, \cdots, y_r^*)$, *select an imputed value of* y_i *by*

$$y_i^{**} = \begin{cases} y_1^* & \text{with probability } 1/r \\ \cdots & \cdots \\ y_r^* & \text{with probability } 1/r \end{cases}$$

independently for each $i = r+1, \cdots, n$.

Using the above ABB imputation m times independently, we can compute the MI point estimator and the MI variance estimator. For the estimation of $\theta = E(Y)$, *we can also establish (4.82) and the asymptotic unbiasedness of the MI variance estimator. Kim (2002) proposed further improvement of the ABB imputation for small sample sizes.*

Example 4.7. *(Regression model imputation)*

Under the linear regression model setup of Example 3.20, multiple imputation can be implemented by applying the steps [P-step]-[I-step] of Example 3.20 independently m times. At each repetition of the imputation $(j = 1, ..., m)$, *we can calculate the imputed version of the full sample estimators*

$$\hat{\beta}_I^{(j)} = \left(\sum_{i=1}^{n} \mathbf{x}_i \mathbf{x}_i'\right)^{-1} \left\{\sum_{i=1}^{r} \mathbf{x}_i y_i + \sum_{i=r+1}^{n} \mathbf{x}_i y_i^{*(j)}\right\}$$

and

$$\hat{V}_I^{(j)} = \left(\sum_{i=1}^{n} \mathbf{x}_i \mathbf{x}_i'\right)^{-1} \hat{\sigma}_I^{(j)2},$$

where

$$\hat{\sigma}_I^{(j)2} = (n-p)^{-1} \left\{\sum_{i=1}^{r} \left(y_i - \mathbf{x}_i' \hat{\beta}_I^{(j)}\right)^2 + \sum_{i=r+1}^{n} \left(y_i^{*(j)} - \mathbf{x}_i' \hat{\beta}_I^{(j)}\right)^2\right\}.$$

The proposed point estimator for the regression coefficient based on m repeated imputations is

$$\hat{\beta}_{MI} = \frac{1}{m} \sum_{j=1}^{m} \hat{\beta}_I^{(j)} \tag{4.84}$$

and the proposed estimator for the variance of $\hat{\beta}_{MI}$ is given by $\hat{V}_{MI}$ in (4.66). Since we can write

$$\begin{aligned}\hat{\beta}_{MI} &= \left(\sum_{i=1}^{n} \mathbf{x}_i \mathbf{x}'\right)^{-1} \left\{\sum_{i=1}^{r} \mathbf{x}_i \left[\mathbf{x}_i'\hat{\beta}_r + \left(y_i - \mathbf{x}_i'\hat{\beta}_r\right)\right]\right\} \\ &+ \left(\sum_{i=1}^{n} \mathbf{x}_i \mathbf{x}'\right)^{-1} \left\{\sum_{i=r+1}^{n} \mathbf{x}_i \left[\mathbf{x}_i'\hat{\beta}_r + m^{-1}\sum_{j=1}^{m}\left(\mathbf{x}_i'\left(\beta^{*(j)} - \hat{\beta}_r\right) + e_i^{*(j)}\right)\right]\right\},\end{aligned}$$

we can decompose it into three independent components as

$$\hat{\beta}_{MI} = \hat{\beta}_r + \frac{1}{m}\sum_{j=1}^{m}\sum_{i=r+1}^{n} \mathbf{h}_i \mathbf{x}_i'\left(\beta^{*(j)} - \hat{\beta}_r\right) + \frac{1}{m}\sum_{j=1}^{m}\sum_{i=r+1}^{n} \mathbf{h}_i e_i^{*(j)}, \tag{4.85}$$

where $\hat{\beta}_r = (X_r'X_r)^{-1}X_r'\mathbf{y}_r$, $\mathbf{h}_i = (X_n'X_n)^{-1}\mathbf{x}_i$ and $\beta^{*(j)}$ is the j-th realization of the parameter values generated from posterior distribution (3.68). The total variance is

$$\begin{aligned}V\left(\hat{\beta}_{MI}\right) &\cong (X_r'X_r)^{-1}\sigma^2 \\ &+m^{-1}(X_n'X_n)^{-1}(X_{n-r}'X_{n-r})(X_r'X_r)^{-1}(X_{n-r}'X_{n-r})(X_n'X_n)^{-1}\sigma^2 \\ &+m^{-1}(X_n'X_n)^{-1}(X_{n-r}'X_{n-r})(X_n'X_n)^{-1}\sigma^2,\end{aligned}$$

which is a special case of the general result in (4.71). Using some matrix algebra similar to (4.75),

$$\begin{aligned}(X_r'X_r)^{-1} &= (X_n'X_n - X_{n-r}'X_{n-r})^{-1} \\ &= (X_n'X_n)^{-1} + (X_n'X_n)^{-1}(X_{n-r}'X_{n-r})(X_n'X_n)^{-1} \\ &\quad + (X_n'X_n)^{-1}(X_{n-r}'X_{n-r})(X_r'X_r)^{-1}(X_{n-r}'X_{n-r})(X_n'X_n)^{-1},\end{aligned}$$

we can write

$$V(\hat{\beta}_{MI}) \cong (X_r'X_r)^{-1}\sigma^2 + m^{-1}\left\{(X_r'X_r)^{-1} - (X_n'X_n)^{-1}\right\}\sigma^2.$$

Also, it can be shown that

$$E(W_m) \cong (X_n'X_n)^{-1}\sigma^2,$$

and using an argument similar to that in (4.73),

$$\begin{aligned}E(B_m) &\cong (X_n'X_n)^{-1}(X_{n-r}'X_{n-r})(X_n'X_n)^{-1} \\ &\quad + (X_n'X_n)^{-1}(X_{n-r}'X_{n-r})(X_r'X_r)^{-1}(X_{n-r}'X_{n-r})(X_n'X_n)^{-1} \\ &= \left\{(X_r'X_r)^{-1} - (X_n'X_n)^{-1}\right\}\sigma^2.\end{aligned}$$

Thus, the asymptotic unbiasedness of the Rubin's variance estimator can be established. Kim (2004) showed that, instead of (3.67), if one uses

$$\sigma^{*2} \mid \mathbf{y}_r \overset{i.i.d.}{\sim} (r-p)\hat{\sigma}_r^2/\chi^2_{r-p+1}, \tag{4.86}$$

then the resulting MI variance estimator can have smaller bias in small sample sizes.

We now discuss an extension under the setup of Example 4.7. Suppose that the parameter of interest is not necessarily the regression coefficient β. Let $\hat{\theta}_n$ be the complete sample estimator of a parameter θ of the form $\hat{\theta}_n = \sum_{i=1}^{n}\alpha_i y_i$ for some coefficients α_i. The MI estimator of θ is computed by

$$\hat{\theta}_{MI} = \frac{1}{m}\sum_{j=1}^{m}\hat{\theta}_I^{(j)},$$

and, for the case of $\hat{\theta}_n = \sum_{i=1}^n \alpha_i y_i$, we can write

$$\hat{\theta}_{MI} = \hat{\theta}_{I,\infty} + \sum_{i=r+1}^{n} \alpha_i \mathbf{x}_i' \left(\bar{\beta}_m^* - \hat{\beta}_r \right) + \sum_{i=r+1}^{n} \alpha_i \bar{e}_i^*,$$

where $\hat{\theta}_{I,\infty} = \sum_{i=1}^r \alpha_i y_i + \sum_{i=r+1}^n \alpha_i \mathbf{x}_i' \hat{\beta}_r$, $\bar{\beta}_m = m^{-1} \sum_{j=1}^m \beta^{*(j)}$, and $\bar{e}_i^* = m^{-1} \sum_{j=1}^m e_i^{*(j)}$. Note that $\hat{\theta}_{I,\infty} = p\lim_{m\to\infty} \hat{\theta}_{MI}$. Thus, the total variance of $\hat{\theta}_{MI}$ is

$$V(\hat{\theta}_{MI}) = V(\hat{\theta}_{I,\infty}) + \frac{1}{m} \left\{ \alpha_{n-r}' X_{n-r} (X_r' X_r)^{-1} X_{n-r}' \alpha_{n-r} + \alpha_{n-r}' \alpha_{n-r} \right\} \sigma^2,$$

where $\alpha_{n-r} = (\alpha_{r+1}, \cdots, \alpha_n)'$.

To discuss variance estimation, first note that, by the same argument for (4.73),

$$E(B_m) = \left\{ \alpha_{n-r}' X_{n-r} (X_r' X_r)^{-1} X_{n-r}' \alpha_{n-r} + \alpha_{n-r}' \alpha_{n-r} \right\} \sigma^2.$$

The following theorem presents the conditions for the asymptotic unbiasedness of the MI variance estimator.

Theorem 4.4. *Assume that $E(\hat{V}_I^{(j)}) \cong V(\hat{\theta}_n)$ holds for each $j = 1, 2, \cdots, m$. Also, assume that*

$$V(\hat{\theta}_{I,\infty}) = V(\hat{\theta}_n) + V(\hat{\theta}_{I,\infty} - \hat{\theta}_n) \tag{4.87}$$

holds. Then, under a linear regression model, the MI variance estimator is asymptotically unbiased for the variance of the MI point estimator.

Proof. By (4.87), the variance of MI point estimator is decomposed into three terms:

$$V(\hat{\theta}_{MI}) = V(\hat{\theta}_n) + V(\hat{\theta}_{I,\infty} - \hat{\theta}_n) + V(\hat{\theta}_{MI} - \hat{\theta}_{I,\infty}).$$

The first term is estimated by W_m by assumption. The third term,

$$V(\hat{\theta}_{MI} - \hat{\theta}_{I,\infty}) = m^{-1} \left\{ \alpha_{n-r}' X_{n-r} (X_r' X_r)^{-1} X_{n-r}' \alpha_{n-r} + \alpha_{n-r}' \alpha_{n-r} \right\} \sigma^2,$$

is estimated by $m^{-1} B_m$. It remains to be shown that the second term is estimated by B_m. Since $\hat{\theta}_{I,\infty} - \hat{\theta}_n = \alpha_{n-r}'(X_{n-r}\hat{\beta}_r - \mathbf{y}_{n-r})$, we have

$$V(\hat{\theta}_{I,\infty} - \hat{\theta}_n) = \left\{ \alpha_{n-r}' X_{n-r} (X_r' X_r)^{-1} X_{n-r}' \alpha_{n-r} + \alpha_{n-r}' \alpha_{n-r} \right\} \sigma^2 = E(B_m),$$

and so the MI variance estimator is asymptotically unbiased. □

Condition (4.87) is crucial for the asymptotic unbiasedness of the MI variance estimator. Meng (1994) called the condition *congeniality*. The congeniality condition is not always achieved. Kim et al. (2006a) discuss sufficient conditions for the congeniality under the linear regression models.

Example 4.8. *Consider the bivariate data (x_i, y_i) of size $n = 200$ where x_i is always observed and y_i is subject to missingness. The sampling distribution of (x_i, y_i) is $x_i \sim N(3,1)$ and $y_i = -2 + x_i + e_i$ with $e_i \sim N(0,1)$. Multiple imputation can be used to estimate $\theta_1 = E(Y)$ and $\theta_2 = Pr(Y < 1)$. To test the performance, a small simulation study was performed. The response mechanism is uniform with response rate 0.6. In estimating θ_2, we used a method-of-moment estimator $\hat{\theta}_2 = n^{-1} \sum_{i=1}^n I(y_i < 1)$ under complete response. An unbiased estimator for the variance of $\hat{\theta}_2$ is then $\hat{V}_2 = (n-1)^{-1} \hat{\theta}_2 (1 - \hat{\theta}_2)$. Multiple imputation with size $m = 50$ was used. After multiple imputation, Rubin's variance formula was used.*

Table 4.1 presents the simulation results for the multiple imputation point estimators. For comparison, we have also computed the complete-sample point estimators. Table 4.2 presents the performance of the multiple imputation variance estimators. The t-statistic is computed to test the

Table 4.1 Simulation results of the MI point estimators

Parameter	Mean	$V(\hat{\theta}_n)$	$V(\hat{\theta}_{MI})$	$V(\hat{\theta}_{MI}-\hat{\theta}_n)$	$Cov(\hat{\theta}_n,\hat{\theta}_{MI}-\hat{\theta}_n)$
θ_1	1.00	0.0100	0.0134	0.0035	0.0000
θ_2	0.50	0.00129	0.00137	0.00046	-0.00019

Table 4.2 Simulation results of the MI variance estimators

Parameter	$E(W_m)$	$E(B_m)$	Rel. Bias (%)	t-statistics
$V(\hat{\theta}_1)$	0.0100	0.0033	-0.24	-0.08
$V(\hat{\theta}_2)$	0.00125	0.000436	23.08	7.48

significance of the Monte Carlo bias of the variance estimator. The MI variance estimator shows significant bias for estimating the variance of $\hat{\theta}_{2,MI}$.

For $\theta=\theta_1$,

$$V(\hat{\theta}_{MI})=V(\hat{\theta}_n)+V(\hat{\theta}_{MI}-\hat{\theta}_n)=\frac{\sigma_y^2}{n}+\left(\frac{1}{r}-\frac{1}{n}\right)\sigma_e^2=\frac{2}{200}+\left(\frac{1}{120}-\frac{1}{200}\right)\cdot 1=0.010+0.0033$$

which is roughly equal to $E(W_m)+E(B_m)$ in the simulation result. However, for $\theta=\theta_2$, we have

$$\begin{aligned} V(\hat{\theta}_{MI}) &= V(\hat{\theta}_n)+V(\hat{\theta}_{MI}-\hat{\theta}_n)+2Cov(\hat{\theta}_n,\hat{\theta}_{MI}-\hat{\theta}_n) \\ &\doteq 0.00129+0.00046+2\cdot(-0.00019)=0.00137, \end{aligned}$$

while

$$E(\hat{V}_{MI})=E(W_m)+(1+m^{-1})E(B_m)=0.00125+1.02\cdot 0.000436=0.00169>0.00137.$$

Thus, the MI variance estimator overestimates the variance because it ignores the covariance term between $\hat{\theta}_n$ and $\hat{\theta}_{MI}-\hat{\theta}_n$. The covariance term is significant because the congeniality condition does not hold when the method-of-moment estimator is used to estimate θ_2.

4.6 Fractional imputation

Fractional imputation was originally proposed by Kalton and Kish (1984) as an imputation method with reduced variance. In fractional imputation (FI), m imputed values are generated for each missing component $\mathbf{y}_{i,\text{mis}}$ of the complete observation $\mathbf{y}_i=(\mathbf{y}_{i,\text{obs}},\mathbf{y}_{i,\text{mis}})$ and m fractional weights are assigned to the imputed values so that the mean score function can be approximated by a weighted sum of the imputed score functions. Let $\mathbf{y}_{ij}^*=(\mathbf{y}_{i,\text{obs}},\mathbf{y}_{i,\text{mis}}^{*(j)})$ be the j-th imputed value of $\mathbf{y}_i$ and let w_{ij}^* be the fractional weight assigned to $\mathbf{y}_{ij}^*$. The fractional weights are constructed to satisfy

$$\sum_{j=1}^{m} w_{ij}^*=1 \tag{4.88}$$

for each $i=1,2,\cdots,n$. Let ψ be the parameter of interest that is consistently estimated by solving

$$\sum_{i=1}^{n} U(\psi;\mathbf{y}_i)=0$$

for ψ under complete response of $\mathbf{y}_i$. In fractional imputation, fractional weights $w_{i1}^*, \cdots, w_{im}^*$ are assigned to $\mathbf{y}_{i1}^*, \cdots, \mathbf{y}_{im}^*$, respectively, such that

$$\sum_{i=1}^{n}\sum_{j=1}^{m} w_{ij}^* U(\psi;\mathbf{y}_{ij}^*) \cong \sum_{i=1}^{n} E\{U(\psi;\mathbf{y}_i) \mid \mathbf{y}_{i,\text{obs}}, \delta_i; \hat{\eta}\}, \tag{4.89}$$

where $\hat{\eta}$ is the maximum likelihood estimator of η in the joint density $f(\mathbf{y},\delta;\eta)$. If $\mathbf{y}_i$ is categorical, then (4.89) can be easily achieved by choosing

$$w_{ij}^* = P\left(\mathbf{y} = \mathbf{y}_{ij}^* \mid \mathbf{y}_{i,\text{obs}}, \delta_i; \hat{\eta}\right).$$

For continuous $\mathbf{y}_i$, we use the following iterative procedure to achieve (4.89) as closely as possible:

[Step 1] Generate m imputed values from some density $h_m(\mathbf{y}_{i,\text{mis}})$ which has the same support as $h(\mathbf{y}_{i,\text{mis}} \mid \mathbf{y}_{i,\text{obs}}, \delta_i; \eta)$ in (3.56). Often, the choice of the proposal density is $h_m(\mathbf{y}_{i,\text{mis}}) = h(\mathbf{y}_{i,\text{mis}} \mid \mathbf{y}_{i,\text{obs}}, \delta_i; \hat{\eta}_p)$, where $\hat{\eta}_p$ is a preliminary estimator of η.

[Step 2] Given $\hat{\eta}_{(t)}$, compute the fractional weights by

$$w_{ij(t)}^* \propto \frac{h(\mathbf{y}_{i,\text{mis}}^{*(j)} \mid \mathbf{y}_{i,\text{obs}}, \delta_i; \hat{\eta}_{(t)})}{h_m(\mathbf{y}_{i,\text{mis}}^{*(j)})} \tag{4.90}$$

with $\sum_{j=1}^{m} w_{ij(t)}^* = 1$.

[Step 3] Given the fractional weights computed from [Step 2], update the parameter $\hat{\eta}_{(t+1)}$ by maximizing

$$Q^*(\eta \mid \hat{\eta}_{(t)}) = \sum_{i=1}^{n}\sum_{j=1}^{m} w_{ij(t)}^* \ln\left\{f\left(\mathbf{y}_{ij}^*, \delta_i; \eta\right)\right\} \tag{4.91}$$

over η, where $f(\mathbf{y}_i, \delta_i; \eta)$ is the joint density of $(\mathbf{y}_i, \delta_i)$

[Step 4] Go to [Step 2] until convergence.

Step 1 can be called the *imputation step*, Step 2 can be called the *weighting step*, and Step 3 can be called the *maximization step* (M-step). The imputation and weighting steps can be combined to implement the E-step of the EM algorithm. Unlike the MCEM method, imputed values are not changed for each EM iteration - only the fractional weights are changed. Thus, the FI method has some computational advantage over the MCEM method.

Note that the fractional weights of the form (4.90) can be written as

$$w_{ij(t)}^* = \frac{h(\mathbf{y}_{i,\text{mis}}^{*(j)} \mid \mathbf{y}_{i,\text{obs}}, \delta_i; \hat{\eta}^{(t)})/h_m(\mathbf{y}_{i,\text{mis}}^{*(j)})}{\sum_{k=1}^{m} h(\mathbf{y}_{i,\text{mis}}^{*(k)} \mid \mathbf{y}_{i,\text{obs}}, \delta_i; \hat{\eta}^{(t)})/h_m(\mathbf{y}_{i,\text{mis}}^{*(k)})}.$$

Since the conditional distribution can be written as

$$h(\mathbf{y}_{i,\text{mis}} \mid \mathbf{y}_{i,\text{obs}}, \delta_i; \hat{\eta}) = \frac{f(\mathbf{y}_i, \delta_i; \hat{\eta})}{\int f(\mathbf{y}_i, \delta_i; \hat{\eta})\, d\mathbf{y}_{i,\text{mis}}} = \frac{f(\mathbf{y}_i, \delta_i; \hat{\eta})}{f_{\text{obs}}(\mathbf{y}_{i,\text{obs}}, \delta_i; \hat{\eta})},$$

where $f_{\text{obs}}(\mathbf{y}_{i,\text{obs}}, \delta_i; \hat{\eta}) = \int f(\mathbf{y}_i, \delta_i; \hat{\eta})\, d\mathbf{y}_{i,\text{mis}}$ is the marginal density of $(\mathbf{y}_{i,\text{obs}}, \delta_i)$, we can express

$$w_{ij(t)}^* = \frac{f(\mathbf{y}_{ij}^*, \delta_i; \hat{\eta}_{(t)})/h_m(\mathbf{y}_{i,\text{mis}}^{*(j)})}{\sum_{k=1}^{m} f(\mathbf{y}_{ik}^*, \delta_i; \hat{\eta}_{(t)})/h_m(\mathbf{y}_{i,\text{mis}}^{*(k)})}. \tag{4.92}$$

Thus, the marginal density in computing the conditional distribution is not needed in computing the fractional weights. Only the joint density is needed.

Given the m imputed values, $\mathbf{y}_{i1}^*, \cdots, \mathbf{y}_{im}^*$, generated from $h_m(\mathbf{y}_{i,\text{mis}})$, the sequence of estimators $\{\hat{\eta}_{(1)}, \hat{\eta}_{(2)}, \ldots\}$ can be constructed using [Step 2]-[Step 3]. The following theorem presents some convergence properties of the sequence of estimators.

Theorem 4.5. *Let $Q^*(\eta \mid \hat{\eta}_{(t)})$ be the weighted log-likelihood function (4.91) based on fractional imputation. If*

$$Q^*(\hat{\eta}_{(t+1)} \mid \hat{\eta}_{(t)}) \geq Q^*(\hat{\eta}_{(t)} \mid \hat{\eta}_{(t)}) \tag{4.93}$$

then

$$l^*_{\text{obs}}(\hat{\eta}_{(t+1)}) \geq l^*_{\text{obs}}(\hat{\eta}_{(t)}), \tag{4.94}$$

where

$$l^*_{\text{obs}}(\eta) = \sum_{i=1}^{n} \ln\{f^*_{obs(i)}(\mathbf{y}_{i,\text{obs}}, \delta_i; \eta)\} \tag{4.95}$$

is the observed log-likelihood constructed from the fractional imputation and

$$f^*_{obs(i)}(\mathbf{y}_{i,\text{obs}}, \delta_i; \eta) = \frac{\sum_{j=1}^{m} f(\mathbf{y}^*_{ij}, \delta_i; \eta)/h_m(\mathbf{y}^{*(j)}_{i,\text{mis}})}{\sum_{j=1}^{m} 1/h_m(\mathbf{y}^{*(j)}_{i,\text{mis}})}.$$

Proof. By (4.92) and using Jensen's inequality,

$$\begin{aligned} l^*_{\text{obs}}(\hat{\eta}_{(t+1)}) - l^*_{\text{obs}}(\hat{\eta}_{(t)}) &= \sum_{i=1}^{n} \ln\left\{\sum_{j=1}^{m} w^*_{ij(t)} \frac{f(\mathbf{y}^*_{ij}, \delta_i; \hat{\eta}_{(t+1)})}{f(\mathbf{y}^*_{ij}, \delta_i; \hat{\eta}_{(t)})}\right\} \\ &\geq \sum_{i=1}^{n}\sum_{j=1}^{m} w^*_{ij(t)} \ln\left\{\frac{f(\mathbf{y}^*_{ij}, \delta_i; \hat{\eta}_{(t+1)})}{f(\mathbf{y}^*_{ij}, \delta_i; \hat{\eta}_{(t)})}\right\} \\ &= Q^*(\hat{\eta}_{(t+1)} \mid \hat{\eta}_{(t)}) - Q^*(\hat{\eta}_{(t)} \mid \hat{\eta}_{(t)}). \end{aligned}$$

Therefore, (4.93) implies (4.94). □

Note that $l^*_{\text{obs}}(\eta)$ is an imputed version of the observed log-likelihood based on the m imputed values, $\mathbf{y}^*_{i1}, \ldots, \mathbf{y}^*_{im}$. By Theorem 4.5, the sequence $l^*_{\text{obs}}(\hat{\eta}_{(t)})$ is monotonically increasing and, under some conditions, the convergence of $\hat{\eta}_{(t)}$ to a stationary point $\hat{\eta}^*_m$ follows for fixed m. The stationary point $\hat{\eta}^*_m$ converges to the MLE of η as $m \to \infty$. Theorem 4.5 does not hold for the sequence obtained from the Monte Carlo EM method for fixed m, because the imputed values are re-generated for each E-step of the Monte Carlo EM method.

In some case, it is desired to create a fractional imputation with small imputation size m, say $m = 10$. If the MLE of η is obtained analytically or computed from a fractional imputation with sufficiently large m, then we can create final weights using smaller m with constraints

$$\sum_{i=1}^{n}\sum_{j=1}^{m} w^*_{ij} S_{\text{com}}\left(\hat{\eta}; \mathbf{y}^*_{ij}, \delta_i\right) = 0, \tag{4.96}$$

where $\hat{\eta}$ is the MLE of η. With this further constraint, the solution to the imputed score equation is equal to the MLE of η even for small m. The fractional imputation satisfying constraints such as (4.96) is called *calibration fractional imputation.*

Finding the fractional weights for calibration fractional imputation can be achieved by the regression weighting technique, by which the fractional weights that satisfy (4.96) and $\sum_{j=1}^{m} w^*_{ij} = 1$ are constructed by

$$w^*_{ij} = w^*_{ij0} + w^*_{ij0}\Delta\left(S^*_{ij} - \bar{S}^*_{i\cdot}\right), \tag{4.97}$$

where w^*_{ij0} is the initial fractional weights defined in (4.90), $S^*_{ij} = S_{\text{com}}(\hat{\eta}; \mathbf{y}^*_{ij}, \delta_i)$, $\bar{S}^*_{i\cdot} = \sum_{j=1}^{m} w^*_{ij0} S^*_{ij}$,

$$\Delta = -\left\{\sum_{i=1}^{n}\sum_{j=1}^{m} w^*_{ij0} S^*_{ij}\right\}' \left[\sum_{i=1}^{n}\sum_{j=1}^{m} w^*_{ij0}\left(S^*_{ij} - \bar{S}^*_{i\cdot}\right)^{\otimes 2}\right]^{-1}.$$

Note that some of the fractional weights computed by (4.97) can take negative values. In this case, some alternative algorithm other than the regression weighting should be used. For example, the fractional weights of the form

$$w_{ij}^* = \frac{w_{ij0}^* \exp\left(\Delta S_{ij}^*\right)}{\sum_{k=1}^m w_{ik0}^* \exp\left(\Delta S_{ik}^*\right)},$$

is approximately equal to the regression fractional weight in (4.97) and is always positive.

In the special case of the exponential family of distributions in (3.45) with $\sum_{i=1}^n \mathbf{T}(\mathbf{y}_i)$ being the complete sufficient statistic for θ. Recall that, under MAR, the M-step of the EM algorithm is implemented by solving (3.47). In this setup, the weighting step in the calibration fractional imputation at the t-th step of the EM algorithm can be expressed as

$$\sum_{i=1}^n \sum_{j=1}^m w_{ij(t)}^* \mathbf{T}\left(\mathbf{y}_{ij}^*\right) = \sum_{i=1}^n E\left\{\mathbf{T}(\mathbf{y}_i) \mid \mathbf{y}_{i,\text{obs}}, \delta_i; \hat{\theta}^{(t)}\right\},$$

with $\sum_{j=1}^m w_{ij(t)}^* = 1$. The M-step remains the same. That is, the parameter is updated by solving

$$\sum_{i=1}^n \sum_{j=1}^m w_{ij(t)}^* \mathbf{T}\left(\mathbf{y}_{ij}^*\right) = \sum_{i=1}^n E_\theta\left\{\mathbf{T}(\mathbf{y}_i)\right\}$$

for θ to get $\hat{\theta}^{(t+1)}$.

Asymptotic properties of the FI estimator are derived as a special case of the general theory in Section 4.2. Variance estimation after fractional imputation is also discussed in Kim (2011).

Example 4.9. *We consider the bivariate normal distribution (2.30) with MAR. Let δ_{1i} and δ_{2i} be the response indicator functions of y_{1i} and y_{2i}, respectively. A set of sufficient statistics for $\theta = (\mu_1, \mu_2, \sigma_{11}, \sigma_{12}, \sigma_{22})$ is $T = \sum_{i=1}^n \left(y_{1i}, y_{2i}, y_{1i}^2, y_{1i}y_{2i}, y_{2i}^2\right)$. Use a property of the normal distribution, and the fractional weights for $\delta_{1i} = 0$ and $\delta_{2i} = 1$ can be constructed by*

$$\sum_{j=1}^m w_{ij(t)}^* \left\{1, y_{1i}^{*(j)}, \left(y_{1i}^{*(j)}\right)^2\right\} = \left\{1, E\left(y_{1i} \mid y_{2i}, \hat{\theta}_{(t)}\right), E\left(y_{1i}^2 \mid y_{2i}, \hat{\theta}_{(t)}\right)\right\},$$

where

$$E\left(y_{1i} \mid y_{2i}, \hat{\theta}\right) = \hat{\mu}_1 + \frac{\hat{\sigma}_{12}}{\hat{\sigma}_{22}}\left(y_{2i} - \hat{\mu}_2\right)$$

and

$$E\left(y_{1i}^2 \mid y_{2i}, \hat{\theta}\right) = \left\{\hat{\mu}_1 + \frac{\hat{\sigma}_{12}}{\hat{\sigma}_{22}}\left(y_{2i} - \hat{\mu}_2\right)\right\}^2 + \hat{\sigma}_{11} - \hat{\sigma}_{12}^2/\hat{\sigma}_{22}.$$

Similarly, the fractional weights for $\delta_{i1} = 1$ and $\delta_{2i} = 0$ are constructed by

$$\sum_{j=1}^m w_{ij(t)}^* \left\{1, y_{2i}^{*(j)}, \left(y_{2i}^{*(j)}\right)^2\right\} = \left\{1, E\left(y_{2i} \mid y_{1i}, \hat{\theta}_{(t)}\right), E\left(y_{2i}^2 \mid y_{1i}, \hat{\theta}_{(t)}\right)\right\}$$

and the fractional weights for $\delta_{i1} = 0$ and $\delta_{2i} = 0$ are constructed by

$$\sum_{j=1}^m w_{ij(t)}^* \left\{1, y_{1i}^{*(j)}, y_{2i}^{*(j)}, \left(y_{1i}^{*(j)}\right)^2, \left(y_{2i}^{*(j)}\right)^2, y_{1i}^{*(j)} y_{2i}^{*(j)}\right\}$$
$$= \left\{1, \hat{\mu}_{1(t)}, \hat{\mu}_{2(t)}, \hat{\mu}_{1(t)}^2 + \sigma_{1(t)}^2, \hat{\mu}_{2(t)}^2 + \sigma_{2(t)}^2, \hat{\mu}_{1(t)}\hat{\mu}_{2(t)} + \hat{\sigma}_{12(t)}\right\}.$$

In the M-step, the parameters are updated by

$$\begin{aligned}
\hat{\mu}_{1(t+1)} &= n^{-1}\sum_{i=1}^{n}\sum_{j=1}^{m} w_{ij(t)}^{*} y_{1i}^{*(j)} \\
\hat{\mu}_{2(t+1)} &= n^{-1}\sum_{i=1}^{n}\sum_{j=1}^{m} w_{ij(t)}^{*} y_{2i}^{*(j)} \\
\hat{\sigma}_{1(t+1)}^{2} &= n^{-1}\sum_{i=1}^{n}\sum_{j=1}^{m} w_{ij(t)}^{*} \left(y_{1i}^{*(j)}\right)^{2} - \hat{\mu}_{1(t+1)}^{2} \\
\hat{\sigma}_{2(t+1)}^{2} &= n^{-1}\sum_{i=1}^{n}\sum_{j=1}^{m} w_{ij(t)}^{*} \left(y_{2i}^{*(j)}\right)^{2} - \hat{\mu}_{2(t+1)}^{2} \\
\hat{\sigma}_{12(t+1)} &= n^{-1}\sum_{i=1}^{n}\sum_{j=1}^{m} w_{ij(t)}^{*} y_{1i}^{*(j)} y_{2i}^{*(j)} - \hat{\mu}_{1(t+1)}\hat{\mu}_{2(t+1)}.
\end{aligned}$$

Note that the parameter estimates are computed by the standard formula for the MLE using fractional weights.

Example 4.10. *We now consider fractional imputation for partially classified categorical data. Let* $\mathbf{y} = (y_1, \cdots, y_p)$ *be the vector of study variables that take categorical values. Let* $\mathbf{y}_i = (y_{i1}, \cdots, y_{ip})$ *be the i-th realization of* $\mathbf{y}$. *Let* δ_{ij} *be the response indicator function for* y_{ij}. *Assume that the response mechanism is missing at random. Based on the realization of* $\delta_i = (\delta_{i1}, \cdots, \delta_{ip})$, *the original observation* $\mathbf{y}_i$ *can decompose into* $(\mathbf{y}_{i,\text{obs}}, \mathbf{y}_{i,\text{mis}})$. *Let* $D_i = \{\mathbf{y}_{i,\text{mis}}^{*(1)}, \cdots, \mathbf{y}_{i,\text{mis}}^{*(M_i)}\}$ *be the set of all possible values of* $\mathbf{y}_{i,\text{mis}}$. *Such enumeration is possible as* $\mathbf{y}_i$ *is categorical with known categories. In this case, the fractional imputation consists of taking all of* M_i *possible values as the imputed values and then assigning them with fractional weights. The fractional weight assigned to* $\mathbf{y}_{i,\text{mis}}^{*(j)}$ *is*

$$w_{ij}^{*} = \frac{\pi(\mathbf{y}_{i,\text{obs}}, \mathbf{y}_{i,\text{mis}}^{*(j)})}{\sum_k \pi(\mathbf{y}_{i,\text{obs}}, \mathbf{y}_{i,\text{mis}}^{*(k)})}, \tag{4.98}$$

where $\pi(\tilde{\mathbf{y}})$ *is the joint probability of* $\tilde{\mathbf{y}}$ *composed of observed and imputed* y_i*'s. If the joint probability is nonparametrically modeled, it is computed by*

$$\pi(\tilde{\mathbf{y}}) = \frac{1}{n}\sum_{i=1}^{n}\sum_{j\in D_i} w_{ij}^{*} I\left\{(\mathbf{y}_{i,\text{obs}}, \mathbf{y}_{i,\text{mis}}^{*(j)}) = \tilde{\mathbf{y}}\right\}. \tag{4.99}$$

Note that (4.98) and (4.99) correspond to the E-step and M-step of the EM algorithm, respectively. The M step (4.99) can be changed if there is a parametric model for the joint probability. For example, if the joint probability can be modeled by a multinomial distribution with parameter θ, *then rather than solving the mean score equation, the M step becomes solving the imputed score equation. The initial values of fractional weights in the EM algorithm can be* $w_{ij}^{*} = 1/M_i$.

Example 4.11. *We now consider fractional imputation under the setup of Example 3.15. Instead of the rejection method, we can consider the following fractional imputation method:*

[Step 1] Generate $y_i^{*(1)}, \cdots, y_i^{*(m)}$ *from* $f\left(y_i \mid x_i; \hat{\theta}_{(0)}\right)$.

[Step 2] Using the m imputed values generated from Step 1, compute the fractional weights by

$$w_{ij(t)}^{*} \propto \frac{f\left(y_i^{*(j)} \mid x_i; \hat{\theta}_{(t)}\right)}{f\left(y_i^{*(j)} \mid x_i; \hat{\theta}_{(0)}\right)} \left\{1 - \pi(x_i, y_i^{*(j)}; \hat{\phi}_{(t)})\right\} \tag{4.100}$$

where

$$\pi(x_i, y_i; \hat{\phi}) = \frac{\exp\left(\hat{\phi}_0 + \hat{\phi}_1 x_i + \hat{\phi}_2 y_i\right)}{1 + \exp\left(\hat{\phi}_0 + \hat{\phi}_1 x_i + \hat{\phi}_2 y_i\right)}.$$

Use the imputed data and the fractional weights and implement the M-step by solving

$$\sum_{i=1}^{n}\sum_{j=1}^{m} w_{ij(t)}^{*} S\left(\theta; x_i, y_i^{*(j)}\right) = 0 \tag{4.101}$$

and

$$\sum_{i=1}^{n}\sum_{j=1}^{m} w_{ij(t)}^{*} \left\{\delta_i - \pi(\phi; x_i, y_i^{*(j)})\right\}\left(1, x_i, y_i^{*(j)}\right) = 0, \tag{4.102}$$

where $S(\theta; x_i, y_i) = \partial \log f(y_i \mid x_i; \theta)/\partial\theta$.

Example 4.12. *We now return to the setup of Example 3.18. The random effects,* a_i, *are to be generated from the conditional distribution in (3.62). Instead of using the Metropolis–Hastings algorithm, which can be computationally heavy, we can use fractional imputation which starts with the generation of m values of* a_i, *denoted by* $a_i^{*(1)}, \cdots, a_i^{*(m)}$, *from some proposal distribution* $h(a_i)$. *The choice of the proposal distribution* $h(a)$ *is somewhat arbitrary, but t-distribution with four degrees of freedom seems to work well in many cases. Then, compute the fractional weights by*

$$w_{ik}^{*} \propto f_1(\mathbf{y}_i \mid \mathbf{x}_i, a_i^{*(k)}; \hat{\beta}) f_2(a_i^{*(k)}; \hat{\sigma})/h(a_i^{*(k)}),$$

where $f_1(\cdot)$ *and* $f_2(\cdot)$ *are defined in (3.62). Given the current parameter values, the M-step updates the parameter estimates by solving*

$$\sum_{i=1}^{n}\sum_{k=1}^{m} w_{ik}^{*} \sum_{j} S_1(\beta; x_{ij}, y_{ij}, a_i^{*(k)}) = 0$$

and

$$\sum_{i=1}^{n}\sum_{k=1}^{m} w_{ik}^{*} S_2(\sigma; a_i^{*(k)}) = 0,$$

where $S_1(\cdot)$ *and* $S_2(\cdot)$ *are the score functions derived from* $f_1(\cdot)$ *and* $f_2(\cdot)$, *respectively.*

Example 4.13. *Suppose that we are interested in estimating parameter* θ *in the conditional distribution* $f(y \mid x; \theta)$. *Instead of observing* (x_i, y_i), *suppose that we observe* (z_i, y_i), *where z is conditionally independent of y given x and the joint distribution of* (x, z) *is known (or estimable from a calibration sample). In this case, we want to create* x_i^* *from the observed values of* (z_i, y_i) *in order to perform regression analysis. This setup is related to the problem of inference with linked data (Lahiri and Larsen, 2005) or measurement error model problem discussed in Example 3.17.*

To apply the FI method in this setup, we first generate m values of x_i, *denoted by* $x_i^{*(1)}, \cdots, x_i^{*(m)}$, *from the conditional distribution* $g(x \mid z_i)$ *obtained from another source and then assign fractional weights computed by*

$$w_{ij}^{*} \propto f\left(x_i^{*(j)} \mid y_i, z_i\right) / g\left(x_i^{*(j)} \mid z_i\right). \tag{4.103}$$

Using (3.61), the above fractional weights can be written

$$w_{ij}^{*} \propto f\left(y_i \mid x_i^{*(j)}\right),$$

which depends on unknown parameters $\theta = (\beta_0, \beta_1, \sigma)$. *Thus, an EM algorithm can be developed for solving*

$$\sum_{i=1}^{n}\sum_{j=1}^{m} w_{ij}^{*}(\theta) S(\theta; x_i^{*(j)}, y_i) = 0,$$

where $S(\theta; x, y) = \partial \ln f(y \mid x; \theta)/\partial\theta$ *and*

$$w_{ij}^{*}(\theta) = \frac{f\left(y_i \mid x_i^{*(j)}; \theta\right)}{\sum_{k=1}^{m} f\left(y_i \mid x_i^{*(k)}; \theta\right)}.$$

Write

$$Q^*(\eta \mid \eta_0) = \sum_{i=1}^{n}\sum_{j=1}^{m} w_{ij}^*(\eta_0)\log f\left(y_{ij}^*, \delta_i; \eta\right), \tag{4.104}$$

where $w_{ij}^*(\eta)$ is the fractional weight associated with y_{ij}^*, denoted by

$$w_{ij}^*(\eta) = \frac{f(\mathbf{y}_{ij}^*, \delta_i; \eta)/h_m(\mathbf{y}_{i,\text{mis}}^{*(j)})}{\sum_{k=1}^{m} f(\mathbf{y}_{ik}^*, \delta_i; \eta)/h_m(\mathbf{y}_{i,\text{mis}}^{*(k)})}, \tag{4.105}$$

the EM algorithm for fractional imputation can be expressed as

$$\hat{\eta}^{(t+1)} \leftarrow \text{argmax } Q^*(\eta \mid \hat{\eta}^{(t)}).$$

Instead of the EM algorithm, a Newton-type algorithm can also be used. Using the Oakes' formula in (3.36), we can obtain the observed information from the Q^* function (4.104) alone, without having to know the observed likelihood function. That is, we have

$$I_{obs}^*(\eta) = -\sum_{i=1}^{n}\sum_{j=1}^{m} w_{ij}^*(\eta)\dot{S}(\eta; \mathbf{y}_{ij}^*, \delta_i) - \sum_{i=1}^{n}\sum_{j=1}^{m} w_{ij}^*(\eta)\left\{S(\eta; \mathbf{y}_{ij}^*, \delta_i) - \bar{S}_i^*(\eta)\right\}^{\otimes 2}, \tag{4.106}$$

where $S(\eta; \mathbf{y}, \delta) = \partial \log f(\mathbf{y}, \delta; \eta)/\partial\eta$, $\dot{S}(\eta; \mathbf{y}, \delta) = \partial S(\eta; \mathbf{y}, \delta)/\partial\eta'$ and

$$\bar{S}_i^*(\eta) = \sum_{j=1}^{m} w_{ij}^*(\eta)S(\eta; \mathbf{y}_{ij}^*, \delta_i).$$

Note that, for $m \to \infty$, (4.106) converges to

$$-\sum_{i=1}^{n} E\left\{\dot{S}(\eta; \mathbf{y}_i, \delta_i) \mid \mathbf{y}_{i,obs}, \delta_i\right\} - \sum_{i=1}^{n} V\left\{S(\eta; \mathbf{y}_i, \delta_i) \mid \mathbf{y}_{i,\text{obs}}, \delta_i\right\},$$

which is equal to the observed information matrix discussed in (2.22). Thus, a Newton-type algorithm for computing the MLE from the fractionally imputed data is given by

$$\hat{\eta}^{(t+1)} = \hat{\eta}^{(t)} + \left\{I_{obs}^*(\hat{\eta}^{(t)})\right\}^{-1} \bar{S}^*(\hat{\eta}^{(t)}), \tag{4.107}$$

with $I_{obs}^*(\hat{\eta})$ defined in (4.106) and

$$\bar{S}^*(\eta) = \sum_{i=1}^{n}\sum_{j=1}^{m} w_{ij}^*(\eta)S(\eta; \mathbf{y}_{ij}^*, \delta_i).$$

We now briefly discuss estimating general parameters Ψ that can be estimated by (4.33) under complete response. The Fractional Imputation (FI) estimator of Ψ is then computed by solving

$$\sum_{i=1}^{n}\sum_{j=1}^{m} w_{ij}^*(\hat{\eta})U(\Psi; \mathbf{y}_{ij}^*) = 0. \tag{4.108}$$

Note that $\hat{\eta}$ is the solution to

$$\sum_{i=1}^{n}\sum_{j=1}^{m} w_{ij}^*(\hat{\eta})S\left(\hat{\eta}; \mathbf{y}_{ij}^*\right) = 0.$$

We can use either the linearization method or the replication method for variance estimation. For the former, we can use (4.41) with $\hat{q}_i^* = \bar{U}_i^* + \hat{\kappa}\bar{S}_i^*$, where $(\bar{U}_i^*, \bar{S}_i^*) = \sum_{j=1}^m w_{ij}^*(U_{ij}^*, S_{ij}^*)$, $U_{ij}^* = U(\hat{\Psi}; \mathbf{y}_{ij}^*)$, $S_{ij}^* = S(\hat{\eta}; \mathbf{y}_{ij}^*)$,

$$\hat{\kappa} = \sum_{i=1}^n \sum_{j=1}^m w_{ij}^*(\hat{\eta}) \left(U_{ij}^* - \bar{U}_i^*\right) S_{ij}^* \{I_{\text{obs}}^*(\hat{\eta})\}^{-1} \tag{4.109}$$

and $I_{\text{obs}}^*(\eta)$ is defined in (4.106). For linearization method, we can use sandwich formula

$$\hat{V}\left(\hat{\Psi}\right) = \hat{\tau}_q^{-1} \hat{\Omega}_q \hat{\tau}_q^{-1'} \tag{4.110}$$

where

$$\begin{aligned}
\hat{\tau}_q &= n^{-1} \sum_{i=1}^n \sum_{j=1}^m w_{ij}^* \dot{U}\left(\hat{\Psi}; \mathbf{y}_{ij}^*\right) \\
\hat{\Omega}_q &= n^{-1}(n-1)^{-1} \sum_{i=1}^n (\hat{q}_i^* - \bar{q}_n^*)^{\otimes 2},
\end{aligned}$$

with $\hat{q}_i^* = \bar{U}_i^* + \hat{\kappa}\bar{S}_i^*$, where $(\bar{U}_i^*, \bar{S}_i^*) = \sum_{j=1}^m w_{ij}^*(U_{ij}^*, S_{ij}^*)$, $U_{ij}^* = U(\hat{\Psi}; \mathbf{y}_{ij}^*)$, $S_{ij}^* = S(\hat{\eta}; \mathbf{y}_{ij}^*)$, and

$$\hat{\kappa} = \sum_{i=1}^n \sum_{j=1}^m w_{ij}^*(\hat{\eta}) \left(U_{ij}^* - \bar{U}_i^*\right) S_{ij}^* \{I_{\text{obs}}^*(\hat{\eta})\}^{-1}.$$

Justification of (4.110) is given in Kim (2011).

For the replication method, we first obtain the k-th replicate $\hat{\eta}^{(k)}$ of $\hat{\eta}$ by solving

$$\bar{S}^{*(k)}(\eta) \equiv \sum_{i=1}^n \sum_{j=1}^m w_i^{(k)} w_{ij}^*(\eta) S\left(\eta; \mathbf{y}_{ij}^*\right) = 0. \tag{4.111}$$

Once $\hat{\eta}^{(k)}$ is obtained then the k-th replicate $\hat{\Psi}^{(k)}$ of $\hat{\Psi}$ is obtained by solving

$$\sum_{i=1}^n \sum_{j=1}^m w_i^{(k)} w_{ij}^*(\hat{\eta}^{(k)}) U(\Psi; \mathbf{y}_{ij}^*) = 0 \tag{4.112}$$

for ψ and the replication variance estimator of $\hat{\Psi}$ from (4.108) is obtained by

$$\hat{V}_{rep}(\hat{\Psi}) = \sum_{k=1}^L c_k \left(\hat{\Psi}^{(k)} - \hat{\Psi}\right)^2.$$

Note that the imputed values are not changed. Only the fractional weights are changed for each replication. Finding the solution $\hat{\eta}^{(k)}$ to (4.111) can require some iterative computation such as the EM algorithm. To avoid iterative computation, one may consider a one-step approximation by applying the following Taylor expansion:

$$\begin{aligned}
0 &= \bar{S}^{*(k)}(\hat{\eta}^{(k)}) \\
&\cong \bar{S}^{*(k)}(\hat{\eta}) + \left\{\frac{\partial}{\partial \eta'} \bar{S}^{*(k)}(\hat{\eta})\right\} \left(\hat{\eta}^{(k)} - \hat{\eta}\right).
\end{aligned}$$

Now, writing $I_{\text{obs}}^{*(k)}(\eta) = \partial \bar{S}^{*(k)}(\eta) / \partial \eta'$, we can obtain, using the argument for (4.106),

$$I_{obs}^{*(k)}(\eta) = -\sum_{i=1}^n w_i^{(k)} \sum_{j=1}^m w_{ij}^*(\eta) \dot{S}(\eta; \mathbf{y}_{ij}^*, \delta_i) - \sum_{i=1}^n w_i^{(k)} \sum_{j=1}^m w_{ij}^*(\eta) \left\{S(\eta; \mathbf{y}_{ij}^*, \delta_i) - \bar{S}_i^*(\eta)\right\}^{\otimes 2}. \tag{4.113}$$

Thus, the one-step approximation of $\hat{\eta}^{(k)}$ to (4.111) is given by

$$\hat{\eta}^{(k)} \cong \hat{\eta} + \left\{ I_{obs}^{*(k)}(\hat{\eta}) \right\}^{-1} \bar{S}^{*(k)}(\hat{\eta}).$$

One-step replicate $\hat{\eta}^{(k)}$ can be used in (4.112) to compute $\hat{\Psi}^{(k)}$.

Example 4.14. *Consider the problem of estimating the p-th quantile $\Psi = F^{-1}(p)$ where $F(y)$ is the marginal CDF of the study variable Y. Under complete response, the estimating equation for Ψ is*

$$U(\Psi) \equiv \sum_{i=1}^{n} \{I(Y_i \leq \Psi) - p\} = 0. \tag{4.114}$$

We may need an interpolation technique to solve (4.114) from the realized sample. Now, using fractionally imputed data, we can solve $\hat{F}_{FI}(\Psi) = p$ for Ψ where

$$\hat{F}_{FI}(\Psi) \equiv n^{-1} \sum_{i=1}^{n} \left\{ \delta_i I(y_i < \Psi) + (1-\delta_i) \sum_{j=1}^{M} w_{ij}^* I\left(y_{ij}^* < \Psi\right) \right\},$$

to obtain FI estimator of Ψ. To estimate the variance of $\hat{\Psi}$ from the FI data, we can simply apply the sandwich formula (4.110) with $U(\Psi; y) = I(y \leq \Psi) - p$ and $\hat{\tau}_q = f\left(\hat{\Psi}; \hat{\eta}\right)$, which is the value of marginal density of Y at $y = \hat{\Psi}$ with parameter value $\hat{\eta}$. If the density is unknown, one can use

$$\hat{f}_h(y) = \frac{1}{2h} \{\hat{F}_{FI}(\hat{\Psi} + h) - \hat{F}_{FI}(\hat{\Psi} - h)\}$$

to estimate the density f around $y = \hat{\Psi}$, where h is the bandwidth. The choice of $h = n^{-1/2}$ can be used.

Example 4.15. *Fractional imputation can be implemented nonparametrically in some simple cases. We assume a bivariate data structure (x_i, y_i) with x_i being fully observed. Assume that the response mechanism is missing at random. Let $K(\cdot)$ be a symmetric density function on the real line and let $h = h_n$ be a smoothing bandwidth such that $h_n \to 0$ and $nh_n \to \infty$ as $n \to \infty$.*

A nonparametric regression estimator of $m(x) = E(y \mid x)$ can be obtained by finding $\hat{m}(x)$ that minimizes

$$\sum_{i=1}^{n} K_h(x_i, x)\, \delta_i \{y_i - m(x)\}^2, \tag{4.115}$$

where $K_h(u, x) = h^{-1} K\{(u-x)/h\}$. The minimizer of (4.115) is

$$\hat{m}_1(x) = \sum_{j=1}^{n} w_{j1}(x)\, y_i, \tag{4.116}$$

where

$$w_{i1}(x) = \frac{\delta_i K_h(x_i, x)}{\sum_{j=1}^{n} \delta_j K_h(x_j, x)}.$$

The weight $w_{i1}(x)$ in (4.116) represents the point mass assigned to y_i when $m_1(x)$ is approximated by $\sum_{i=1}^{n} w_{i1}(x)\, y_i$. If we write

$$\hat{\theta}_1 = \frac{1}{n} \sum_{i=1}^{n} \{\delta_i y_i + (1-\delta_i) \hat{m}_1(x_i)\} = \frac{1}{n} \sum_{i=1}^{n} \left\{ \delta_i y_i + (1-\delta_i) \sum_{j=1}^{n} w_{j1}(x_i) y_j \right\}, \tag{4.117}$$

the weight $w_{ij}^* = w_{j1}(x_i)$ is essentially the fractional weight assigned to the j-th imputed value for missing unit i. Cheng (1994) proved, under some regularity conditions,

$$\sqrt{n}\left(\hat{\theta}_1 - \theta\right) \to N\left(0, \sigma_1^2\right),$$

where $\sigma_1^2 = V\{m(X)\} + E\left[\{\pi(X)\}^{-1} V(Y \mid X)\right]$ and $\pi(X) = P(\delta = 1 \mid X)$.

We now consider the *fractional hot deck imputation* in which m imputed values are taken from the set of respondents. For simplicity, we consider a bivariate data structure (x_i, y_i) with x_i fully observed. Let $\{y_1, \cdots, y_r\}$ be the set of respondents and we want to obtain m imputed values, $y_i^{*(1)}, \cdots, y_i^{*(m)}$, for $i = r+1, \cdots, n$ from the respondents. Let w_{ij}^* be the fractional weights assigned to $y_i^{*(j)}$ for $j = 1, 2, \cdots, m$. We consider the special case of $m = r$. In this case, the fractional weight represents the point mass assigned to each responding y_i. Thus, it is desirable to compute the fractional weights $w_{i1}^*, \cdots, w_{ir}^*$ such that $\sum_{j=1}^r w_{ij}^* = 1$ and

$$\sum_{j=1}^{r} w_{ij}^* I(y_j < y) \cong Pr(y_i < y \mid x_i, \delta_i = 0). \tag{4.118}$$

Note that for the special case of $w_{ij}^* = 1/r$, the left side of the above equality estimates $Pr(y_i < y \mid \delta_i = 1)$.

If we can assume a parametric model $f(y \mid x; \theta)$ for the conditional distribution of y on x and the response probability model is given by $Pr(\delta_i = 1 \mid x_i, y_i) = \pi(x_i, y_i; \phi)$, then the fractional weights satisfying (4.118) are given by

$$\begin{aligned} w_{ij}^* &\propto f\left(y_j \mid x_i, \delta_i = 0; \hat{\theta}, \hat{\phi}\right) / f(y_j \mid \delta_j = 1) \\ &\propto f\left(y_j \mid x_i, \hat{\theta}\right)\left\{1 - \pi\left(x_i, y_j; \hat{\phi}\right)\right\} / f(y_j \mid \delta_j = 1). \end{aligned}$$

Since

$$\begin{aligned} f(y_j \mid \delta_j = 1) &\propto \int \pi(x, y_j) f(y_j \mid x) f(x) dx \\ &\cong \frac{1}{n} \sum_{i=1}^{n} \pi(x_i, y_j) f(y_j \mid x_i), \end{aligned}$$

we can compute

$$w_{ij}^* \propto \frac{f\left(y_j \mid x_i, \hat{\theta}\right)\left\{1 - \pi\left(x_i, y_j; \hat{\phi}\right)\right\}}{\sum_{k=1}^{n} \pi\left(x_k, y_j; \hat{\phi}\right) f\left(y_j \mid x_k; \hat{\theta}\right)}.$$

Under MAR, the fractional weight is

$$w_{ij}^* \propto \frac{f(y_j \mid x_i; \hat{\theta})}{\sum_{k; \delta_k = 1} f(y_j \mid x_k; \hat{\theta})}$$

with $\sum_{j; \delta_j = 1} w_{ij}^* = 1$. Once the fractional imputation is created, the imputed estimating equation for η is computed by

$$\sum_{i=1}^{n} \left\{ \delta_i U(\eta; x_i, y_i) + (1 - \delta_i) \sum_{j; \delta_j = 1} w_{ij}^* U(\eta; x_i, y_j) \right\} = 0 \tag{4.119}$$

where

$$w_{ij}^* = \frac{f(y_j \mid x_i; \hat{\theta}) / \{\sum_{k; \delta_k = 1} f(y_j \mid x_k; \hat{\theta})\}}{\sum_{j; \delta_j = 1} \left[f(y_j \mid x_i; \hat{\theta}) / \{\sum_{k; \delta_k = 1} f(y_j \mid x_k; \hat{\theta})\} \right]}. \tag{4.120}$$

The fractional weights in (4.119) leads to robust estimation in the sense that certain level of misspecification in $f(y \mid x)$ can still provide a consistent estimator of η. The fractional weights can be further adjusted to satisfy

$$\sum_{i=1}^{n} \{\delta_i S(\hat{\theta}; x_i, y_i) + (1 - \delta_i) \sum_{j; \delta_j = 1} w_{ij}^* S(\hat{\theta}; x_i, y_j)\} = 0, \tag{4.121}$$

where $S(\theta; x_i, y_i)$ is the score function of θ and $\hat{\theta}$ is the MLE of θ.

Example 4.16. *Consider the setup of Example 4.8, except that* $e_i = u_i - 1$ *where* u_i *is the exponential distribution with mean 1. Suppose that the imputation model for the error term is* $e_i \sim N(0, \sigma^2)$*. Thus, the imputation model is not correct because the true sampling distribution is* $e_i \sim \exp(1) - 1$*. We are interested in estimating* $\theta_1 = E(Y)$ *and* $\theta_2 = Pr(Y < 1)$*. A simulation study was performed to compare the three imputation methods: multiple imputation (MI), parametric fractional imputation (PFI) of Kim (2011), and fractional hot deck imputation (FHDI), with* $m = 50$ *for all methods. Table 4.3 presents the result from the simulation study. For estimation of* $\theta_1 = E(Y)$*, all imputation methods provide unbiased point estimators because* $E(y_i^*) = E(y_i)$*, where the expectation is taken under the true model. However, the imputation methods considered here provide biased estimates for* θ_2 *because* $E\{I(y_i^* < 1)\} \neq E\{I(y_i < 1)\}$*. The simulation results in Table 4.3 show that the bias of the FHDI estimator is the smallest for* θ_2 *in the setup considered.*

Table 4.3 Monte Carlo biases and standard errors of the point estimators under model misspecification for imputation model

Parameter	Method	Bias	Standard Error
θ_1	MI	0.00	0.084
	PFI	0.00	0.084
	FHDI	0.00	0.084
θ_2	MI	-0.014	0.026
	PFI	-0.014	0.026
	FHDI	-0.001	0.029

Fractional imputation also provides a likelihood-based inference. Since we can construct the observed log-likelihood function, which is given by (4.95), we can establish some asymptotic results for the likelihood ratio statistics from the observed likelihood. That is, under some regularity condition, we can show that

$$-2\{l_{\text{obs}}^*(\eta_0) - l_{\text{obs}}^*(\hat{\eta})\} \xrightarrow{d} \chi^2(p), \tag{4.122}$$

where $\hat{\eta}$ is the MLE of η and p is the dimension of η. To show (4.122), we use the second-order Taylor expansion to obtain

$$\begin{aligned} l_{\text{obs}}^*(\eta_0) &\cong l_{\text{obs}}^*(\hat{\eta}) + \frac{\partial}{\partial \eta'} l_{\text{obs}}^*(\hat{\eta})(\eta_0 - \hat{\eta}) \\ &+ \frac{1}{2}(\eta_0 - \hat{\eta})' \left\{ \frac{\partial^2}{\partial \eta \partial \eta'} l_{\text{obs}}^*(\hat{\eta}) \right\} (\eta_0 - \hat{\eta}). \end{aligned}$$

Note that, by the definition of $\hat{\eta}$, we have

$$\frac{\partial}{\partial \eta'} l_{\text{obs}}^*(\hat{\eta}) = 0.$$

Also, after some algebra, it can be shown that

$$-\frac{\partial^2}{\partial \eta \partial \eta'} l_{\text{obs}}^*(\eta) = I_{\text{obs}}^*(\eta)$$

where $I_{\text{obs}}^*(\eta)$ is defined in (4.106). Thus, we have

$$-\frac{\partial^2}{\partial \eta \partial \eta'} l_{\text{obs}}^*(\hat{\eta}) \xrightarrow{p} \mathcal{I}_{\text{obs}}^*(\eta_0) = \{V(\hat{\eta})\}^{-1}$$

and result (4.122) follows. Likelihood-ratio (LR) test can be constructed from (4.122). Further details can be found in Yang and Kim (2013).

Exercises

1. Show (4.4).
2. Prove (4.45).
3. Derive (4.106).
4. Derive (4.109).
5. Under the setup of Example 4.1, let $\hat{y}_i = \hat{\beta}_0 + \hat{\beta}_1 x_i$ be the (deterministically) imputed value of y_i. Let $y_i^* = \hat{y}_i + \hat{e}_i^*$, where $\hat{e}_i^*$ is randomly selected from $\hat{e}_i = y_i - \hat{y}_i$ among the respondents of y.
   (a) Prove that the estimator of σ_y^2 using the deterministic imputation

$$\hat{\sigma}_{y,d}^2 = \frac{1}{n}\sum_{i=1}^{n}\left\{\delta_i y_i^2 + (1-\delta_i)\hat{y}_i^2 - (\bar{y}_{Id})^2\right\},$$

   where $\bar{y}_{Id} = n^{-1}\sum_{i=1}^{n}\{\delta_i y_i + (1-\delta_i)\hat{y}_i\}$, satisfies

$$E\left(\hat{\sigma}_{y,d}^2\right) = \sigma_y^2 - \frac{n-r}{n}\sigma_y^2\left(1-\rho^2\right) < \sigma_y^2.$$

   (b) Prove that the estimator of σ_y^2 using the stochastic imputation

$$\hat{\sigma}_{y,s}^2 = \frac{1}{n}\sum_{i=1}^{n}\left\{\delta_i y_i^2 + (1-\delta_i)y_i^{*2} - (\bar{y}_{Is})^2\right\},$$

   where $\bar{y}_{Is} = n^{-1}\sum_{i=1}^{n}\{\delta_i y_i + (1-\delta_i)y_i^*\}$, satisfies $E\left(\hat{\sigma}_{y,s}^2\right) \doteq \sigma_y^2$.
6. Let $y_1, \cdots, y_n$ be IID samples from a Bernoulli distribution with probability of $y_i = 1$ being equal to

$$p_i = \frac{\exp(\beta_0 + \beta_1 x_i)}{1 + \exp(\beta_0 + \beta_1 x_i)}$$

   and let δ_i be the response indicator function of y_i. Assume that x_i is always observed and the response mechanism for y is MAR. In this case, the maximum likelihood estimator of the probability p_i is $\hat{p}_i = p_i(\hat{\beta})$ with $\hat{\beta} = (\hat{\beta}_0, \hat{\beta}_1)$ computed by

$$\sum_{i=1}^{n}\delta_i\{y_i - p_i(\beta)\}(1, x_i)' = (0,0)'.$$

   The parameter of interest is $\theta = E(Y)$.
   (a) If the imputation method is deterministic in the sense that we use $\hat{y}_i = \hat{p}_i$ as the imputed value for missing y_i, derive the variance estimation formula for the imputed estimator of θ.
   (b) If the imputation method is stochastic in the sense that we use $y_i^* \sim Bernoulli(\hat{p}_i)$ as the imputed value for missing y_i, derive the variance estimation formula for the imputed estimator of θ.
7. A random sample of size $n = 100$ is drawn from an infinite population composed of two cells, where the distribution of y is

$$y_i \sim \begin{cases} N\left(\mu_1, \sigma_1^2\right), & \text{if } i \in \text{cell 1} \\ N\left(\mu_2, \sigma_2^2\right), & \text{if } i \in \text{cell 2}, \end{cases}$$

   where μ_g and σ_g^2 are unknown parameters and subscript g denotes cells, $g = 1, 2$. In this sample, we have n_1 elements from cell 1 and n_2 elements from cell 2. Among the n_1 elements, r_1 elements are observed and $m_1 = n_1 - r_1$ elements are missing in y. In cell 2, we have r_2 respondents and

$m_2 = n_2 - r_2$ nonrespondents. We assume that the response mechanism is missing at random in the sense that the respondents are independently and identically distributed within the same cell. We use a with-replacement hot deck imputation in the sense that, for each cell g, m_g imputed values are randomly selected from r_g respondents with replacement. Let y_i^* be the imputed value of y_i obtained from the hot deck imputation. The parameter of interest is the population mean of y.

(a) Under the above setup, compute the variance of the imputed point estimator.

(b) Show that the expected value of the naive variance estimator $\hat{V}_I$ (that is obtained by treating the imputed values as if observed and applying the standard variance estimation formula to the imputed data) is equal to

$$E\left\{\hat{V}_I\right\} = V\left\{n^{-1}\sum_{g=1}^{2} n_g \mu_g\right\} + \frac{1}{n(n-1)} E\left\{\sum_{g=1}^{2}\left(n_g - \frac{n_g}{n} - 2\frac{m_g}{n} - \frac{m_g(m_g-1)}{n r_g}\right)\right\}\sigma_g^2.$$

(c) We use the following bias-corrected estimator

$$\hat{V} = \hat{V}_I + \sum_{g=1}^{2} c_g S_{Rg}^2$$

to estimate the variance of the imputed point estimator. Find c_1 and c_2.

(d) Show that the Rao and Shao (1992) jackknife variance estimator defined below is approximately unbiased.

$$\hat{V}_{JK} = \frac{n-1}{n}\sum_{k=1}^{n}\left(\hat{\theta}_I^{(k)} - \hat{\theta}_I\right)^2,$$

where

$$w_i^{(k)} = \begin{cases} 1/(n-1) & \text{if } i \neq k \\ 0 & \text{if } i = k, \end{cases}$$

$$\hat{\theta}_I^{(k)} = \sum_{i=1}^{n} w_i^{(k)}\left(\delta_i y_i + (1-\delta_i) y_i^{*(k)}\right)$$

and

$$y_i^{*(k)} = \bar{y}_g^{(k)} - \bar{y}_g + y_i^*$$

for $i \in A_g$ and A_g is the index set of samples in cell g. Here, $\bar{y}_g$ is the sample mean of y_i (with $\delta_i = 1$) in cell g and $\bar{y}_g^{(k)} = \left(\sum_{i\in A_c} w_i^{(k)}\delta_i y_i\right) / \left(\sum_{i\in A_c} w_i^{(k)}\delta_i\right)$.

8. Assume that we have a random sample of $(\mathbf{x}_i, y_i)$ where only y_i is subject to missingness. Let $\hat{y}_i = m(\mathbf{x}_i;\hat{\beta})$ be the imputed value of y_i, with $m(\cdot)$, $q_i(\cdot)$ being known functions, and $\hat{\beta}$ obtained by solving

$$\sum_{i=1}^{n}\delta_i\left\{y_i - m(\mathbf{x}_i;\beta)\right\} q_i(\beta)\mathbf{x}_i = 0. \tag{4.123}$$

for β. In this case, the deterministic imputation for $\theta = E(Y)$ is given by

$$\hat{\theta}_{dI} = n^{-1}\sum_{i=1}^{n}\left\{\delta_i y_i + (1-\delta_i) m(\mathbf{x}_i;\hat{\beta})\right\}.$$

(a) Under the assumption of $y_i \mid \mathbf{x}_i \sim (m(\mathbf{x}_i;\beta_0), v_i)$ for some $v_i > 0$ and MAR, discuss the optimal choice of q_i in (4.123) that minimizes the variance of $\hat{\beta}$.

(b) Find the linearization variance estimator of $\hat{\theta}_{dI}$.

9. Under the setup of Example 4.8, suppose that we are only interested in estimating $\theta_2 = Pr(Y < c)$. Suppose that a fractional imputation of size m is used where the imputed values are generated from $N(\hat{\beta}_0 + \hat{\beta}_1 x_i, \hat{\sigma}_e^2)$, where $(\hat{\beta}_0, \hat{\beta}_1, \hat{\sigma}_e^2)$ is obtained by the ordinary linear regression from the set of respondents. Discuss how to obtain a consistent variance estimator of the fractionally imputed estimator of θ_2.

10. Assume that (x_1, x_2, y) is a vector of random variables from a multivariate normal distribution. Let the sample be partitioned into three sets, H_1, H_2 and H_3, where (x_{1i}, x_{2i}, y_i) are observed in H_1, (x_{1i}, x_{2i}) are observed in H_2, and only x_{1i} are observed in H_3. That is, x_{1i} are observed throughout the sample, x_{2i} are observed in $H_1 \cup H_2$, and y_i are observed only in H_1. Suppose that we are interested in estimating $\theta = E(Y)$ from the sample. We consider the following imputation estimator of θ:

$$\hat{\theta}_1 = n^{-1}\left\{\sum_{i \in H_1} y_i + \sum_{i \in H_2} \hat{y}_{2i} + \sum_{i \in H_3} \hat{y}_{1i}\right\},$$

where $\hat{y}_{2i} = \hat{\beta}_0 + \hat{\beta}_1 x_{1i} + \hat{\beta}_2 x_{2i}$, $\hat{y}_{1i} = \hat{\beta}_0 + \hat{\beta}_1 x_{1i} + \hat{\beta}_2 \hat{x}_{2i}$, and $\hat{x}_{2i} = \hat{\gamma}_0 + \hat{\gamma}_1 x_{1i}$. The regression coefficients are computed by

$$\begin{pmatrix} \hat{\beta}_0 \\ \hat{\beta}_1 \\ \hat{\beta}_2 \end{pmatrix} = \left\{\sum_{i \in H_1} \begin{pmatrix} 1 \\ x_{1i} \\ x_{2i} \end{pmatrix} \begin{pmatrix} 1 \\ x_{1i} \\ x_{2i} \end{pmatrix}'\right\}^{-1} \sum_{i \in H_1} \begin{pmatrix} 1 \\ x_{1i} \\ x_{2i} \end{pmatrix} y_i$$

and

$$\begin{pmatrix} \hat{\gamma}_0 \\ \hat{\gamma}_1 \end{pmatrix} = \left\{\sum_{i \in H_1 \cup H_2} \begin{pmatrix} 1 \\ x_{1i} \end{pmatrix} \begin{pmatrix} 1 \\ x_{1i} \end{pmatrix}'\right\}^{-1} \sum_{i \in H_1 \cup H_2} \begin{pmatrix} 1 \\ x_{1i} \end{pmatrix} y_i.$$

(a) Compute the variance of $\hat{\theta}$ in terms of the model parameters in the joint distribution of (x_1, x_2, y).

(b) Derive the linearized variance estimator of $\hat{\theta}$.

11. Under the setup in Example 4.9, discuss how to compute the observed information using the formula in (4.106).

Chapter 5

Propensity scoring approach

5.1 Introduction

Assume that the parameter of interest θ_0 is defined implicitly by $E\{U(\theta;Z)\}=0$ and $U(\theta;z)$ is the estimating function for θ_0. Under complete response, a consistent estimator of θ can be obtained by solving

$$\hat{U}_n(\theta) \equiv n^{-1}\sum_{i=1}^{n} U(\theta;z_i)=0 \tag{5.1}$$

for θ. Note that by Lemma 4.1, the solution $\hat{\theta}$ is asymptotically unbiased for θ_0 if $\hat{U}(\theta_0)$ is asymptotically unbiased for zero. We assume that the solution to (5.1) is unique. Under some regularity conditions, $\hat{\theta}$ converges to θ_0 in probability and the limiting distribution of $\hat{\theta}$ is normal.

Suppose that we can write $z_i=(x_i,y_i)$, where x_i are always observed and y_i are subject to missingness. Let δ_i be the response indicator function for y_i that takes the value one if and only if y_i is observed. The complete sample estimating equation (5.1) cannot be directly computed under the existence of missing data. To resolve this problem, we can consider an expected estimating equation approach that computes the conditional expectation of $U(\theta;Z)$ given the observations, which requires correct specification of the conditional distribution of y_i on x_i. That is, we can consider

$$n^{-1}\sum_{i=1}^{n}\left[\delta_i U(\theta;x_i,y_i)+(1-\delta_i)E\{U(\theta;x_i,y_i)\mid x_i,\delta_i=0\}\right]=0$$

and apply the imputation approach discussed in Chapter 4. Such an approach, called the *prediction model approach*, requires correct specification of the prediction model, the model for the conditional distribution of y on x and δ. While the prediction model approach provides efficient estimates when the prediction model is true, correct specification of the prediction model and parameter estimation can be difficult especially when y is vector-valued. In this chapter, we consider an alternative approach based on modeling δ_i using all available information. Note that, even when y_i is a vector, we may have a scalar δ_i for unit nonresponse and so the modeling of δ_i may be easier than the modeling of y_i. This alternative approach based on the model for δ_i is called *response probability model approach*.

Under the existence of missing data, the complete case (CC) method, defined by solving

$$\sum_{i=1}^{n}\delta_i U(\theta;z_i)=0,$$

can lead to a biased estimator of θ unless $Cov(\delta_i,U_i)=0$, where $U_i=U(\theta_0;z_i)$. So, unless the missing mechanism is missing completely at random (MCAR), the CC method results in biased estimation. Furthermore, the CC method does not make use of the observed information of x_i for $\delta_i=0$. Thus, it is not fully efficient.

To correct for the bias, a weighted complete case (WCC) estimator can be obtained by solving

$$\hat{U}_W(\theta) \equiv \frac{1}{n}\sum_{i=1}^{n}\delta_i\frac{1}{\pi_i}U(\theta;z_i)=0, \tag{5.2}$$

where $\pi_i = Pr(\delta_i = 1 \mid z_i)$. The response probability π_i is often called the *propensity score*, termed by Rosenbaum (1983). When the true propensity score is known, as in survey sampling, asymptotic properties of the WCC estimator can be obtained using the sandwich formula. Let $\hat{\theta}_W$ be the (unique) solution to (5.2) with π_i known. The solution is also called the *Horvitz–Thompson estimator* in survey sampling. Using Lemma 4.1, we have

$$\hat{\theta}_W - \theta_0 \cong -\left[E\left\{\dot{U}(\theta_0;Z)\right\}\right]^{-1}\hat{U}_W(\theta_0),$$

where $\dot{U}(\theta;z) = \partial U(\theta;z)/\partial\theta'$. Thus,

$$E\left\{\hat{U}_W(\theta_0)\right\} = E\left[E\left\{\hat{U}_W(\theta_0) \mid \mathbf{z}\right\}\right] = E\left\{\hat{U}_n(\theta_0)\right\} = 0,$$

and the WCC estimator is asymptotically unbiased. Also, the asymptotic variance of $\hat{\theta}_W$ is computed by the sandwich formula

$$V\left(\hat{\theta}_W\right) \cong \tau^{-1}V\left\{\hat{U}_W(\theta_0)\right\}\tau^{-1'},$$

where $\tau = E\left\{\dot{U}(\theta_0;Z)\right\}$ and, assuming that $Cov(\delta_i,\delta_j) = 0$ for $i \neq j$,

$$V\left\{\hat{U}_W(\theta_0)\right\} = V\left\{\hat{U}_n(\theta_0)\right\} + E\left\{n^{-2}\sum_{i=1}^{n}\left(\pi_i^{-1}-1\right)U(\theta;z_i)^{\otimes 2}\right\}.$$

If $z_1,\cdots,z_n$ are independently and identically distributed, then

$$V\left\{\hat{U}_n(\theta_0)\right\} \cong E\left\{\frac{1}{n^2}\sum_{i=1}^{n}U(\theta_0;z_i)^{\otimes 2} - \frac{1}{n}\bar{U}_n(\theta_0)^{\otimes 2}\right\},$$

where $\bar{U}_n(\theta_0) = n^{-1}\sum_{i=1}^{n}U(\theta_0;z_i)$. Thus,

$$\begin{aligned} V\left\{\hat{U}_W(\theta_0)\right\} &= n^{-1}E\left\{n^{-1}\sum_{i=1}^{n}\pi_i^{-1}U(\theta;z_i)^{\otimes 2} - \bar{U}_n(\theta_0)^{\otimes 2}\right\} \\ &\cong E\{n^{-2}\sum_{i=1}^{n}\pi_i^{-1}U(\theta_0;z_i)^{\otimes 2}\}, \end{aligned} \tag{5.3}$$

and a consistent estimator for the variance of $\hat{\theta}_W$ is computed by

$$\hat{V}\left(\hat{\theta}_W\right) = \hat{\tau}^{-1}\hat{V}_u\hat{\tau}^{-1'},$$

where $\hat{\tau} = n^{-1}\sum_{i=1}^{n}\delta_i\pi_i^{-1}\dot{U}(\hat{\theta}_W;z_i)$ and $\hat{V}_u = n^{-2}\sum_{i=1}^{n}\delta_i\pi_i^{-2}U(\hat{\theta}_W;z_i)^{\otimes 2}$.

Example 5.1. *Let the parameter of interest be $\theta = E(Y)$ and we use $U(\theta;z) = (y-\theta)$ to compute θ. The WCC estimator of θ can be written*

$$\hat{\theta}_W = \frac{\sum_{i=1}^{n}\delta_i y_i/\pi_i}{\sum_{i=1}^{n}\delta_i/\pi_i}. \tag{5.4}$$

The asymptotic variance of $\hat{\theta}_W$ in (5.4) is equal to, by (5.3),

$$E\left\{n^{-2}\sum_{i=1}^{n}\pi_i^{-1}(y_i-\theta)^2\right\}. \tag{5.5}$$

In the context of survey sampling, where the population size is equal to N and δ_i corresponds to the sampling indicator function, the estimator (5.4) is called the Hájek estimator. *The asymptotic variance in (5.5) represents the asymptotic variance of the Hájek estimator under Poisson sampling when the finite population is a random sample from an infinite population, called* superpopulation, *and the parameter θ is the superpopulation parameter. See Godambe and Thompson (1986).*

5.2 Regression weighting method

In practice, the propensity score is usually unknown and the theory in Section 5.1 cannot be directly applied. In this section, we introduce a useful technique for obtaining a weighted estimator when the propensity score is unknown. Assume that auxiliary variables $\mathbf{x}_i$ are observed throughout the sample and the response probability satisfies

$$\frac{1}{\pi_i} = \mathbf{x}_i'\lambda \tag{5.6}$$

for all unit i in the sample, where λ is unknown. We assume that an intercept is included in $\mathbf{x}_i$. Under these conditions, according to Fuller et al. (1994), the regression estimator defined by

$$\hat{\theta}_{reg} = \sum_{i=1}^{n} \delta_i w_i y_i, \tag{5.7}$$

where

$$w_i = \left(\frac{1}{n} \sum_{i=1}^{n} \mathbf{x}_i \right)' \left(\sum_{i=1}^{n} \delta_i \mathbf{x}_i \mathbf{x}_i' \right)^{-1} \mathbf{x}_i$$

is asymptotically unbiased for $\theta = E(Y)$. To show this, note that we can write

$$\hat{\theta}_{reg} = \bar{\mathbf{x}}_n' \hat{\beta}_r,$$

where

$$\hat{\beta}_r = \left(\sum_{i=1}^{n} \delta_i \mathbf{x}_i \mathbf{x}_i' \right)^{-1} \sum_{i=1}^{n} \delta_i \mathbf{x}_i y_i.$$

Because an intercept term is included in $\mathbf{x}_i$, we have

$$\hat{\theta}_n \equiv \bar{y}_n = \bar{\mathbf{x}}_n' \hat{\beta}_n, \tag{5.8}$$

where

$$\hat{\beta}_n = \left(\sum_{i=1}^{n} \mathbf{x}_i \mathbf{x}_i' \right)^{-1} \sum_{i=1}^{n} \mathbf{x}_i y_i.$$

Thus, we can write

$$\hat{\theta}_{reg} - \hat{\theta}_n = \bar{\mathbf{x}}_n' \left(\sum_{i=1}^{n} \delta_i \mathbf{x}_i \mathbf{x}_i' \right)^{-1} \sum_{i=1}^{n} \delta_i \mathbf{x}_i \left(y_i - \mathbf{x}_i' \hat{\beta}_n \right)$$

and so

$$E\left(\hat{\theta}_{reg} - \hat{\theta}_n \mid \mathbf{X}, \mathbf{Y} \right) \cong \bar{\mathbf{x}}_n' \left(\sum_{i=1}^{n} \pi_i \mathbf{x}_i \mathbf{x}_i' \right)^{-1} \sum_{i=1}^{n} \pi_i \mathbf{x}_i \left(y_i - \mathbf{x}_i' \hat{\beta}_n \right),$$

where the expectation is taken with respect to the response mechanism. Thus, to show that $\hat{\theta}_{reg}$ is asymptotically unbiased, we have only to show that

$$\sum_{i=1}^{n} \pi_i \mathbf{x}_i \left(y_i - \mathbf{x}_i' \hat{\beta}_n \right) = 0 \tag{5.9}$$

holds. By (5.6) and (5.8), we have

$$0 = \sum_{i=1}^{n} \left(y_i - \mathbf{x}_i \hat{\beta}_n \right) = \sum_{i=1}^{n} \pi_i \left(\lambda' \mathbf{x}_i \right) \left(y_i - \mathbf{x}_i' \hat{\beta}_n \right) = \lambda' \sum_{i=1}^{n} \pi_i \mathbf{x}_i \left(y_i - \mathbf{x}_i' \hat{\beta}_n \right),$$

which implies that (5.9) holds.

Example 5.2. *Assume that the sample is partitioned into G exhaustive and mutually exclusive groups, denoted by $A_1, \cdots, A_G$, where $|A_g| = n_g$ with g being the group indicator. Assume a uniform response mechanism for each group. Thus, we assume that $\pi_i = p_g$ for some $p_g \in (0,1]$ if $i \in A_g$. Let*

$$x_{ig} = \begin{cases} 1 & \text{if } i \in A_g \\ 0 & \text{otherwise.} \end{cases}$$

Then, $\mathbf{x}_i = (x_{i1}, \cdots, x_{iG})$ satisfies (5.6). The regression estimator (5.7) of $\theta = E(Y)$ can be written as

$$\hat{\theta}_{reg} = \frac{1}{n}\sum_{g=1}^{G}\frac{n_g}{r_g}\sum_{i \in A_g}\delta_i y_i = \frac{1}{n}\sum_{g=1}^{G} n_g \bar{y}_{Rg},$$

where $r_g = \sum_{i \in A_g}\delta_i$ is the realized size of respondents in group g and $\bar{y}_{Rg} = \left(\sum_{i \in A_g}\delta_i\right)^{-1}\sum_{i \in A_g}\delta_i y_i$. Because the covariate satisfies (5.6), the regression estimator is asymptotically unbiased and the asymptotic variance of $\hat{\theta}_{reg}$ is

$$V\left(\hat{\theta}_{reg}\right) = V\left(\hat{\theta}_n\right) + E\left\{n^{-2}\sum_{g=1}^{G}\left(\frac{n_g}{r_g} - 1\right)\sum_{i \in A_g}(y_i - \bar{y}_{ng})^2\right\}.$$

If we write

$$V\left(\hat{\theta}_n\right) = E\left\{\frac{1}{n}S_n^2\right\} \doteq E\left\{\frac{1}{n^2}\sum_{i=1}^{n}(y_i - \bar{y}_n)^2\right\},$$

we also have

$$V\left(\hat{\theta}_{reg}\right) \doteq E\left[n^{-2}\sum_{g=1}^{G}\sum_{i \in A_g}\left\{\frac{n_g}{r_g}(y_i - \bar{y}_{ng})^2 + (\bar{y}_{ng} - \bar{y}_n)^2\right\}\right].$$

This form shows that compared with the complete sample case, the between-group variations are unchanged but the within-group variances are increased by n_g/r_g.

To discuss variance estimation of the regression estimator in (5.7) with the covariates $\mathbf{x}_i$ satisfying (5.6), write

$$\begin{aligned} \bar{\mathbf{x}}_n'\hat{\beta}_r &= \bar{\mathbf{x}}_n'\beta + \bar{\mathbf{x}}_n'\left(\hat{\beta}_r - \beta\right) \\ &= \bar{\mathbf{x}}_n'\beta + \bar{\mathbf{x}}_n'\left(\sum_{i=1}^{n}\delta_i\mathbf{x}_i\mathbf{x}_i'\right)^{-1}\sum_{i=1}^{n}\delta_i\mathbf{x}_i(y_i - \mathbf{x}_i'\beta) \\ &\cong \bar{\mathbf{x}}_n'\beta + \bar{\mathbf{x}}_n'\left(\sum_{i=1}^{n}\pi_i\mathbf{x}_i\mathbf{x}_i'\right)^{-1}\sum_{i=1}^{n}\delta_i\mathbf{x}_i(y_i - \mathbf{x}_i'\beta), \end{aligned}$$

where β is the probability limit of $\hat{\beta}_r$. By the fact that 1 is included in $\mathbf{x}_i$ (indicating the existence of an intercept) and by (5.6), it can be shown that

$$\bar{\mathbf{x}}_n'\left(\sum_{i=1}^{n}\pi_i\mathbf{x}_i\mathbf{x}_i'\right)^{-1}\sum_{i=1}^{n}\delta_i\mathbf{x}_i(y_i - \mathbf{x}_i'\beta) = \frac{1}{n}\sum_{i=1}^{n}\frac{\delta_i}{\pi_i}(y_i - \mathbf{x}_i'\beta) \quad (5.10)$$

and

$$V\left(\hat{\theta}_{reg}\right) \doteq V\left(\frac{1}{n}\sum_{i=1}^{n}d_i\right), \quad (5.11)$$

where $d_i = \mathbf{x}_i'\beta + \delta_i\pi_i^{-1}(y_i - \mathbf{x}_i'\beta)$. Variance estimation can be implemented by using a standard variance estimation formula applied to $\hat{d}_i = \mathbf{x}_i'\hat{\beta}_r + \delta_i n w_i(y_i - \mathbf{x}_i'\hat{\beta}_r)$.

5.3 Propensity score method

We now consider a more realistic situation of the response probability being unknown. Suppose the true response probability is parametrically modeled by

$$\pi_i = \pi(z_i; \phi_0)$$

for some $\phi_0 \in \Omega$. If the maximum likelihood estimator of ϕ_0, denoted by $\hat{\phi}$, is available, then the propensity score adjusted (PSA) estimator of θ, denoted by $\hat{\theta}_{PSA}$, can be computed by solving

$$\hat{U}_{PSA}(\theta) \equiv \frac{1}{n}\sum_{i=1}^{n} \delta_i \frac{1}{\hat{\pi}_i} U(\theta; z_i) = 0, \tag{5.12}$$

where $\hat{\pi}_i = \pi(z_i; \hat{\phi})$. Strictly speaking, the PSA estimator in (5.12) is also a function of $\hat{\phi}$. Thus, we can write $(\hat{\theta}_{PSA}, \hat{\phi})$ as the solution to $\hat{U}_1(\theta, \phi) = 0$ and $S(\phi) = 0$, where $\hat{U}_1(\theta, \hat{\phi}) = \hat{U}_{PSA}(\theta)$, with $\hat{U}_{PSA}(\theta)$ defined as in (5.12), and $S(\phi)$ is the score function for ϕ.

The following lemma presents some results on the relationship between the PSA estimating function $\hat{U}_{PSA}(\theta)$ and the score function $S(\phi)$.

Lemma 5.1. *Let*

$$U_1(\theta, \phi) = \sum_{i=1}^{n} u_{i1}(\theta, \phi),$$

where $u_{i1}(\theta, \phi) = u_{i1}(\theta, \phi; z_i, \delta_i)$, be an estimating equation satisfying

$$E\{U_1(\theta_0, \phi_0)\} = 0.$$

Let $\pi_i = \pi_i(\phi)$ be the probability of response. Then,

$$E\{-\partial U_1 / \partial \phi'\} = Cov(U_1, S), \tag{5.13}$$

where S is the score function of ϕ.

Proof. Since $E\{U_1(\theta_0, \phi_0)\} = 0$, we have

$$\begin{aligned} 0 &= \partial E\{U_1(\theta_0, \phi_0)\} / \partial \phi' \\ &= \sum_{i=1}^{n} \frac{\partial}{\partial \phi'} \int u_{i1}(\theta_0, \phi_0) f(\delta_i \mid z_i, \phi_0) f(z_i) d\delta_i dz_i \\ &= \sum_{i=1}^{n} \int \left[\frac{\partial}{\partial \phi'} u_{i1}(\theta_0, \phi_0)\right] f(\delta_i \mid z_i, \phi_0) f(z_i) d\delta_i dz_i \\ &+ \sum_{i=1}^{n} \int u_{i1}(\theta_0, \phi_0) \frac{\partial}{\partial \phi'} [f(\delta_i \mid z_i, \phi_0)] f(z_i) d\delta_i dz_i \\ &= E\{\partial U / \partial \phi'\} + E\left\{U(\theta_0, \phi_0) S(\phi_0)'\right\}, \end{aligned}$$

which proves (5.13). □

If we set $U_1(\theta, \phi) = S(\phi)$, then (5.13) reduces to $E\{-\partial S(\phi)/\partial \phi'\} = E\left\{S(\phi)^{\otimes 2}\right\}$, which is already presented in (2.3).

Under some regularity conditions, the solution $(\hat{\theta}_{PSA}, \hat{\phi})$ to

$$\begin{aligned} \hat{U}_{PSA}(\theta, \phi) &= 0 \\ S(\phi) &= \mathbf{0} \end{aligned}$$

is asymptotically normal with mean $(\theta_0, \phi_0)'$ and variance $A^{-1}BA'^{-1}$, where

$$
\begin{aligned}
A &= \begin{bmatrix} E\{-\partial \hat{U}_{PSA}/\partial \theta'\} & E\{-\partial U_{PSA}/\partial \phi'\} \\ E\{-\partial S/\partial \theta'\} & E\{-\partial S/\partial \phi'\} \end{bmatrix} = \begin{bmatrix} A_{11} & A_{12} \\ 0 & A_{22} \end{bmatrix} \\
B &= \begin{bmatrix} V(\hat{U}_1) & C(\hat{U}_1, S) \\ C(S, \hat{U}_1) & V(S) \end{bmatrix} = \begin{bmatrix} B_{11} & B_{12} \\ B_{21} & B_{22} \end{bmatrix}.
\end{aligned}
$$

Then, using

$$
A^{-1} = \begin{bmatrix} A_{11}^{-1} & -A_{11}^{-1}A_{12}A_{22}^{-1} \\ 0 & A_{22}^{-1} \end{bmatrix},
$$

we have

$$
Var\left(\hat{\theta}_{PSA}\right) \cong A_{11}^{-1}\left[B_{11} - A_{12}A_{22}^{-1}B_{21} - B_{12}A_{22}^{-1}A'_{12} + A_{12}A_{22}^{-1}B_{22}A_{22}^{-1}A'_{12}\right]A_{11}'^{-1}. \tag{5.14}
$$

By Lemma 5.1, $B_{22} = A_{22}$ and $B_{12} = A_{12}$. Thus,

$$
V\left(\hat{\theta}_{PSA}\right) \cong A_{11}^{-1}\left[B_{11} - B_{12}B_{22}^{-1}B_{21}\right]A_{11}'^{-1}. \tag{5.15}
$$

Note that $\hat{\theta}_W = \hat{\theta}_W(\phi_0)$ computed from (5.2) with known π_i satisfies

$$
V\left(\hat{\theta}_W\right) \cong A_{11}^{-1}B_{11}A_{11}^{-1'}.
$$

Ignoring the smaller order terms, we have

$$
V\left(\hat{\theta}_W\right) \geq V\left(\hat{\theta}_{PSA}\right). \tag{5.16}
$$

The result of (5.16) means that the PSA estimator with estimated π_i is more efficient than the PSA estimator with known π_i. Such contradictory phenomenon has been discussed in Rosenbaum (1987), Robins et al. (1994), and Kim and Kim (2007). See also Henmi and Eguchi (2004).

Write $\hat{\theta}_{PSA} = \hat{\theta}_W(\hat{\phi})$, and another way of understanding (5.15) is

$$
V\left(\hat{\theta}_{PSA}\right) \cong E\left\{V\left(\hat{\theta}_W \mid \hat{\phi}^{\perp}\right)\right\} \cong E\left\{V\left(\hat{\theta}_W \mid S^{\perp}\right)\right\}, \tag{5.17}
$$

where

$$
V\left(Y \mid X^{\perp}\right) = V(Y) - C(Y,X)\{V(X)\}^{-1}C(X,Y),
$$

and $S = S(\phi)$ is the score function of ϕ. Here, under the joint asymptotic normality of $(\hat{\theta}_W, \hat{\phi})$, we have

$$
\begin{aligned}
V\left(\hat{\theta}_W \mid \hat{\phi}\right) &= V(\hat{\theta}_W) - C(\hat{\theta}_W, \hat{\phi})\left\{V(\hat{\phi})\right\}^{-1}C(\hat{\phi}, \hat{\theta}_W) \\
&\cong V(\hat{\theta}_W) - C(\hat{\theta}_W, S)\{\mathcal{I}(\phi_0)\}C(S, \hat{\theta}_W) \\
&\cong V(\hat{\theta}_W) - C(\hat{\theta}_W, S)\{V(S)\}^{-1}C(S, \hat{\theta}_W),
\end{aligned}
$$

where the second equality follows from $\hat{\phi} - \phi_0 \cong \{\mathcal{I}(\phi_0)\}^{-1}S(\phi_0)$ and $V(\hat{\phi}) \cong \{\mathcal{I}(\phi_0)\}^{-1}$. Thus, the PSA estimator $\hat{\theta}_{PSA}$ with $\hat{\pi}_i = \pi_i(\hat{\phi})$, where $\hat{\phi}$ is from the maximum likelihood method, can be viewed as a Rao-Blackwellization of $\hat{\theta}_W$ by making use of the score function $S(\phi)$. That is, we can express

$$
\begin{aligned}
\hat{\theta}_{PSA} &\cong \hat{\theta}_W - C(\hat{\theta}_W, \hat{\phi})\left\{V(\hat{\phi})\right\}^{-1}(\hat{\phi} - \phi_0) \\
&\cong \hat{\theta}_W - C(\hat{\theta}_W, S)\{V(S)\}^{-1}S(\phi_0).
\end{aligned}
$$

If $\hat{\phi}$ is computed from some estimating equation other than the score equation, this property does not hold and the variance should be computed directly from the sandwich formula in (5.14).

The variance formula is also useful for deriving a variance estimator for the PSA estimators. If the response mechanism is ignorable, then we can further write

$$\pi_i = \pi(x_i; \phi_0) \tag{5.18}$$

for some $\phi_0 \in \Omega$, where x_i is completely observed in the sample. In this case, the propensity score can be estimated by the maximum likelihood method that solves

$$S(\phi) \equiv \sum_{i=1}^{n} \{\delta_i - \pi(x_i; \phi)\} \frac{1}{\pi(x_i; \phi)\{1 - \pi(x_i; \phi)\}} \dot{\pi}(x_i; \phi) = 0, \tag{5.19}$$

where $\dot{\pi}(x_i; \phi) = \partial \pi(x_i; \phi) / \partial \phi$. Under the logistic regression model

$$\pi(x_i; \phi) = \frac{\exp(\phi_0 + x_i\phi_1)}{1 + \exp(\phi_0 + x_i\phi_1)},$$

we have $\dot{\pi}(x_i; \phi) = \pi(x_i; \phi)\{1 - \pi(x_i; \phi)\}(1, x_i)$.

Using (5.15), a plug-in variance estimator of the PSA estimator is computed by

$$\hat{V}\left(\hat{\theta}_{PSA}\right) = \hat{A}_{11}^{-1}\left[\hat{B}_{11} - \hat{B}_{12}\hat{B}_{22}^{-1}\hat{B}_{21}\right]\hat{A}_{11}^{\prime -1},$$

where $\hat{A}_{11} = n^{-1}\sum_{i=1}^{n} \delta_i \pi_i^{-1} \dot{U}(\hat{\theta}; z_i)$ and

$$\begin{aligned}
\hat{B}_{11} &= n^{-2}\sum_{i=1}^{n} \delta_i \hat{\pi}_i^{-2} U(\hat{\theta}; z_i)^{\otimes 2} \\
\hat{B}_{12} &= n^{-2}\sum_{i=1}^{n} \delta_i \hat{\pi}_i^{-1}(\hat{\pi}_i^{-1} - 1) U(\hat{\theta}; z_i)\mathbf{h}_i \\
\hat{B}_{22} &= n^{-2}\sum_{i=1}^{n} \delta_i \hat{\pi}_i^{-1}(\hat{\pi}_i^{-1} - 1)\mathbf{h}_i\mathbf{h}_i',
\end{aligned}$$

where $\hat{\theta} = \hat{\theta}_{PSA}$ and $\mathbf{h}_i = \dot{\pi}_i/(1 - \pi_i)$.

Example 5.3. *We consider the case of $\theta = E(Y)$ in Example 5.1 when the true response probability is unknown and is assumed to follow a parametric model (5.18). There are two types of PSA estimators for θ. The first one is*

$$\hat{\theta}_{PSA1} = \frac{1}{n}\sum_{i=1}^{n} \frac{\delta_i}{\hat{\pi}_i} y_i$$

and the second one is a Hájek-type estimator of the form

$$\hat{\theta}_{PSA2} = \frac{\sum_{i=1}^{n} \delta_i y_i / \hat{\pi}_i}{\sum_{i=1}^{n} \delta_i / \hat{\pi}_i}.$$

For the variance of $\hat{\theta}_{PSA1}$, use (5.15) with $\hat{U}_1 = n^{-1}\sum_{i=1}^{n}(\delta_i y_i/\hat{\pi}_i - \theta)$, then

$$\begin{aligned}
V\left(\hat{\theta}_{PSA1}\right) &\cong V\left(\hat{\theta}_{W1}\right) - C\left(\hat{\theta}_{W1}, S\right)\{V(S)\}^{-1} C\left(S, \hat{\theta}_{W1}\right) \\
&= V\left\{\hat{\theta}_{W1} - B^*(\phi)' S(\phi)\right\},
\end{aligned}$$

where $\hat{\theta}_{W1} = n^{-1}\sum_{i=1}^{n} \delta_i y_i/\pi_i$ and $B^*(\phi) = [V\{S(\phi)\}]^{-1} C\{S(\phi), \hat{\theta}_{W1}\}$. If the score function is of the form

$$S(\phi) = \sum_{i=1}^{n}\left\{\frac{\delta_i}{\pi_i(\phi)} - 1\right\}\mathbf{h}_i(\phi),$$

then

$$B^*(\phi) = \left\{ \sum_{i=1}^{n} (\pi_i^{-1} - 1)\mathbf{h}_i\mathbf{h}_i' \right\}^{-1} \sum_{i=1}^{n} (\pi_i^{-1} - 1)\mathbf{h}_i y_i \tag{5.20}$$

and the asymptotic variance of $\hat{\theta}_{PSA1}$ *is equal to the variance of*

$$\hat{\theta}_{PSA1,l} = \frac{1}{n}\sum_{i=1}^{n} B^*(\phi)'\mathbf{h}_i(\phi) + \frac{1}{n}\sum_{i=1}^{n} \frac{\delta_i}{\pi_i}\left\{y_i - B^*(\phi)'\mathbf{h}_i(\phi)\right\}, \tag{5.21}$$

with the subscript l denoting linearization. Also, note that the variance $\hat{\theta}_{PSA1,l}$ *can be written as*

$$V\left(\hat{\theta}_{PSA1,l}\right) = V\left(\frac{1}{n}\sum_{i=1}^{n} y_i\right) + E\left\{\frac{1}{n^2}\sum_{i=1}^{n}\frac{1-\pi_i}{\pi_i}\left\{y_i - B^*(\phi)'\mathbf{h}_i(\phi)\right\}^2\right\}$$

and $B^*(\phi)'$ *in (5.20) minimizes the second term of the variance. For variance estimation, we can write*

$$\hat{\theta}_{PSA1,l} = \frac{1}{n}\sum_{i=1}^{n} d_i(\phi),$$

where

$$d_i(\phi) = B^*(\phi)'\mathbf{h}_i(\phi) + \frac{\delta_i}{\pi_i}\left\{y_i - B^*(\phi)'\mathbf{h}_i(\phi)\right\}.$$

With the linearized pseudo values $d_i(\phi)$ *directly used for variance estimation, the variance of* $\hat{\theta}_{PSA1,l}$ *is consistently estimated by*

$$\frac{1}{n}\frac{1}{n-1}\sum_{i=1}^{n}\left(\hat{d}_i - \frac{1}{n}\sum_{j=1}^{n}\hat{d}_j\right)^2,$$

with $\hat{d}_i = d_i(\hat{\phi})$.

For the Hájek-type estimator $\hat{\theta}_{PSA2}$*, we can apply the same argument to show that* $\hat{\theta}_{PSA2}$ *is asymptotically equivalent to*

$$\hat{\theta}_{PSA2,l} = \frac{1}{n}\sum_{i=1}^{n} B_2^*(\phi)'\mathbf{h}_i(\phi) + \frac{1}{n}\sum_{i=1}^{n}\frac{\delta_i}{\pi_i}\left\{y_i - \hat{\theta}_W - B_2^*(\phi)'\mathbf{h}_i(\phi)\right\}, \tag{5.22}$$

where

$$B_2^* = \left\{\sum_{i=1}^{n}(\pi_i^{-1} - 1)\mathbf{h}_i\mathbf{h}_i'\right\}^{-1}\sum_{i=1}^{n}(\pi_i^{-1} - 1)\mathbf{h}_i(y_i - \hat{\theta}_W).$$

The pseudo value for variance estimation is

$$d_i(\phi) = B_2^*(\phi)'\mathbf{h}_i(\phi) + \frac{\delta_i}{\pi_i}\left\{y_i - \hat{\theta}_W(\phi) - B_2^*(\phi)'\mathbf{h}_i(\phi)\right\}.$$

To improve the efficiency of the PSA estimator in (5.12), one can consider a class of estimating equations of the form

$$\sum_{i=1}^{n}\delta_i\frac{1}{\hat{\pi}_i}\left\{U(\theta; x_i, y_i) - b(\theta; x_i)\right\} + \sum_{i=1}^{n} b(\theta; x_i) = 0, \tag{5.23}$$

where $b(\theta; x_i)$ is to be determined. Assume that the estimated response probability is computed by $\hat{\pi}_i = \pi(x_i; \hat{\phi})$, where $\hat{\phi}$ is computed by

$$\sum_{i=1}^{n}\left(\frac{\delta_i}{\pi(x_i;\phi)} - 1\right)\mathbf{h}_i(\phi) = \mathbf{0}$$

for some $\mathbf{h}_i(\phi) = \mathbf{h}(x_i;\phi)$. The score equation in (5.19) uses $\mathbf{h}_i(\phi) = \dot{\pi}_i(\phi)/\{1-\pi_i(\phi)\}$. The following theorem, which was originally proved by Robins et al. (1994), presents asymptotic properties of the solution to the augmented estimating equation (5.23).

Theorem 5.1. *Assume that the response probability $Pr(\delta = 1 \mid x, y) = \pi(x)$ does not depend on the value of y. Let $\hat{\theta}_b$ be the solution to (5.23) given $b(\theta; x_i)$. Under some regularity conditions, $\hat{\theta}_b$ is consistent and its asymptotic variance satisfies*

$$V(\hat{\theta}_b) \geq n^{-1}\tau^{-1}\left[V\{E(U \mid X)\} + E\left\{\pi^{-1}V(U \mid X)\right\}\right](\tau^{-1})', \tag{5.24}$$

where $\tau = E(\partial U/\partial \theta')$ and the equality holds when $b(\theta;x_i) = E\{U(\theta;x_i,y_i) \mid x_i\}$.

Proof. We first consider the case when the true response probability $\pi(x) = Pr(\delta = 1 \mid x)$ is known. Let

$$U_b(\theta) = \frac{1}{n}\sum_{i=1}^{n} b(\theta;x_i) + \frac{1}{n}\sum_{i=1}^{n}\frac{\delta_i}{\pi_i}\{U(\theta;x_i,y_i) - b(\theta;x_i)\},$$

where $\pi_i = \pi(x_i)$ and $b(\theta;\mathbf{x}_i)$ is to be determined. Since $E\{U_b(\theta_0)\} = E\{U(\theta_0;x,y)\} = 0$, by Lemma 4.1, the solution $\hat{\theta}_b$ to $U_b(\theta) = 0$ is asymptotically unbiased for θ_0 and

$$V(\hat{\theta}_b) \cong \tau^{-1}V\{U_b(\theta_0)\}(\tau^{-1})',$$

where $\tau = E\{\partial U_b(\theta_0)/\partial\theta'\} = E\{\partial U(\theta_0;x,y)/\partial\theta'\}$. Now, define $b^*(\theta;x_i) = E\{U(\theta;x_i,y_i) \mid x_i\}$ and writing

$$U_b(\theta) = U_{b^*}(\theta) + D_b(\theta), \tag{5.25}$$

where

$$\begin{aligned} U_{b^*}(\theta) &= \frac{1}{n}\sum_{i=1}^{n} b^*(\theta;x_i) + \frac{1}{n}\sum_{i=1}^{n}\frac{\delta_i}{\pi_i}\{U(\theta;x_i,y_i) - b^*(\theta;x_i)\} \\ D_b(\theta) &= \frac{1}{n}\sum_{i=1}^{n}\left(\frac{\delta_i}{\pi_i} - 1\right)\{b^*(\theta;x_i) - b(\theta;x_i)\} \end{aligned}$$

and $b^*(\theta;x_i) = E\{U(\theta;x_i,y_i) \mid x_i\}$, we have

$$V\{U_b(\theta)\} = V\{U_{b^*}(\theta)\} + V\{D_b(\theta)\} + 2Cov\{U_{b^*}(\theta), D_b(\theta)\}.$$

Note that

$$Cov\{U_{b^*}(\theta), D_b(\theta)\} = E\left\{n^{-2}\sum_{i=1}^{n}\left(\frac{1}{\pi_i} - 1\right)(U_i - b_i^*)(b_i^* - b_i)\right\},$$

where $U_i = U(\theta;x_i,y_i)$, $b_i = b(\theta;x_i)$, and $b_i^* = b^*(\theta;x_i)$. Because $E(U_i \mid x_i) = b_i^*$, the above covariance term is equal to zero and

$$V(\hat{\theta}_b) \cong \tau^{-1}[V\{U_{b^*}(\theta_0)\} + V\{D_b(\theta_0)\}](\tau^{-1})' \geq \tau^{-1}V\{U_{b^*}(\theta_0)\}(\tau^{-1})'.$$

Since

$$\begin{aligned} V\{U_{b^*}(\theta)\} &= V\left\{\frac{1}{n}\sum_{i=1}^{n} U(\theta;x_i,y_i)\right\} + E\left\{\frac{1}{n^2}\sum_{i=1}^{n}\left(\frac{1}{\pi_i} - 1\right)(U_i - b_i^*)^2\right\} \\ &= n^{-1}V(U) + n^{-1}E\left\{(\pi^{-1} - 1)V(U \mid X)\right\} \\ &= n^{-1}V\{E(U \mid X)\} + n^{-1}E\left\{\pi^{-1}V(U \mid X)\right\}, \end{aligned}$$

result (5.24) holds when the true response probability is known.

Now, to discuss the case with the response probability unknown, let $\hat{\pi}_i = \pi(x_i; \hat{\phi})$, where $\hat{\phi}$ is estimated by solving $U_2(\phi) = 0$. Write

$$U_{1b}(\theta,\phi) = \frac{1}{n}\sum_{i=1}^{n} b(\theta;x_i) + \frac{1}{n}\sum_{i=1}^{n}\frac{\delta_i}{\pi(x_i;\phi)}\left\{U(\theta;x_i,y_i) - b(\theta;x_i)\right\},$$

then the solution $(\hat{\theta}_b, \hat{\phi})$ to $U_{1b}(\theta,\phi) = 0$ and $U_2(\phi) = 0$ is consistent for (θ_0, ϕ_0). Using the same argument for deriving (5.14), we can show

$$V(\hat{\theta}_b) \cong \tau^{-1} V\{U_{1b}(\theta_0,\phi_0) - CU_2(\phi_0)\}(\tau^{-1})', \tag{5.26}$$

where $C = (\partial U_{1b}/\partial\phi')(\partial U_2/\partial\phi')^{-1}$. Now, similar to (5.25), we can write

$$U_{1b}(\theta,\phi) - CU_2(\phi) = U_{b^*}(\theta) + D_{2b}(\theta,\phi),$$

where $D_{2b}(\theta,\phi) = D_b(\theta) - CU_2(\phi)$. Because $U_2(\phi)$ does not depend on y, we can show that

$$Cov(U_{b^*}, D_{2b}) = E\{n^{-2}\sum_{i=1}^{n}(\pi_i^{-1} - 1)(U_i - b_i^*)(b_i^* - b_i - \mathbf{C}\mathbf{h}_i)\} = 0$$

and result (5.24) follows. □

Note that, for the choice of $b^*(\theta;x_i) = E\{U(\theta;x_i,y_i) \mid x_i\}$, the resulting estimator achieves the lower bound (semiparametric variance lower bound) in (5.24) regardless of whether $\pi_i = Pr(\delta_i = 1 \mid x_i)$ is known or estimated. For the case of known π_i, the lower bound in (5.24) was also discussed by Godambe and Joshi (1965) and Isaki and Fuller (1982) in the context of survey sampling.

Example 5.4. *Consider the sample from a linear regression model*

$$y_i = \mathbf{x}_i'\beta + e_i, \tag{5.27}$$

where e_i's are independent with $E(e_i \mid \mathbf{x}_i) = 0$. Assume that $\mathbf{x}_i$ is available from the full sample and y_i is observed only when $\delta_i = 1$. The response propensity model follows from the logistic regression model with $logit(\pi_i) = \mathbf{x}_i'\phi$. We are interested in estimating $\theta = E(Y)$ from the partially observed data.

To construct the optimal estimator that achieves the minimum variance in (5.24), we can use $U_i(\theta) = y_i - \theta$ and $b_i^*(\theta) = \mathbf{x}_i'\beta - \theta$. Thus, the optimal estimator using $\hat{b}_i^*(\theta) = \mathbf{x}_i'\hat{\beta} - \theta$ in (5.23) is given by

$$\hat{\theta}_{opt}(\hat{\beta}) = \frac{1}{n}\sum_{i=1}^{n}\frac{\delta_i}{\hat{\pi}_i}y_i + \frac{1}{n}\left(\sum_{i=1}^{n}\mathbf{x}_i - \sum_{i=1}^{n}\frac{\delta_i}{\hat{\pi}_i}\mathbf{x}_i\right)'\hat{\beta}, \tag{5.28}$$

where $\hat{\beta}$ is any estimator of β satisfying the $\sqrt{n}$-consistency, which means $\sqrt{n}(\hat{\beta} - \beta) = O_p(1)$, with $X_n = O_p(1)$ denoting that X_n is bounded in probability. Note that the choice of $\hat{\beta}$ does not play any leading role in the asymptotic variance of $\hat{\theta}_{opt}(\hat{\beta})$. This is because

$$\hat{\theta}_{opt}(\hat{\beta}) \cong \hat{\theta}_{opt}(\beta_0) + E\left\{\frac{\partial}{\partial\beta'}\hat{\theta}_{opt}(\beta_0)\right\}\left(\hat{\beta} - \beta_0\right) \tag{5.29}$$

and, under the correct response model,

$$E\left\{\frac{\partial}{\partial\beta'}\hat{\theta}_{opt}(\beta_0)\right\} = E\left\{\frac{1}{n}\left(\sum_{i=1}^{n}\mathbf{x}_i - \sum_{i=1}^{n}\frac{\delta_i}{\hat{\pi}_i}\mathbf{x}_i\right)\right\} \cong \mathbf{0}$$

and so the second term of (5.29) becomes negligible. Furthermore, it can be shown that the choice of $\hat{\phi}$ in $\hat{\pi}_i = \pi_i(\hat{\phi})$ does not matter much either as long as the regression model holds. See Exercise 5.3.

5.4 Optimal estimation

In Example 5.4, the optimal PSA estimator is discussed under the assumption that the prediction model, which is the regression model (5.27) in this example, is correctly specified. If the regression model is not true, then optimality is not achieved. We now discuss optimal estimation with the propensity score method without introducing the prediction model. We assume that the propensity score is computed by (5.19). In general, the PSA estimator $\hat{\theta}$ applied to $\theta = E(X)$ is not equal to the complete sample estimator $\hat{\theta}_n = n^{-1}\sum_{i=1}^{n} x_i$. Thus, the complete sample estimator $\bar{x}_n$ can be used to improve the efficiency of the PSA estimator.

To discuss efficient estimation, we consider a more general problem of minimizing the objective function

$$Q = \begin{pmatrix} \hat{X}_1 - \mu_x \\ \hat{X}_2 - \mu_x \\ \hat{Y} - \mu_y \end{pmatrix}' \begin{pmatrix} V(\hat{X}_1) & C(\hat{X}_1, \hat{X}_2) & C(\hat{X}_1, \hat{Y}) \\ C(\hat{X}_1, \hat{X}_2) & V(\hat{X}_2) & C(\hat{X}_2, \hat{Y}) \\ C(\hat{X}_1, \hat{Y}) & C(\hat{X}_2, \hat{Y}) & V(\hat{Y}) \end{pmatrix}^{-1} \begin{pmatrix} \hat{X}_1 - \mu_x \\ \hat{X}_2 - \mu_x \\ \hat{Y} - \mu_y \end{pmatrix}, \tag{5.30}$$

where $\hat{X}_1$ and $\hat{X}_2$ are two unbiased estimators of μ_x and $\hat{Y}$ is an unbiased estimator of μ_y. The solution to the minimization is called the *GLS estimator* or simply the *optimal estimator.* The following lemma presents the optimal estimator in the sense that it has minimal asymptotic variance among linear estimators.

Lemma 5.2. *The optimal (GLS) estimator of (μ_x, μ_y) that minimizes Q in (5.30) is given by*

$$\hat{\mu}_x^* = \alpha^* \hat{X}_1 + (1 - \alpha^*) \hat{X}_2 \tag{5.31}$$

and

$$\hat{\mu}_y^* = \hat{Y} + B_1 \left(\hat{\mu}_x^* - \hat{X}_1\right) + B_2 \left(\hat{\mu}_x^* - \hat{X}_2\right), \tag{5.32}$$

where

$$\alpha^* = \frac{V(\hat{X}_2) - C(\hat{X}_1, \hat{X}_2)}{V(\hat{X}_1) + V(\hat{X}_2) - 2C(\hat{X}_1, \hat{X}_2)}$$

and

$$\begin{pmatrix} B_1 \\ B_2 \end{pmatrix} = \begin{pmatrix} V(\hat{X}_1) & C(\hat{X}_1, \hat{X}_2) \\ C(\hat{X}_1, \hat{X}_2) & V(\hat{X}_2) \end{pmatrix}^{-1} \begin{pmatrix} C(\hat{X}_1, \hat{Y}) \\ C(\hat{X}_2, \hat{Y}) \end{pmatrix}. \tag{5.33}$$

Proof. First, do a mental partition of the matrix component of Q located in the middle of (5.30). Using the inverse of the partitioned matrix, we can write

$$Q = Q_1 + Q_2$$

where

$$Q_1 = \begin{pmatrix} \hat{X}_1 - \mu_x \\ \hat{X}_2 - \mu_x \end{pmatrix}' \begin{pmatrix} V(\hat{X}_1) & C(\hat{X}_1, \hat{X}_2) \\ C(\hat{X}_1, \hat{X}_2) & V(\hat{X}_2) \end{pmatrix}^{-1} \begin{pmatrix} \hat{X}_1 - \mu_x \\ \hat{X}_2 - \mu_x \end{pmatrix},$$

$$Q_2 = \left\{\hat{Y} - \mu_y - B_1(\hat{X}_1 - \mu_x) - B_2(\hat{X}_2 - \mu_x)\right\}' V_{ee}^{-1} \left\{\hat{Y} - \mu_y - B_1(\hat{X}_1 - \mu_x) - B_2(\hat{X}_2 - \mu_x)\right\},$$

and $V_{ee} = V(\hat{Y}) - (B_1, B_2)\{V(\hat{X}_1, \hat{X}_2)\}^{-1}(B_1, B_2)'$. Minimizing Q_1 with respect to μ_x yields $\hat{\mu}_x^*$ in (5.31) and minimizing Q_2 with respect to μ_y given $\hat{\mu}_x^*$ yields $\hat{\mu}_y^*$ in (5.32). □

The optimal estimators in (5.31) and (5.32) are unbiased with minimum variance in the class of linear estimators of $\hat{X}_1$, $\hat{X}_2$, and $\hat{Y}$. The optimal estimator of μ_y takes the form of a regression estimator with $\hat{\mu}_x^*$ as the control. That is, the optimal estimator is the predicted value of $\hat{y} = b_0 + b_1 x$ evaluated at $x = \hat{\mu}_x^*$. Since $\hat{\mu}_x^* - \hat{X}_1 = (1 - \alpha^*)(\hat{X}_2 - \hat{X}_1)$ and $\hat{\mu}_x^* - \hat{X}_2 = -\alpha^*(\hat{X}_2 - \hat{X}_1)$, we can express

$$\begin{aligned} \hat{\mu}_y^* &= \hat{Y} + \{B_1(1 - \alpha^*) - B_2\alpha^*\}(\hat{X}_2 - \hat{X}_1) \\ &= \hat{Y} - C\left(\hat{Y}, \hat{X}_2 - \hat{X}_1\right)\left\{V\left(\hat{X}_2 - \hat{X}_1\right)\right\}^{-1}(\hat{X}_2 - \hat{X}_1). \end{aligned}$$

Under the missing data setup where $\mathbf{x}_i$ is always observed and y_i is subject to missingness, if we know π_i, then we can use $\hat{X}_1 = n^{-1}\sum_{i=1}^n \mathbf{x}_i = \hat{X}_n$, $\hat{X}_2 = n^{-1}\sum_{i=1}^n \delta_i \mathbf{x}_i/\pi_i = \hat{X}_W$, and $\hat{Y} = n^{-1}\sum_{i=1}^n \delta_i y_i/\pi_i = \hat{Y}_W$. In this case, $C(\hat{X}_2, \hat{X}_1) = V(\hat{X}_1)$ and so $\alpha^* = 1$, which leads to $\hat{\mu}_x^* = \bar{X}_1$. In this case, the optimal estimator of μ_y reduces to

$$\begin{aligned} \hat{\mu}_y^* &= \hat{Y} + C\left(\hat{Y}, \hat{X}_2 - \hat{X}_1\right)\left\{V\left(\hat{X}_2 - \hat{X}_1\right)\right\}^{-1}(\hat{X}_1 - \hat{X}_2) \\ &= \hat{Y}_W + (\hat{X}_n - \hat{X}_W)' B^*, \end{aligned}$$

where

$$B^* = E\left(\sum_{i=1}^n \frac{1-\pi_i}{\pi_i}\mathbf{x}_i\mathbf{x}_i'\right)^{-1} E\left(\sum_{i=1}^n \frac{1-\pi_i}{\pi_i}\mathbf{x}_i y_i\right).$$

In practice, we cannot estimate B^* and have to resort to a plug-in estimator

$$\hat{\theta}_{opt} = \hat{Y}_W + (\hat{X}_n - \hat{X}_W)'\hat{B}^*, \tag{5.34}$$

where

$$\hat{B}^* = \left(\sum_{i=1}^n \delta_i \frac{1-\pi_i}{\pi_i^2}\mathbf{x}_i\mathbf{x}_i'\right)^{-1}\left(\sum_{i=1}^n \delta_i \frac{1-\pi_i}{\pi_i^2}\mathbf{x}_i y_i\right).$$

The estimator (5.34) is the asymptotically optimal estimator among the class of linear unbiased estimators. Furthermore, it satisfies the calibration constraint that establishes the identity between the optimal estimator and the full sample estimator under the special case of $y_i = \mathbf{x}_i'\alpha$ for some α. To see this, note that $y_i = \mathbf{x}_i'\alpha$ for all i implies $\hat{B}^* = \alpha$ and

$$\begin{aligned} \hat{\theta}_{opt} &= n^{-1}\sum_{i=1}^n \pi_i^{-1}\delta_i \mathbf{x}_i'\alpha + \left\{n^{-1}\sum_{i=1}^n \mathbf{x}_i - n^{-1}\sum_{i=1}^n \pi_i^{-1}\delta_i\mathbf{x}_i\right\}'\alpha \\ &= n^{-1}\sum_{i=1}^n \mathbf{x}_i'\alpha = n^{-1}\sum_{i=1}^n y_i. \end{aligned}$$

The property $\hat{\theta}_{opt} = \hat{\theta}_n$ for $y_i = \mathbf{x}_i'\alpha$ can be called the *external consistency property*.

We now discuss the case when π_i is unknown and is estimated by $\hat{\pi}_i = \pi_i(\hat{\phi})$, where $\hat{\phi}$ is the maximum likelihood estimator of ϕ in the response probability model. In this case, the optimal estimator of μ_x is still equal to $\bar{x}_n$, but the optimal estimator of $\theta = E(Y)$ in (5.34) using $\hat{\pi}_i$ instead of π_i is not optimal because the covariance between $\hat{Y}_{PSA}$ and $(\hat{X}_{PSA}, \hat{X}_n)$ is different from the covariance between $\hat{Y}_W$ and $(\hat{X}_W, \hat{X}_n)$. To construct the optimal estimator, we can consider an estimator of the form

$$\hat{\theta}_B = \hat{Y}_{PSA} + (\hat{X}_n - \hat{X}_{PSA})B,$$

indexed by B, and find the optimal coefficient B^* that minimizes the variance of $\hat{\theta}_B$. The solution is

$$B^* = \left\{V(\hat{X}_{PSA} - \hat{X}_n)\right\}^{-1} C\left(\hat{X}_{PSA} - \hat{X}_n, \hat{Y}_{PSA}\right).$$

Using the argument for (5.17), we can write

$$B^* = \left\{V(\hat{X}_W - \hat{X}_n \mid S^{\perp})\right\}^{-1} C\left(\hat{X}_W - \hat{X}_n, \hat{Y}_W \mid S^{\perp}\right). \tag{5.35}$$

Note that the optimal estimator from minimizing Q in (5.30) with $\hat{X}_1 = \hat{X}_n$, $\hat{X}_2 = \hat{X}_{PSA}$, and $\hat{Y} = \hat{Y}_{PSA}$ can be obtained by minimizing

$$Q = \left(\hat{\mathbf{Z}} - \mu_z\right)'\left\{V\left(\hat{\mathbf{Z}}_0 \mid S^{\perp}\right)\right\}^{-1}\left(\hat{\mathbf{Z}} - \mu_z\right), \tag{5.36}$$

where $\hat{\mathbf{Z}} = (\hat{X}_n, \hat{X}_{PSA}, \hat{Y}_{PSA})'$, $\hat{\mathbf{Z}}_0 = (\hat{X}_n, \hat{X}_W, \hat{Y}_W)'$ and $\mu_z = (\mu_x, \mu_x, \mu_y)'$. The optimal estimator minimizing Q in (5.36) can also be obtained by minimizing the augmented Q given by

$$Q^*(\mu_z, \phi) = \begin{pmatrix} \hat{\mathbf{Z}}_0 - \mu_z \\ S(\phi) \end{pmatrix}' \left\{ \begin{array}{cc} V(\hat{\mathbf{Z}}_0) & C\{\hat{\mathbf{Z}}_0, S(\phi)\} \\ C\{S(\phi), \hat{\mathbf{Z}}_0\} & V\{S(\phi)\} \end{array} \right\}^{-1} \begin{pmatrix} \hat{\mathbf{Z}}_0 - \mu_z \\ S(\phi) \end{pmatrix}. \tag{5.37}$$

To see the equivalence, note that we can establish

$$Q^*(\mu_z, \phi) = Q_1(\mu_z \mid \phi) + Q_2(\phi), \tag{5.38}$$

where

$$Q_1(\mu_z \mid \phi) = \left\{\hat{\mathbf{Z}}_0 - \mu_z - BS(\phi)\right\}' \left\{V\left(\hat{\mathbf{Z}}_0 \mid S^{\perp}\right)\right\}^{-1} \left\{\hat{\mathbf{Z}}_0 - \mu_z - BS(\phi)\right\},$$

with $B = C\left\{\hat{\mathbf{Z}}_0, S(\phi)\right\} [V\{S(\phi)\}]^{-1}$ and

$$Q_2(\phi) = S(\phi)' \{V(S)\}^{-1} S(\phi).$$

If $\hat{\phi}$ satisfies $S(\hat{\phi}) = 0$ then $Q_2(\hat{\phi}) = 0$ and $Q_1\left(\mu_z \mid \hat{\phi}\right) = Q$ in (5.36). Thus, the effect of using an estimated propensity score can be easily taken into account by simply adding the score function for ϕ into the Q term. The optimization method based on the augmented Q in (5.37) is especially useful for handling longitudinal missing data where the missing pattern is more complicated. See Zhou and Kim (2012).

Example 5.5. *Under the response model where the score function for ϕ is*

$$S(\phi) = \frac{1}{n} \sum_{i=1}^{n} \left\{ \frac{\delta_i}{\pi_i(\phi)} - 1 \right\} \mathbf{h}_i(\phi),$$

the optimal coefficient in (5.35) can be written as

$$B_1^* = \left\{V_{xx} - V_{xs} V_{ss}^{-1} V_{sx}\right\}^{-1} \left\{V_{xy} - V_{xs} V_{ss}^{-1} V_{sy}\right\}$$

where

$$\begin{pmatrix} V_{xx} & V_{xy} & V_{xs} \\ V_{yx} & V_{yy} & V_{ys} \\ V_{sx} & V_{sy} & V_{ss} \end{pmatrix} = \begin{pmatrix} V(\hat{X}_d) & C(\hat{X}_d, \hat{Y}_W) & C(\hat{X}_d, S) \\ C(\hat{Y}_W, \hat{X}_d) & V(\hat{Y}_W) & C(\hat{Y}_W, S) \\ C(S, \hat{X}_d) & C(S, \hat{Y}_W) & V(S) \end{pmatrix}$$

and $\hat{X}_d = \hat{X}_W - \hat{X}_n$. Thus, a consistent estimator of B_1^* is

$$\hat{B}_1^* = (I_p, O_q) \left\{ \sum_{i=1}^{n} \delta_i b_i \begin{pmatrix} \mathbf{x}_i \\ \hat{\mathbf{h}}_i \end{pmatrix} \begin{pmatrix} \mathbf{x}_i \\ \hat{\mathbf{h}}_i \end{pmatrix}' \right\}^{-1} \sum_{i=1}^{n} \delta_i b_i \begin{pmatrix} \mathbf{x}_i \\ \hat{\mathbf{h}}_i \end{pmatrix} y_i, \tag{5.39}$$

where $b_i = \hat{\pi}_i^{-2}(1 - \hat{\pi}_i)$, $\hat{\mathbf{h}}_i = \mathbf{h}_i(\hat{\phi})$, p is the dimension of $\mathbf{x}_i$, and q is the dimension of ϕ.

The optimal PSA estimator of $\theta = E(Y)$ is then given by

$$\hat{\theta}_{opt} = \hat{Y}_{PSA} + \left(\hat{X}_n - \hat{X}_{PSA}\right)' \hat{B}_1^*, \tag{5.40}$$

which was originally proposed in Cao et al. (2009). For variance estimation, we can use a linearization method. Since $S(\hat{\phi}) = 0$, we can write (5.40) as

$$\hat{\theta}_{opt} = \hat{Y}_{PSA} + \left(\hat{X}_n - \hat{X}_{PSA}\right)' \hat{B}_1^* + S(\hat{\phi}) \hat{B}_2^*, \tag{5.41}$$

where

$$\begin{pmatrix} \hat{B}_1^* \\ \hat{B}_2^* \end{pmatrix} = \left\{ \sum_{i=1}^{n} \delta_i b_i \begin{pmatrix} \mathbf{x}_i \\ \hat{\mathbf{h}}_i \end{pmatrix} \begin{pmatrix} \mathbf{x}_i \\ \hat{\mathbf{h}}_i \end{pmatrix}' \right\}^{-1} \sum_{i=1}^{n} \delta_i b_i \begin{pmatrix} \mathbf{x}_i \\ \hat{\mathbf{h}}_i \end{pmatrix} y_i.$$

The pseudo values for variance estimation can take the form

$$\eta_i = \mathbf{x}_i'\hat{B}_1^* + \hat{\mathbf{h}}_i'\hat{B}_2^* + \frac{\delta_i}{\hat{\pi}_i}\left\{y_i - \mathbf{x}_i'\hat{B}_1^* - \hat{\mathbf{h}}_i'\hat{B}_2^*\right\}$$

and the final variance estimator of $\hat{\theta}_{opt}$ is

$$n^{-1}(n-1)^{-1}\sum_{i=1}^{n}(\eta_i - \bar{\eta}_n)^2.$$

5.5 Doubly robust method

In this section, we consider some means of protection against the failure of the assumed model. The optimal PSA estimator in (5.40) is asymptotically unbiased and optimal under the assumed response model. If the response model does not hold, then the validity of the optimal PSA estimator is no longer guaranteed. An estimator is called *doubly robust (DR)* if it remains consistent if either model (outcome regression model or response model) is true. The DR procedure offers some protection against misspecification of one model or the other. In this sense, it can be called the *doubly protected procedure*, as termed by Kim and Park (2006).

To discuss DR estimators, consider the following outcome regression (OR) model

$$E(y_i \mid \mathbf{x}_i) = m(\mathbf{x}_i; \beta_0),$$

for some $m(\mathbf{x}_i; \beta_0)$ known up to β_0 and assume the MAR. For the response probability (RP) model, we can assume (5.18). Under these models, we can consider the following class of doubly robust estimators, with the subscript "DR" denoting "doubly robust":

$$\hat{\theta}_{DR} = \frac{1}{n}\sum_{i=1}^{n}\left\{\hat{y}_i + \frac{\delta_i}{\hat{\pi}_i}(y_i - \hat{y}_i)\right\}, \tag{5.42}$$

where $\hat{y}_i = m(\mathbf{x}_i; \hat{\beta})$ for some function $m(\mathbf{x}_i; \beta_0)$ known up to β_0 and $\hat{\beta}$ is an estimator of β_0. The predicted value $\hat{y}_i$ is derived under the OR model while $\hat{\pi}_i$ is obtained from the RP model. Writing $\hat{\theta}_n = n^{-1}\sum_{i=1}^{n} y_i$, we have

$$\hat{\theta}_{DR} - \hat{\theta}_n = n^{-1}\sum_{i=1}^{n}\left(\frac{\delta_i}{\hat{\pi}_i} - 1\right)(y_i - \hat{y}_i). \tag{5.43}$$

Taking an expectation of the above, we note that the first term has approximate zero expectation if the RP model is true. The second term has approximate zero expectation if the OR model is true. Thus, $\hat{\theta}_{DR}$ is approximately unbiased when either RP model or OR model is true. DR estimation has been considered by Robins et al. (1994), Bang and Robins (2005), Tan (2006), Kang and Schafer (2007), Cao et al. (2009), and Kim and Haziza (2013). In particular, the optimal PSA estimator in (5.40) is doubly robust with $E(y_i \mid \mathbf{x}_i) = \mathbf{x}_i'B_0$ and $E(\delta_i \mid \mathbf{x}_i) = \pi_i(\phi_0)$ as long as $\hat{B}^*$ in (5.39) is consistent for B_0 under the OR model or $\hat{\phi}$ is consistent for ϕ_0 under the response probability model. Because the optimal PSA estimator in (5.40) is obtained by minimizing the variance of the PSA estimator under the response probability model, it is optimal when the assumed response probability model holds.

We now consider optimal estimation in the context of doubly robust estimation with a general form of the conditional expectation $m(\mathbf{x}_i; \beta_0)$ in the OR model. The optimality criteria for doubly robust estimation is somewhat unclear since there are two models involved. We consider the approach used in Cao et al. (2009) with the main objective of minimizing the variance of the DR estimator under the response probability model while maintaining the consistency of the point estimator under the OR model. In the class of the estimators of the form

$$\hat{\theta}_{DR}(\hat{\beta}, \hat{\phi}) = n^{-1}\sum_{i=1}^{n}\left[m(\mathbf{x}_i; \hat{\beta}) + \frac{\delta_i}{\pi_i(\hat{\phi})}\left(y_i - m(\mathbf{x}_i; \hat{\beta})\right)\right], \tag{5.44}$$

where $\hat{\phi}$ is the MLE and $\hat{\beta}$ is to be determined, Rubin and der Laan (2008) considered obtaining $\hat{\beta}$ that minimizes

$$\sum_{i=1}^{n} \frac{\delta_i}{\pi_i(\hat{\phi})} \left\{ \frac{1}{\pi_i(\hat{\phi})} - 1 \right\} \{y_i - m(\mathbf{x}_i;\beta)\}^2,$$

which can be justified when the response probability π_i is known, rather than estimated. To correctly account for the effect of estimating ϕ_0, one can use the linearization technique in Section 4.3 to obtain the optimal estimator. To be specific, we can account for the effect of π_i being estimated by writing (5.44) as

$$\hat{\theta}_{DR}(\hat{\beta},\hat{\phi},\mathbf{k}) = n^{-1}\sum_{i=1}^{n}\left[\mathbf{k}'\mathbf{h}_i(\hat{\phi}) + m(\mathbf{x}_i;\hat{\beta}) + \frac{\delta_i}{\pi_i(\hat{\phi})}\left\{y_i - m(\mathbf{x}_i;\hat{\beta}) - \mathbf{k}'\mathbf{h}_i(\hat{\phi})\right\}\right], \quad (5.45)$$

where $\mathbf{h}_i = (\partial \pi_i/\partial \phi)/(1-\pi_i)$ and $(\hat{\beta},\mathbf{k})$ is to be determined. Under the response probability model, the variance of the DR estimator is minimized by finding $(\hat{\beta},\mathbf{k})$ that minimizes

$$\sum_{i=1}^{n} \frac{\delta_i}{\pi_i(\hat{\phi})} \left\{ \frac{1}{\pi_i(\hat{\phi})} - 1 \right\} \{y_i - m(\mathbf{x}_i;\beta) - \mathbf{k}'\mathbf{h}_i(\hat{\phi})\}^2. \quad (5.46)$$

Note that in this case there is no guarantee that the resulting estimator is efficient under the OR model. Also, the computation for minimizing (5.46) can be cumbersome.

Tan (2006) slightly changed the class of estimators to be

$$\hat{\theta}_{DR}(\mathbf{k}) = n^{-1}\sum_{i=1}^{n}\left\{\mathbf{k}_1'\hat{\mathbf{m}}_i + \frac{\delta_i}{\pi_i(\hat{\phi})}(y_i - \mathbf{k}_1'\hat{\mathbf{m}}_i)\right\}, \quad (5.47)$$

where $\mathbf{k}_1 = (k_0,k_1)'$, $\hat{\mathbf{m}}_i = (1,\hat{m}_i)$, $\hat{m}_i = m(\mathbf{x}_i;\hat{\beta})$, and $\hat{\beta}$ is optimally computed in advance under the OR model. Therefore, if we have some extra information about $V(y_i \mid \mathbf{x}_i)$, then we can incorporate it to obtain $\hat{\beta}$ in (5.47) while the DR estimator of Cao et al. (2009) does not use the information. The optimal estimator among the class in (5.47) can be obtained by minimizing

$$\sum_{i=1}^{n} \frac{\delta_i}{\pi_i(\hat{\phi})} \left\{ \frac{1}{\pi_i(\hat{\phi})} - 1 \right\} \{y_i - \mathbf{k}_1'\hat{\mathbf{m}}_i - \mathbf{k}_2'\mathbf{h}_i(\hat{\phi})\}^2. \quad (5.48)$$

The solution can be written as

$$\begin{pmatrix} \hat{\mathbf{k}}_1^* \\ \hat{\mathbf{k}}_2^* \end{pmatrix} = \left\{\sum_{i=1}^{n} \delta_i b_i \begin{pmatrix} \hat{\mathbf{m}}_i \\ \hat{\mathbf{h}}_i \end{pmatrix}\begin{pmatrix} \hat{\mathbf{m}}_i \\ \hat{\mathbf{h}}_i \end{pmatrix}'\right\}^{-1} \sum_{i=1}^{n} \delta_i b_i \begin{pmatrix} \hat{\mathbf{m}}_i \\ \hat{\mathbf{h}}_i \end{pmatrix} y_i, \quad (5.49)$$

where $b_i = \hat{\pi}_i^{-1}(\hat{\pi}_i^{-1} - 1)$ and $\hat{\mathbf{h}}_i = \mathbf{h}_i(\hat{\phi})$. Note that the expected value of $\hat{\mathbf{k}}_1^*$ is approximately equal to $(0,1)'$ under the OR model. Thus, under the OR model, the optimal DR estimator of Tan is asymptotically equivalent to (5.44). Furthermore, if $m(x_i;\beta) = \beta_0 + \beta_1 x_i$, then the optimal estimator is equal to the optimal PSA estimator in (5.40). In fact, if both the OR model and the response probability model are correct, then the choice of $\hat{\beta}$ is not critical. This phenomenon, called the *local efficiency* of the DR estimator, was first discussed by Robins et al. (1994).

Kim and Riddles (2012) considered an augmented propensity model of the form

$$\hat{\pi}_i^* = \pi_i^*(\hat{\phi},\hat{\lambda}) = \frac{\pi_i(\hat{\phi})}{\pi_i(\hat{\phi}) + \{1-\pi_i(\hat{\phi})\}\exp(\hat{\lambda}_0 + \hat{\lambda}_1\hat{m}_i)}, \quad (5.50)$$

where $\pi_i(\hat{\phi})$ is the estimated response probability under the response probability model and $(\hat{\lambda}_0,\hat{\lambda}_1)$ satisfies

$$\sum_{i=1}^{n} \frac{\delta_i}{\pi_i^*(\hat{\phi},\hat{\lambda})}(1,\hat{m}_i) = \sum_{i=1}^{n}(1,\hat{m}_i). \quad (5.51)$$

According to Kim and Riddles (2012), the augmented PSA estimator, defined by $\hat{\theta}^*_{PSA} = n^{-1}\sum_{i=1}^n \delta_i y_i/\hat{\pi}_i^*$, based on the augmented propensity in (5.50) satisfies, under the assumed response probability model,

$$\hat{\theta}^*_{PSA} \cong \frac{1}{n}\sum_{i=1}^n \left\{ \hat{b}_0 + \hat{b}_1\hat{m}_i + \frac{\delta_i}{\hat{\pi}_i}\left(y_i - \hat{b}_0 - \hat{b}_1\hat{m}_i\right)\right\}, \tag{5.52}$$

where

$$\begin{pmatrix} \hat{b}_0 \\ \hat{b}_1 \end{pmatrix} = \left\{\sum_{i=1}^n \delta_i\left(\frac{1}{\hat{\pi}_i}-1\right)\begin{pmatrix} 1 \\ \hat{m}_i \end{pmatrix}\begin{pmatrix} 1 \\ \hat{m}_i \end{pmatrix}'\right\}^{-1} \sum_{i=1}^n \delta_i\left(\frac{1}{\hat{\pi}_i}-1\right)\begin{pmatrix} 1 \\ \hat{m}_i \end{pmatrix} y_i.$$

Thus, if we use another augmented propensity model of the form

$$\hat{\pi}_i^* = \pi_i^*(\hat{\phi},\hat{\lambda}) = \frac{\pi_i(\hat{\phi})}{\pi_i(\hat{\phi}) + \{1-\pi_i(\hat{\phi})\}\exp\{\hat{\lambda}_0/\pi_i(\hat{\phi}) + \hat{\lambda}_1\hat{m}_i/\pi_i(\hat{\phi})\}}, \tag{5.53}$$

where $(\hat{\lambda}_0, \hat{\lambda}_1)$ satisfies (5.51), the resulting augmented PSA estimator is asymptotically equivalent to Tan's estimator in (5.47) with the optimal coefficients in (5.49) under the response probability model.

Remark 5.1. *We can construct a fractional imputation method that is doubly robust. In fractional imputation, several imputed values are used for each missing value and fractional weights are assigned to the imputed values. Let* $y_{ij}^* = m(\mathbf{x}_j;\hat{\beta}) + \hat{e}_i$ *be the imputed value for unit* j *using donor* i *where* $\hat{e}_j = y_j - m(\mathbf{x}_j;\hat{\beta})$. *The FI estimator can be written as*

$$\hat{\theta}_{FI} = n^{-1}\sum_{j=1}^n \left\{\delta_j y_j + (1-\delta_j)\sum_{i=1}^n w_{ij}^*\delta_i y_{ij}^*\right\}, \tag{5.54}$$

where w_{ij}^* *are the fractional weights attached to unit* j *such that* $\sum_{i=1}^n w_{ij}^*\delta_i = 1$. *Note that (5.54) can be alternatively written as*

$$\hat{\theta}_{FI} = n^{-1}\sum_{i=1}^n m(\mathbf{x}_i;\hat{\beta}) + n^{-1}\sum_{i=1}^n \delta_i\left\{1 + \sum_{j=1}^n (1-\delta_j)w_{ij}^*\right\}\hat{e}_i. \tag{5.55}$$

Compare (5.55) with (5.42), and it follows that the FI estimator (5.54) is doubly robust if

$$1 + \sum_{j=1}^n (1-\delta_j)w_{ij}^* = \frac{1}{\pi_i(\hat{\phi})}. \tag{5.56}$$

If the estimated propensity score satisfies

$$n^{-1}\sum_{i=1}^n \frac{\delta_i}{\pi_i(\hat{\phi})} = 1,$$

then the choice

$$w_{ij}^* = \frac{\{1/\pi_i(\hat{\phi}) - 1\}}{\sum_{k=1}^n \delta_k\{1/\pi_k(\hat{\phi}) - 1\}} \tag{5.57}$$

satisfies (5.56) and the FI estimator is doubly robust.

5.6 Empirical likelihood method

The empirical likelihood (EL) method, proposed by Owen (1988), has become a very powerful tool for nonparametric inference in statistics. It uses a likelihood-based approach without having to make a parametric distributional assumption about the data observed, often resulting in efficient estimation.

To discuss the empirical likelihood method under missing data, consider a multivariate random variable (X,Y) with distribution function $F(x,y)$, which is completely unspecified except that $E\{U(\theta_0;X,Y)\}=0$ for some θ_0. If (x_i,y_i), $i=1,2,\ldots,n$, are n independent realizations of the random variable (X,Y), a consistent estimator of θ_0 can be obtained by solving

$$\sum_{i=1}^{n} U(\theta;x_i,y_i)=0.$$

Assume that x_i is always observed and y_i is subject to missingness. Let $\delta_i=1$ if y_i is observed and $\delta_i=0$ otherwise.

Note that the joint density of the observed data can be written as

$$p^{n_r}(1-p)^{n-n_r}\times\prod_{\delta_i=1} f(x_i,y_i|\delta_i=1)\prod_{r_i=0} f(x_i|\delta_i=0),$$

where n_r is the respondents' sample size, $p=Pr(\delta=1)$, $f(x,y|\delta)$ is the conditional density of (X,Y) given δ, and $f(x_i|\delta_i=0)=\int f(x_i,y_i|\delta_i=0)dy_i$ is the marginal density of X among $\delta=0$.

In the empirical likelihood approach, the distribution is assumed to have the support on the sample observation. Let $F_1(x,y)=Pr(X\le x,Y\le y|\delta=1)$ and $F_0(x,y)=Pr(X\le x,Y\le y|\delta=0)$. Under the empirical likelihood approach, we can express

$$F_1(x,y)=\sum_{\delta_i=1}\omega_i I(x_i\le x,y_i\le y), \tag{5.58}$$

where $\sum_{\delta_i=1}\omega_i=1$, ω_i is the point mass assigned to (x_i,y_i) in the nonparametric distribution of $F_1(x,y)$, and $I(B)$ is an indicator function for event B. To express $F_0(x,y)$ using ω_i, note that we can write

$$f(x_i,y_i|\delta_i=0)=f(x_i,y_i|\delta_i=1)\times\frac{Odd(x_i,y_i)}{E\{Odd(x_i,y_i)|\delta_i=1\}},$$

where

$$Odd(x,y)=\frac{Pr(\delta=0\mid x,y)}{Pr(\delta=1\mid x,y)}.$$

Thus, we can express $F_0(x,y)=Pr(X\le x,Y\le y|\delta=0)$ by

$$F_0(x,y)=\frac{\sum_{\delta_i=1}\omega_i O_i I(x_i\le x,y_i\le y)}{\sum_{\delta_i=1}\omega_i O_i}, \tag{5.59}$$

where $O_i=Odd(x_i,y_i)$. Note that $F_0(x,y)$ is completely determined by two factors: ω_i and O_i. The factor ω_i is determined by the distribution $F_1(x,y)$ and the factor O_i is determined by the response mechanism. If $Odd(x,y)$ is a known function of (x,y), then we only have to determine ω_i.

From (5.59), the joint distribution of (x,y) can be written as

$$\begin{aligned} F_w(x,y) &= p\times\sum_{\delta_i=1}\omega_i I(x_i\le x,y_i\le y)+(1-p)\times\left\{\frac{\sum_{\delta_i=1}\omega_i O_i I(x_i\le x,y_i\le y)}{\sum_{\delta_i=1}\omega_i O_i}\right\} \\ &= p\times\left\{\sum_{\delta_i=1}\omega_i I(x_i\le x,y_i\le y)+(1/p-1)\frac{\sum_{\delta_i=1}\omega_i O_i I(x_i\le x,y_i\le y)}{\sum_{\delta_i=1}\omega_i O_i}\right\}. \end{aligned}$$

Note that (5.58) implies

$$\begin{aligned}\sum_{\delta_i=1} \omega_i(O_i+1) &= E\left\{\frac{1}{\pi(X,Y)}|\delta=1\right\} \\ &= \int \frac{1}{\pi(x,y)} f(x,y|\delta=1)dxdy \\ &= \int \frac{1}{\pi(x,y)} \frac{\pi(x,y)f(x,y)}{p} dxdy = 1/p.\end{aligned}$$

Thus, we have $\sum_{\delta_i=1} \omega_i O_i = 1/p - 1$ and

$$F_w(x,y) = \frac{\sum_{\delta_i=1} \omega_i(1+O_i)I(x_i \le x, y_i \le y)}{\sum_{\delta_i=1} \omega_i(O_i+1)}.$$

Thus, the empirical likelihood approach can be formulated as maximizing

$$l_e(\theta) = \sum_{\delta_i=1} \log(\omega_i), \tag{5.60}$$

subject to

$$\sum_{\delta_i=1} \omega_i = 1, \quad \sum_{\delta_i=1} \omega_i(1+O_i)U(\theta; x_i, y_i) = 0. \tag{5.61}$$

Note that, in constraint (5.61), the observed values of x_i with $\delta_i = 0$ are not used. To incorporate the partial information, we can impose

$$\frac{\sum_{\delta_i=1} \omega_i(1+O_i)h(x_i;\theta)}{\sum_{\delta_i=1} \omega_i(1+O_i)} = n^{-1}\sum_{i=1}^{n} h(x_i;\theta) \tag{5.62}$$

as an additional constraint for some $h(x_i;\theta)$.

If the response probability $\pi_i = Pr(\delta_i = 1 \mid x_i, y_i)$ is known, then $1+O_i = \pi_i^{-1}$ and the empirical likelihood estimator of θ can be obtained by maximizing (5.60) subject to

$$\sum_{\delta_i=1} \omega_i = 1, \quad \sum_{\delta_i=1} \omega_i \pi_i^{-1}\left\{h_i(\theta) - n^{-1}\sum_{i=1}^{n} h_i(\theta)\right\} = 0, \quad \sum_{\delta_i=1} \omega_i \pi_i^{-1} U_i(\theta) = 0. \tag{5.63}$$

For $\theta = E(Y)$, a popular choice of $h(\theta)$ is $h(\theta) = x$. In this case, the EL estimator of θ is obtained by $\hat{\theta}_{h1} = \sum_{\delta_i=1} w_i^* \pi_i^{-1} y_i / (\sum_{\delta_i=1} w_i^* \pi_i^{-1})$ where $w_i^* = n_r^{-1}\{1+\hat{\lambda}\pi_i^{-1}(x_i - \bar{x}_n)\}^{-1}$, $\bar{x}_n = n^{-1}\sum_{i=1}^{n} x_i$ and $\hat{\lambda}$ is constructed to satisfy $\sum_{\delta_i=1} w_i^* \pi_i^{-1}(x_i - \bar{x}_n) = 0$.

Using a Taylor linearization, it can be shown (Chen, 2013) that the solution $\hat{\theta}_h$ satisfies

$$\sqrt{n}\left(\hat{\theta}_h - \theta_0\right) \to^d N(0, V_h), \tag{5.64}$$

where $\to^d$ denotes convergence in distribution, $V_h = \tau^{-1}\Omega_h(\tau^{-1})'$, $\tau = E(\partial U/\partial\theta')$,

$$\Omega_h = V\left\{\frac{\delta}{\pi}(U - Bh) + Bh\right\} = E\left\{(\frac{1}{\pi} - 1)(U - Bh)^{\otimes 2}\right\} + V(U), \tag{5.65}$$

$B = E(Uh'/\pi)\{E(hh'/\pi)\}^{-1}$, and $A^{\otimes 2} = AA'$. Furthermore, the asymptotic variance of $\hat{\theta}_h$ is minimized when $h \propto h^* = E(U|X)$. The asymptotic variance satisfies

$$V_h \ge \tau^{-1}\left\{E\left(\frac{UU'}{\pi}\right) - E\left(\frac{1-\pi}{\pi}h^*U'\right)\right\}(\tau^{-1})'. \tag{5.66}$$

The lower bound in (5.66) is the same as the semi-parametric lower bound in (5.24) for the asymptotic variance discussed in Robins et al. (1994).

For the special case of $\theta = E(Y)$ and $h = x$, after some algebra, we have

$$\hat{\theta}_h \cong \hat{\bar{y}}_d - \hat{B}(\hat{\bar{x}}_d - \bar{x}_n),$$

where

$$(\hat{\bar{y}}_d, \hat{\bar{x}}_d) = (\sum_{\delta_i=1} \pi_i^{-1})^{-1}(\sum_{\delta_i=1} \pi_i^{-1} y_i, \sum_{\delta_i=1} \pi_i^{-1} x_i),$$

and

$$\hat{B} = \{\sum_{\delta_i=1} \pi_i^{-2}(x_i - \hat{\bar{x}}_d)^2\}^{-1} \sum_{\delta_i=1} \pi_i^{-2}(x_i - \hat{\bar{x}}_d)(y_i - \hat{\bar{y}}_d).$$

In practice, we do not know the true response probability π_i and so we use $\hat{\pi}_i$ instead of π_i in computing the empirical likelihood estimator. The asymptotic variance using $\hat{\pi}_i$ will remain unchanged for $h^* = E(U \mid X)$ and the same lower bound of the asymptotic variance will be achieved. Double robustness can also be established for the choice of $h^* = E(U \mid X)$. See Chen (2013) for details.

Remark 5.2. *The empirical likelihood function in (5.60) can be called a* partial likelihood function *in the sense that we only consider the likelihood function corresponding to* $\sum_{\delta_i=1} \log f(x_i, y_i \mid \delta_i = 1)$. *Instead of considering the partial likelihood, Qin et al. (2009) propose an alternative approach that uses the full likelihood to maximize*

$$l(\theta, \phi) = \sum_{i=1}^{n} \log(\omega_i),$$

subject to

$$\sum_{i=1}^{n} \omega_i = 1, \ \sum_{i=1}^{n} \omega_i \left(\frac{\delta_i}{\pi_i} - 1\right) h(x_i; \theta) = 0, \ \sum_{i=1}^{n} \omega_i \frac{\delta_i}{\pi_i} U(\theta; x_i, y_i) = 0.$$

Thus, full sample instead of partial likelihood can still be used when constraints are properly imposed. The full-sample empirical likelihood estimator $\hat{\theta}_{hf}$ *satisfies*

$$\sqrt{n}\left(\hat{\theta}_{hf} - \theta_0\right) \to^d N(0, V_h),$$

where $\to^d$ *denotes convergence in distribution,* $V_{hf} = \tau^{-1}\Omega_{hf}(\tau^{-1})'$, $\tau = E(\partial U / \partial \theta')$,

$$\Omega_{hf} = V\left\{\frac{r}{\pi}(U - B_f h) + B_f h\right\} = E\left\{(\frac{1}{\pi} - 1)(U - B_f h)^{\otimes 2}\right\} + V(U),$$

and $B_f = E((\pi^{-1} - 1)Uh')\{E((\pi^{-1} - 1)hh')\}^{-1}$. *Note that the choice of* $B = B_f$ *minimizes the variance term* $E\{(\pi^{-1} - 1)(U - Bh)^{\otimes 2}\}$ *among different choices of B. Thus, we have* $V_{hf} \le V_h$, *where* V_h *is defined in (5.64), with equality at* $h \propto h^* = E(U \mid X)$. *That is, the full-sample EL estimator is more efficient because it uses the full likelihood for maximization.*

5.7 Nonparametric method

Instead of using parametric models for propensity scores, nonparametric approaches can also be used. We assume a bivariate data structure (x_i, y_i) with x_i fully observed. We assume that the response mechanism is missing at random. We assume that $\pi(x) = Pr(\delta = 1 \mid x)$ is completely unspecified, except that it is a smooth function of x with bounded partial derivatives of order 2. Using the argument discussed in Example 4.15, a nonparametric regression estimator of $\pi(x) = E(\delta \mid x)$ can be obtained by

$$\hat{\pi}_h(x) = \frac{\sum_{i=1}^{n} \delta_i K_h(x_i, x)}{\sum_{i=1}^{n} K_h(x_i, x)}, \tag{5.67}$$

where K_h is the kernel function which satisfies certain regularity conditions and h is the bandwidth. Use of nonparametric propensity scores has been considered by Hirano et al. (2003) and Cattaneo (2010). Once a nonparametric estimator of $\pi(x)$ is obtained, the nonparametric PSA estimator $\hat{\theta}_{NPS}$ of $\theta_0 = E(Y)$ is given by

$$\hat{\theta}_{NPS} = \frac{1}{n}\sum_{i=1}^{n} \frac{\delta_i}{\hat{\pi}_h(x_i)} y_i. \tag{5.68}$$

To discuss asymptotic properties of the nonparametric PSA estimator in (5.68), assume the following regularity conditions:

(C1) The marginal density of X, denoted by $f(x)$, and the unknown response probability $\pi(x) = E(\delta \mid x)$ have bounded partial derivatives with respect to x up to order 2 almost surely.

(C2) The Kernel function $K(s)$ satisfies the following regularity conditions:

1. It is bounded and has compact support.
2. It is symmetric and $\sigma_K^2 = \int s^2 K(s) ds < \infty$.
3. $K(s) \geq c$ for some $c > 0$ in some closed interval centered at zero.

(C3) The bandwidth h satisfies $nh^2 \to \infty$ and $nh^4 \to 0$.

(C4) $E\{Y^2\} < \infty$ and the density of X decays exponentially fast.

(C5) $1 > \pi(x) > d > 0$ almost surely.

The following theorem, originally proved by Hirano et al. (2003), establishes the $\sqrt{n}$-consistency of the PSA estimator of $\theta = E(Y)$. Notation $X_n = o_p(1)$ denotes that X_n converges to zero in probability.

Theorem 5.2. *Under the regularity conditions (C1)-(C5), we have*

$$\hat{\theta}_{NPS} = \frac{1}{n}\sum_{i=1}^{n}\left[m(x_i) + \frac{\delta_i}{\pi(x_i)}\{y_i - m(x_i)\}\right] + o_p(n^{-1/2}), \tag{5.69}$$

where $m(x) = E(Y \mid x)$ and $\pi(x) = P(\delta = 1 \mid x)$. Furthermore, we have

$$\sqrt{n}\left(\hat{\theta}_{NPS} - \theta\right) \to N\left(0, \sigma_1^2\right),$$

where $\sigma_1^2 = V\{m(X)\} + E\left[\{\pi(X)\}^{-1} V(Y \mid X)\right]$.

Proof. By the standard arguments in Kernel smoothing, we have

$$E\left\{\frac{1}{n}\sum_{j=1}^{n} K_h(X_i, X_j)\right\} = f(X_i) + O(h^2) \tag{5.70}$$

and

$$E\left\{\frac{1}{n}\sum_{j=1}^{n} \delta_j K_h(X_i, X_j)\right\} = \pi(X_i) f(X_i) + O(h^2). \tag{5.71}$$

By (5.70) and (5.71), we can use Taylor expansion to get

$$\begin{aligned}
\frac{\sum_{j=1}^{n} K_h(X_i,X_j)}{\sum_{j=1}^{n}\delta_j K_h(X_i,X_j)} &= \frac{1}{\pi(X_i)} + \frac{1}{\pi(X_i)f(X_i)}\left\{\frac{1}{n}\sum_{j=1}^{n} K_h(X_i,X_j) - f(X_i)\right\} \\
&\quad - \frac{1}{\{\pi(X_i)\}^2 f(X_i)}\left\{\frac{1}{n}\sum_{j=1}^{n}\delta_j K_h(X_i,X_j) - \pi(X_i)f(X_i)\right\} + O(h^2) \\
&= \frac{1}{\pi(X_i)} + \frac{1}{n}\sum_{j=1}^{n}\frac{K_h(X_i,X_j)}{\pi(X_i)f(X_i)}\left\{1 - \frac{\delta_j}{\pi(X_i)}\right\} + O_p(h^2).
\end{aligned}$$

By (C3), we can write

$$\begin{aligned}
\hat{\theta}_{NPS} &= \frac{1}{n}\sum_{i=1}^{n} r_i \left\{ \frac{\sum_{j=1}^{n} K_h(X_i,X_j)}{\sum_{j=1}^{n} \delta_j K_h(X_i,X_j)} \right\} y_i \\
&= \frac{1}{n}\sum_{i=1}^{n} \frac{\delta_i}{\pi(X_i)} Y_i + \frac{1}{n^2}\sum_{i=1}^{n}\sum_{j=1}^{n} \delta_i Y_i \frac{K_h(X_i,X_j)}{\pi(X_i)f(X_i)} \left\{1 - \frac{\delta_j}{\pi(X_i)}\right\} + O_p(h^2) \\
&= \frac{1}{n}\sum_{i=1}^{n} \frac{\delta_i}{\pi(X_i)} Y_i + \frac{1}{n^2}\sum_{i=1}^{n} \delta_i Y_i \frac{K(0)}{\pi(X_i)f(X_i)} \left\{1 - \frac{\delta_i}{\pi(X_i)}\right\} \\
&+ \frac{1}{n^2}\sum_{i=1}^{n}\sum_{j\neq i} \delta_i Y_i \frac{K_h(X_i,X_j)}{\pi(X_i)f(X_i)} \left\{1 - \frac{\delta_j}{\pi(X_i)}\right\} + O_p(h^2) \\
&= \frac{1}{n}\sum_{i=1}^{n} \frac{\delta_i}{\pi(X_i)} Y_i + \frac{1}{n(n-1)}\sum_{i=1}^{n}\sum_{j\neq i} \delta_i Y_i \frac{K_h(X_i,X_j)}{\pi(X_i)f(X_i)} \left\{1 - \frac{\delta_j}{\pi(X_i)}\right\} + o_p(n^{-1/2}).
\end{aligned}$$

So, we can express

$$\hat{\theta}_{NPS} = \frac{1}{n}\sum_{i=1}^{n} \frac{\delta_i y_i}{\pi(x_i)} + \frac{1}{n(n-1)}\sum_{i\neq j} h(Z_i,Z_j) + o_p(n^{-1/2}), \tag{5.72}$$

where $Z_i = (X_i, Y_i, \delta_i)$ and

$$\begin{aligned}
h(Z_i,Z_j) &= \frac{1}{2}\left[\delta_i Y_i \frac{K_h(X_i,X_j)}{\pi(X_i)f(X_i)} \left\{1 - \frac{\delta_j}{\pi(X_i)}\right\} + \delta_j Y_j \frac{K_h(X_j,X_i)}{\pi(X_j)f(X_j)} \left\{1 - \frac{\delta_i}{\pi(X_j)}\right\}\right] \\
&=: \frac{1}{2}(\zeta_{ij} + \zeta_{ji}).
\end{aligned}$$

Thus, $\sum_{j\neq i} h(Z_i,Z_j)/\{n(n-1)\}$ is a U-statistics. Let $s = (X_i - X_j)/h$ and by a Taylor expansion again, we have

$$\begin{aligned}
E(\zeta_{ij} \mid Z_i) &= \frac{\delta_i Y_i}{\pi(X_i)f(X_i)} \frac{1}{h} \int K\left(\frac{X_i - X_j}{h}\right) \left\{1 - \frac{\pi(X_j)}{\pi(X_i)}\right\} f(X_j) dX_j \\
&= \frac{\delta_i Y_i}{\pi(X_i)f(X_i)} \int K(s) \left\{1 - \frac{\pi(X_i + hs)}{\pi(X_i)}\right\} f(X_i + hs) dX_j \\
&= O_p(h^2)
\end{aligned} \tag{5.73}$$

and

$$\begin{aligned}
E(\zeta_{ji} \mid Z_i) &= \frac{1}{h}\int \frac{Y_j}{f(X_j)} K\left(\frac{X_i - X_j}{h}\right) \left\{1 - \frac{\delta_i}{\pi(X_j)}\right\} f(X_j, Y_j) dX_j dY_j \\
&= \int \frac{Y_j}{f(X_i + hs)} K(s) \left\{1 - \frac{\delta_i}{\pi(X_i + hs)}\right\} f(X_i + hs, Y_j) ds dY_j \\
&= \left\{1 - \frac{\delta_i}{\pi(X_i)}\right\} m(X_i) + O_p(h^2).
\end{aligned} \tag{5.74}$$

By (5.72), (5.73), (5.74), and by the theory of U-statistics (Serfling, 1980, Chapter 5), we have (5.69). □

Note that the asymptotic variance of the nonparametric PSA estimator is the same as that of the nonparametric fractional imputation estimator discussed in Example 4.15. The asymptotic variance is equal to the lower bound in (5.24).

Exercises

1. Show (5.10).
2. Show (5.26).
3. Under the setup of Example 5.4, answer the following questions.
   (a) Write the optimal estimator $\hat{\theta}_{opt}(\hat{\beta})$ as $\hat{\theta}_{opt}(\hat{\beta}, \hat{\phi})$, where $\hat{\phi}$ is the estimator for computing $\hat{\pi}_i = \pi_i(\hat{\phi})$ in (5.28), and show that

$$E\left\{\frac{\partial}{\partial \phi}\hat{\theta}_{opt}(\beta, \phi)\right\} = 0$$

   (b) Prove that, under the regression model in (5.27), $\hat{\theta}_{opt}(\hat{\beta}, \hat{\phi})$ is asymptotically equivalent to $\hat{\theta}_{opt}(\hat{\beta}, \phi_0)$ as long as $(\hat{\beta}, \hat{\phi})$ is consistent for (β, ϕ).
   (c) Prove that if the propensity scores are constructed to satisfy

$$\sum_{i=1}^{n} \frac{\delta_i}{\hat{\pi}_i}\mathbf{x}_i = \sum_{i=1}^{n} \mathbf{x}_i,$$

   then the PSA estimator $\hat{\theta}_{PSA}$ is optimal in the sense that it achieves the lower bound in (5.24).
4. Under the setup of Example 5.4 again, suppose that we are going to use

$$\hat{\theta}_p = \frac{1}{n}\sum_{i=1}^{n} \mathbf{x}_i'\hat{\beta}_c, \tag{5.75}$$

   where

$$\hat{\beta}_c = \left\{\sum_{i=1}^{n} \delta_i c_i \mathbf{x}_i \mathbf{x}_i'\right\}^{-1} \sum_{i=1}^{n} \delta_i c_i \mathbf{x}_i y_i$$

   and c_i is to be determined. Answer the following questions:
   (a) Show that $\hat{\theta}_p$ is asymptotically unbiased for $\theta = E(Y)$ regardless of the choice of c_i.
   (b) If $\mathbf{x}_i$ contains an intercept term then the choice of $c_i = 1/\hat{\pi}_i$ makes the resulting estimator optimal in the sense that it achieves the lower bound in (5.24).
   (c) Instead of $\hat{\theta}_p$ in (5.75), suppose that an alternative estimator

$$\hat{\theta}_{p2} = \frac{1}{n}\sum_{i=1}^{n}\{\delta_i y_i + (1-\delta_i)\mathbf{x}_i'\hat{\beta}_c\}$$

   is used, where $\hat{\beta}_c$ is as defined in (5.75). Find a set of conditions for $\hat{\theta}_{p2}$ to be optimal.
5. Prove (5.38).
6. Suppose that the response probability is parametrically modeled by

$$\pi_i = \Phi(\phi_0 + \phi_1 x_i)$$

   for some (ϕ_0, ϕ_1), where $\Phi(\cdot)$ is the cumulative distribution function of the standard normal distribution. Assume that x_i is completely observed and y_i is observed only when $\delta_i = 1$, where δ_i follows from the Bernoulli distribution with parameter π_i.
   (a) Find the score equation for (ϕ_0, ϕ_1).
   (b) Discuss asymptotic variance of the PSA estimator of $\theta = E(Y)$ using the MLE of (ϕ_0, ϕ_1).

7. Prove that minimizing (5.30) is algebraically equivalent to minimizing

$$Q_2 = \begin{pmatrix} \hat{X}_1 - \hat{X}_2 \\ \hat{Y} - \mu_y \end{pmatrix}' \begin{pmatrix} V(\hat{X}_1 - \hat{X}_2) & C(\hat{X}_1 - \hat{X}_2, \hat{Y}) \\ C(\hat{X}_1 - \hat{X}_2, \hat{Y}) & V(\hat{Y}) \end{pmatrix}^{-1} \begin{pmatrix} \hat{X}_1 - \hat{X}_2 \\ \hat{Y} - \mu_y \end{pmatrix}$$

   and the resulting optimal estimator minimizing Q_2 is given by

$$\hat{\mu}_y^* = \hat{Y} - \frac{C(\hat{X}_1 - \hat{X}_2, \hat{Y})}{V(\hat{X}_1 - \hat{X}_2)} (\hat{X}_1 - \hat{X}_2).$$

   Show that it is equal to the solution in (5.32).

8. Prove (5.52).

9. Devise a linearization variance estimator for the doubly robust fractional imputation estimator in (5.54).

10. Let $\hat{\pi}_i = \pi(x_i; \hat{\phi})$ be the estimated response probability. Consider a regression estimator of the form

$$\hat{\theta}_{reg} = \sum_{i \in A} w_i y_i,$$

   where

$$w_i = \left(n^{-1} \sum_{i=1}^{n} \mathbf{z}_i \right)' \left(\sum_{i=1}^{n} \mathbf{z}_i \mathbf{z}_i' \right)^{-1} \mathbf{z}_i$$

   and $\mathbf{z}_i' = (\hat{\pi}_i^{-1}, \mathbf{x}_i')$.

   (a) Show that $\hat{\theta}_{reg}$ is asymptotically unbiased under the response model $Pr(\delta = 1 \mid x) = \pi(x_i; \phi)$ and $\hat{\phi}$ is a consistent estimator of ϕ.

   (b) Construct a consistent variance estimator of $\hat{\theta}_{reg}$.

Chapter 6

Nonignorable missing data

6.1 Nonresponse instrument

We now consider the case of nonignorable missing data. This occurs when the probability of response depends on the variable that is not always observed. Let $\mathbf{x}_i$ be the variables that are always observed and y_i be the variable that is subject to missingness. Let δ_i be the response indicator function of y_i. In this case, the observed likelihood, conditional on $\mathbf{x}_i$'s, is

$$L_{\text{obs}}(\theta,\phi) = \prod_{\delta_i=1} f(y_i \mid \mathbf{x}_i;\theta)\, g(\delta_i \mid \mathbf{x}_i, y_i;\phi) \prod_{\delta_i=0} \int f(y_i \mid \mathbf{x}_i;\theta)\, g(\delta_i \mid \mathbf{x}_i, y_i;\phi)\, dy_i, \tag{6.1}$$

where $g(\delta_i \mid \mathbf{x}_i, y_i;\phi)$ is the conditional distribution of δ_i given $(y_i, \mathbf{x}_i)$ and ϕ is an unknown parameter. If $\mathbf{x}$ is null, then (6.1) becomes (2.10). If the response mechanism is ignorable in the sense that

$$g(\delta_i \mid \mathbf{x}_i, y_i;\phi) = g(\delta_i \mid \mathbf{x}_i;\phi)$$

then the observed likelihood in (6.1) can be written as

$$L_{\text{obs}}(\theta,\phi) = \prod_{\delta_i=1} f(y_i \mid \mathbf{x}_i;\theta) \times \prod_{i=1}^{n} g(\delta_i \mid \mathbf{x}_i;\phi) = L_1(\theta) \times L_2(\phi)$$

and the maximum likelihood estimator of θ can be obtained by maximizing $L_1(\theta)$. Otherwise one needs to maximize the full likelihood (6.1) directly.

There are several problems in maximizing the full likelihood (6.1). First, the parameters in the full likelihood are not always identifiable. Second, the integrals in (6.1) are not easy to handle. Finally, inference with nonignorable missing data is sensitive to the failure of the assumed parametric model. The identifiability issue is illustrated in the following example.

Example 6.1. *Consider the case where there is no covariate, and $f(y;\theta)$ is normal with unknown mean μ and variance σ^2. Also, consider the logistic model $g(\delta = 1|y;\phi) = [1+\exp(\alpha+\beta y)]^{-1}$ with unknown real-valued α and β. Missing is ignorable if and only if $\beta = 0$. Note that*

$$g(\delta = 1|y;\phi)f(y;\theta) = \frac{\exp[-(y-\mu)^2/2\sigma^2]}{\sqrt{2\pi}\sigma[1+\exp(\alpha+\beta y)]}.$$

The parameters are not identifiable if two different sets of parameters, $(\alpha,\beta,\mu,\sigma)$ and $(\alpha',\beta',\mu',\sigma')$,

$$\frac{\exp[-(y-\mu)^2/2\sigma^2]}{\sigma[1+\exp(\alpha+\beta y)]} = \frac{\exp[-(y-\mu')^2/2\sigma'^2]}{\sigma'[1+\exp(\alpha'+\beta' y)]} \quad \text{for all } y \in \mathcal{R}, \tag{6.2}$$

where $\mathcal{R}$ denotes the real line. It can be easily verified that (6.2) holds if $\sigma = \sigma'$, $\alpha' = -\alpha$, $\beta' = -\beta$, $\alpha = (\mu'^2 - \mu^2)/2\sigma^2$, and $\beta = (\mu' - \mu)/\sigma^2$. Hence, the parameters are not identifiable unless $\beta = \beta' = 0$ (ignorable missing).

In Example 6.1, if there is a covariate z such that the conditional distribution of y given z depends on the value of z, and $g(\delta|y,z)$ does not depend on z, then all parameters are identifiable. This is a special case of the following result, which was discussed in Wang et al. (2013).

Lemma 6.1. *Suppose that we can decompose the covariate vector* $\mathbf{x}$ *into two parts,* $\mathbf{u}$ *and* $\mathbf{z}$*, such that*

$$g(\delta|y,\mathbf{x}) = g(\delta|y,\mathbf{u}) \tag{6.3}$$

and, for any given $\mathbf{u}$*, there exist* $z_{\mathbf{u},1}$ *and* $z_{\mathbf{u},2}$ *such that*

$$f(y|\mathbf{u},\mathbf{z}=z_{\mathbf{u},1}) \neq f(y|\mathbf{u},\mathbf{z}=z_{\mathbf{u},2}), \tag{6.4}$$

then under some other minor conditions, all the parameters in f *and* g *are identifiable.*

In the literature of measurement error, where a covariate $\mathbf{x}^*$ associated with a study variable y^* is measured with error, valid estimators of regression parameters can be obtained by utilizing an instrument $\mathbf{z}$ that is correlated with $\mathbf{x}^*$ but independent of y^* conditioned on $\mathbf{x}^*$. In (6.3)-(6.4), we decompose the covariate vector $\mathbf{x}$ into two parts, $\mathbf{u}$ and $\mathbf{z}$, such that $\mathbf{z}$ plays the same role as an instrument, i.e., $\mathbf{z}$ is correlated with $\mathbf{x}^* = (y,\mathbf{u})$, and $\mathbf{z}$ is independent of $y^* = \delta$ conditioned on $\mathbf{x}^* = (y,\mathbf{u})$. Unconditionally, $\mathbf{z}$ may still be related to δ. Since y is subject to nonresponse, not measurement error, we name $\mathbf{z}$ as a *nonresponse instrument*. The nonresponse instrument $\mathbf{z}$ helps to identify the unknown quantities. Condition (6.3) can also be written as

$$Cov(\delta,\mathbf{z}\mid y,\mathbf{u}) = 0.$$

Once identifiability is guaranteed, the observed likelihood has a unique maximum and one can obtain the MLE that maximizes the observed likelihood (6.1). This is called the full likelihood or the full parametric likelihood approach. To deal with the integral in the observed likelihood, numerical methods such as the EM algorithm have to be used to compute the MLE; see Example 3.15. Baker and Laird (1988) discussed the EM method for a categorical y. Ibrahim et al. (1999) considered continuous y variable using a Monte Carlo EM method of Wei and Tanner (1990). Chen and Ibrahim (2006) extend the method to generalized additive models under a parametric assumption on the response model. Kim and Kim (2012) describes the use of fractional imputation to handle parameter estimation with nonignorable missing data.

Example 6.2. *We now revisit the parametric fractional imputation in Example 4.9. The parametric fractional imputation can be described as follows:*

[Step 1] Generate $y_i^{*(1)},\cdots,y_i^{*(m)}$ *from* $h(y_i\mid x_i)$*.*

[Step 2] Using the m imputed values generated from [Step 1], compute the fractional weights by

$$w_{ij(t)}^{*} \propto \frac{f\left(y_i^{*(j)}\mid x_i;\hat{\theta}_{(t)}\right)}{h\left(y_i^{*(j)}\mid x_i\right)}\left\{1-\pi(x_i,y_i^{*(j)};\hat{\phi}_{(t)})\right\}, \tag{6.5}$$

where $\pi(x_i,y_i;\hat{\phi})$ *is the estimated response probability evaluated at* $\hat{\phi}$*.*

[Step 3] Using the imputed data and the fractional weights, the M-step can be implemented by using (4.101) and (4.102).

[Step 4] Set $t=t+1$ *and go to [Step 2]. Continue until convergence.*

We now discuss the choice of the proposal density $h(y_i\mid x_i)$ *in [Step 1]. Often, it is possible to specify a "working" model, denoted by* $f(y_i\mid\mathbf{x}_i,\delta_i=1)$*, for the conditional distribution of* y_i *given* $\mathbf{x}_i$ *among* $\delta_i=1$ *and then estimate the conditional distribution by* $\hat{f}(y_i\mid\mathbf{x}_i,\delta_i=1)$*. Once* $\hat{f}(y_i\mid\mathbf{x}_i,\delta_i=1)$ *is computed, one can generate imputed values from*

$$\hat{f}(y_i\mid\mathbf{x}_i,\delta_i=0) = \frac{\hat{f}(y_i\mid\mathbf{x}_i,\delta_i=1)\left\{1/\pi(\mathbf{x}_i,y_i;\hat{\phi}_{(0)})-1\right\}}{\int\hat{f}(y_i\mid\mathbf{x}_i,\delta_i=1)\left\{1/\pi(\mathbf{x}_i,y_i;\hat{\phi}_{(0)})-1\right\}dy_i}. \tag{6.6}$$

In this case, we can use $h(y_i \mid \mathbf{x}_i) = \hat{f}(y_i \mid \mathbf{x}_i, \delta_i = 0)$ so that the fractional weights in (6.5) become

$$w^*_{ij(t)} \propto \frac{f(y_i^{*(j)} \mid \mathbf{x}_i; \hat{\theta}_{(t)})\pi(\mathbf{x}_i, y_i^{*(j)}; \hat{\phi}_{(0)})}{\hat{f}(y_i^{*(j)} \mid \mathbf{x}_i, \delta_i = 1)} \times \frac{1 - \pi\left(\mathbf{x}_i, y_i^{*(j)}; \hat{\phi}_{(t)}\right)}{1 - \pi\left(\mathbf{x}_i, y_i^{*(j)}; \hat{\phi}_{(0)}\right)}$$

with $\sum_{j=1}^{M} w^*_{ij(t)} = 1$. Generating imputed values from (6.6) can be implemented by applying the Monte Carlo sampling methods described in Chapter 3.

Such fully parametric approach in the nonignorable missing data case is known to be sensitive to the failure of the assumed parametric model (Kenward, 1998). Park and Brown (1994) used a Bayesian method to avoid the instability of the maximum likelihood estimators in the analysis of categorial missing data. Sensitivity analysis for nonignorable missingness can be a useful tool for addressing the issue associated with the nonignorable missingness. See Little (1995), Copas and Li (1997), Scharfstein et al. (1999), and Copas and Eguchi (2001) for some examples of the sensitivity analysis of missing data inference under nonignorable nonresponse.

6.2 Conditional likelihood approach

To avoid complicated computation, we now consider a likelihood-based approach of estimating parameters using only a part of the observed sample data. Recall that, if the parameter of interest is θ in $f(y \mid \mathbf{x}; \theta)$ and the response mechanism is ignorable, then the maximum likelihood method that maximizes

$$l_c(\theta) = \sum_{\delta_i=1} \log\{f(y_i \mid \mathbf{x}_i, \delta_i = 1; \theta)\} \tag{6.7}$$

is consistent. The likelihood function in (6.7) is a conditional likelihood because it is based on the conditional distribution given $\delta = 1$. The conditional likelihood is very close to the partial likelihood in survival analysis, which is very popular in analyzing censored data under Cox's proportional hazard model (Cox, 1972).

Following the decomposition below

$$f(y_i \mid \mathbf{x}_i) g(\delta_i \mid \mathbf{x}_i, y_i) = f_1(y_i \mid \mathbf{x}_i, \delta_i) g_1(\delta_i \mid \mathbf{x}_i),$$

the observed likelihood can be expressed as

$$L_{\text{obs}}(\theta, \phi) = \prod_{\delta_i=1} f_1(y_i \mid \mathbf{x}_i, \delta_i = 1) \times \prod_{i=1}^{n} g_1(\delta_i \mid \mathbf{x}_i). \tag{6.8}$$

The first component on the right hand side of (6.8) is the conditional likelihood

$$L_c(\theta) = \prod_{\delta_i=1} f_1(y_i \mid \mathbf{x}_i, \delta_i = 1) = \prod_{\delta_i=1} \left\{ \frac{f(y_i \mid \mathbf{x}_i; \theta)\, \pi(\mathbf{x}_i, y_i)}{\int f(y \mid \mathbf{x}_i; \theta)\, \pi(\mathbf{x}_i, y) dy} \right\}, \tag{6.9}$$

where

$$\pi(\mathbf{x}_i, y_i) = \pi_i = P(\delta_i = 1 \mid \mathbf{x}_i, y_i). \tag{6.10}$$

Unlike the observed likelihood (6.1), the conditional likelihood can be used even when $\mathbf{x}_i$'s associated with the missing y-values are not observed.

The score function derived from the conditional likelihood is

$$\begin{aligned} S_c(\theta) &= \frac{\partial}{\partial \theta} \ln L_c(\theta) \\ &= \sum_{i=1}^{n} \delta_i [S_i(\theta) - E\{S_i(\theta) \mid \mathbf{x}_i, \delta_i = 1; \theta\}] \\ &= \sum_{i=1}^{n} \delta_i \left[S_i(\theta) - \frac{E\{S_i(\theta)\pi_i \mid \mathbf{x}_i; \theta\}}{E(\pi_i \mid \mathbf{x}_i; \theta)} \right], \end{aligned}$$

where $S_i(\theta) = \partial \ln f(y_i \mid \mathbf{x}_i;\theta)/\partial\theta$. On the other hand, the score function derived from the observed likelihood (6.1) is

$$S_{\text{obs}}(\theta) = \frac{\partial}{\partial\theta}\ln L_{\text{obs}}(\theta) = \sum_{i=1}^n \delta_i S_i(\theta) + \sum_{i=1}^n (1-\delta_i)\frac{E\{S_i(\theta)(1-\pi_i) \mid \mathbf{x}_i;\theta\}}{E\{(1-\pi_i) \mid \mathbf{x}_i;\theta\}}.$$

If the response mechanism is ignorable such that $\pi_i = \pi(\mathbf{x}_i)$, then the score functions reduce to

$$S_c(\theta) = \sum_{i=1}^n \delta_i \{S_i(\theta) - E\{S_i(\theta) \mid \mathbf{x}_i;\theta\}\}$$

and

$$S_{\text{obs}}(\theta) = \sum_{i=1}^n \delta_i S_i(\theta) + \sum_{i=1}^n (1-\delta_i) E\{S_i(\theta) \mid \mathbf{x}_i;\theta\},$$

which are the same since $E\{S_i(\theta) \mid \mathbf{x}_i;\theta\} = 0$.

Assume that π_i is known. Then maximizing the conditional likelihood (6.9) is to solve $S_c(\theta) = 0$, and we can apply the Fisher-scoring method. Note that

$$\frac{\partial}{\partial\theta'}S_c(\theta) = \sum_{i=1}^n \delta_i \left[\frac{\partial}{\partial\theta'}S_i(\theta) - \frac{\partial}{\partial\theta'}\left\{\frac{E\{S_i(\theta)\pi_i \mid \mathbf{x}_i;\theta\}}{E(\pi_i \mid \mathbf{x}_i;\theta)}\right\}\right].$$

Writing $\dot{S}_i(\theta) = \partial S_i(\theta)/\partial\theta'$ and using

$$\frac{\partial}{\partial\theta'}E\{S_i(\theta)\pi_i \mid \mathbf{x}_i;\theta\} = E\{\dot{S}_i\pi_i \mid \mathbf{x}_i;\theta\} + E\{S_i S_i'\pi_i \mid \mathbf{x}_i;\theta\}$$

and

$$\frac{\partial}{\partial\theta'}E\{\pi_i \mid \mathbf{x}_i;\theta\} = E\{S_i\pi_i \mid \mathbf{x}_i;\theta\},$$

we have

$$\frac{\partial}{\partial\theta'}S_c(\theta) = \sum_{i=1}^n \delta_i\dot{S}_i(\theta) - \sum_{i=1}^n \delta_i\frac{E\{\dot{S}_i\pi_i \mid \mathbf{x}_i;\theta\} + E\{S_iS_i'\pi_i \mid \mathbf{x}_i;\theta\}}{E(\pi_i \mid \mathbf{x}_i;\theta)} + \sum_{i=1}^n \delta_i\frac{\{E(S_i\pi_i \mid \mathbf{x}_i;\theta)\}^{\otimes 2}}{\{E(\pi_i \mid \mathbf{x}_i;\theta)\}^2}.$$

Hence,

$$\mathcal{I}_c(\theta) = -E\left\{\frac{\partial}{\partial\theta'}S_c(\theta) \mid \mathbf{x}_i;\theta\right\} = \sum_{i=1}^n \left[E\{S_iS_i'\pi_i \mid \mathbf{x}_i;\theta\} - \frac{\{E(S_i\pi_i \mid \mathbf{x}_i;\theta)\}^{\otimes 2}}{E(\pi_i \mid \mathbf{x}_i;\theta)}\right]. \tag{6.11}$$

The Fisher-scoring method for obtaining the MLE from the conditional likelihood is then given by

$$\hat{\theta}^{(t+1)} = \hat{\theta}^{(t)} + \left\{\mathcal{I}_c(\hat{\theta}^{(t)})\right\}^{-1} S_c(\hat{\theta}^{(t)}), \qquad t = 0,1,2,\ldots$$

Furthermore, under some regularity conditions, it can be shown that the solution $\hat{\theta}_c$ to $S_c(\theta) = 0$ satisfies

$$\mathcal{I}_c^{1/2}(\hat{\theta}_c - \theta_0) \to_d N(0,I), \tag{6.12}$$

where $\to_d$ denotes convergence in distribution, $\mathcal{I}_c = \mathcal{I}_c(\theta_0)$ in (6.11), and I is the identity matrix.

In practice, the response probability $\pi_i = \pi(x_i, y_i)$ is generally unknown. To apply the conditional likelihood method, π_i can be replaced by a consistent estimator $\hat{\pi}_i$. Such a consistent estimator cannot be obtained from $(y_i, \mathbf{x}_i)$ with $\delta_i = 1$ only. This will be further studied in Sections 6.3 and 6.5.

The following example, originally presented in Chambers et al. (2012), indicates how to obtain estimators from both the full and conditional likelihood when π_i is unknown but it depends on θ only.

Example 6.3. *Assume that the original sample is a random sample from an exponential distribution with mean $\mu = 1/\theta$. That is, the probability density function of y is $f(y;\theta) = \theta\exp(-\theta y)I(y > 0)$. Suppose that we observe y_i only when $y_i > K$ for a known $K > 0$. Thus, the response indicator function is defined by $\delta_i = 1$ if $y_i > K$ and $\delta_i = 0$ otherwise. To compute the maximum likelihood estimator from the observed likelihood, note that*

$$S_{\text{obs}}(\theta) = \sum_{\delta_i=1}\left(\frac{1}{\theta} - y_i\right) + \sum_{\delta_i=0}\left\{\frac{1}{\theta} - E(y_i \mid \delta_i = 0)\right\}.$$

Since

$$E(Y \mid y \le K) = \frac{1}{\theta} - \frac{K\exp(-\theta K)}{1-\exp(-\theta K)},$$

the maximum likelihood estimator of θ can be obtained by the following iteration equation:

$$\left\{\hat{\theta}^{(t+1)}\right\}^{-1} = \bar{y}_r - \frac{n-r}{r}\left\{\frac{K\exp(-K\hat{\theta}^{(t)})}{1-\exp(-K\hat{\theta}^{(t)})}\right\}, \tag{6.13}$$

where $r = \sum_{i=1}^n \delta_i$ and $\bar{y}_r = r^{-1}\sum_{i=1}^n \delta_i y_i$. Similarly, we can derive the maximum conditional likelihood estimator. Note that $\pi_i = Pr(\delta_i = 1 \mid y_i) = I(y_i > K)$ and $E(\pi_i) = E\{I(y_i > K)\} = \exp(-K\theta)$. Thus, the conditional likelihood in (6.9) reduces to

$$\prod_{\delta_i=1}\theta\exp\{-\theta(y_i - K)\}.$$

The maximum conditional likelihood estimator of θ is

$$\hat{\theta}_c = \frac{1}{\bar{y}_r - K}.$$

Since $E(y \mid y > K) = \mu + K$, the maximum conditional likelihood estimator of μ, which is $\hat{\mu}_c = 1/\hat{\theta}_c$, is unbiased for μ.

6.3 Generalized method of moments (GMM) approach

In the previous section, $\pi(\mathbf{x}, y)$ given by (6.10) is usually unknown and has to be estimated. To estimate $\pi(\mathbf{x}, y)$, we assume (6.3)-(6.4) and the following parametric response model

$$\pi(\mathbf{x}, y) = \pi(\phi; \mathbf{u}, y), \tag{6.14}$$

where ϕ is an unknown parameter vector not depending on $\mathbf{z}$ and π is a known strictly monotone and twice differentiable function from $\mathcal{R}$ to $(0,1]$.

To estimate parameter ϕ, we consider the generalized method of moment (GMM) approach (Hansen (1982); Hall (2005)). The key idea of the GMM is to construct a set of L estimating functions

$$g_l(\phi, \mathbf{d}), \quad l = 1, ..., L, \quad \phi \in \Psi,$$

where $\mathbf{d}$ is a vector of observations, Ψ is the parameter space containing the true parameter value ϕ, $L \ge$ the dimension of Ψ, g_l's are non-constant functions with $E[g_l(\phi, \mathbf{d})] = 0$ for all l, and g_l's are not linearly dependent, i.e., the $L \times L$ matrix whose (l, l')th element being $E[g_l(\phi, \mathbf{d}) g_{l'}(\phi, \mathbf{d})]$ is positive definite, which can usually be achieved by eliminating redundant functions when g_l's are linearly dependent. Let $\mathbf{d}_1, ..., \mathbf{d}_n$ be n independent vectors distributed as $\mathbf{d}$ and

$$G(\varphi) = \left(\frac{1}{n}\sum_i g_1(\phi, \mathbf{d}_i), ..., \frac{1}{n}\sum_i g_L(\phi, \mathbf{d}_i)\right)^T, \quad \phi \in \Psi, \tag{6.15}$$

where a^T denotes the transpose of the vector a. If L has the same dimension as ϕ, then we may be able to find a $\hat{\phi}$ such that $G(\hat{\phi}) = 0$. If L is larger than the dimension of ϕ, however, a solution to $G(\varphi) = 0$ may not exist. In any case, a GMM estimator of ϕ can be obtained using the following two-step algorithm:

1. Obtain $\hat{\phi}^{(1)}$ by minimizing $G^T(\phi)G(\phi)$ over $\phi \in \Psi$.
2. Let $\hat{W}$ be the inverse of the $L \times L$ matrix whose (l, l')-th element is equal to

$$n^{-1} \sum_i g_l(\hat{\phi}^{(1)}, \mathbf{d}_i) g_{l'}(\hat{\phi}^{(1)}, \mathbf{d}_i).$$

The GMM estimator $\hat{\phi}$ is obtained by minimizing $G^T(\phi)\hat{W}G(\phi)$ over $\phi \in \Psi$.

To explain the use of GMM for parameter estimation with nonignorable missing data, we assume the following response model

$$P(\delta = 1|\mathbf{x}, y) = \pi(\phi_0 + \phi_1' \mathbf{u} + \phi_2 y), \tag{6.16}$$

where π is defined in (6.14), and $\phi = (\phi_0, \phi_1, \phi_2)$ is a $(k+2)$-dimensional unknown parameter vector not depending on values of $\mathbf{x}$. A similar assumption to (6.16) was made in Qin et al. (2002) and Kott and Chang (2010).

We assume that $\mathbf{z}$ has both continuous and discrete components and $\mathbf{u}$ is a continuous k-dimensional covariate. Let $\mathbf{z} = (z_d, \mathbf{z}_c)$, where $\mathbf{z}_c$ is m-dimensional continuous vector and z_d is discrete taking values $1, \ldots, J$. To estimate ϕ, the GMM can be applied with the following $L = k+m+J$ estimating functions:

$$\mathbf{g}(\phi, y, \mathbf{z}, \mathbf{u}, \delta) = \boldsymbol{\xi}[\delta \omega(\phi) - 1], \tag{6.17}$$

where $\boldsymbol{\xi} = (\boldsymbol{\zeta}^T, \mathbf{z}_c^T, \mathbf{u}^T)^T$, $\boldsymbol{\zeta}$ is a J-dimensional row vector whose lth component is $I(z_d = l)$, $I(A)$ is the indicator function of A, and $\omega(\phi) = [\pi(\phi_0 + \phi_1 \mathbf{u} + \phi_2 y)]^{-1}$.

The estimating function $\mathbf{g}$ is motivated by the fact that, when ϕ is the true parameter value,

$$\begin{aligned} E[\mathbf{g}(\phi, y, \mathbf{z}, \mathbf{u}, \delta)] &= E\{\boldsymbol{\xi}[\delta w(\phi) - 1]\} \\ &= E(E\{\boldsymbol{\xi}[\delta w(\phi) - 1]|y, \mathbf{z}, \mathbf{u}\}) \\ &= E\left\{\boldsymbol{\xi}\left[\frac{E(\delta|y, \mathbf{z}, \mathbf{u})}{P(\delta = 1|y, \mathbf{z}, \mathbf{u})} - 1\right]\right\} \\ &= 0. \end{aligned}$$

Let G be defined by (6.15) with g_l being the lth function of $\mathbf{g}$ and $\mathbf{d} = (y, \mathbf{z}, \mathbf{u}, \delta)$, $\hat{W}$ be the $\hat{W}$ in the previously described two-step algorithm and $\hat{\phi}$ be the two-step GMM estimator of ϕ in (6.16). Under some regularity conditions, as discussed in Wang et al. (2013), we can establish that

$$\sqrt{n}(\tilde{\phi} - \phi) \rightarrow_d N\left(0, (\Gamma^T \Sigma^{-1} \Gamma)^{-1}\right), \tag{6.18}$$

where $\rightarrow_d$ denotes convergence in distribution,

$$\Gamma = \begin{bmatrix} E[\delta \boldsymbol{\zeta}^T \omega'(\phi)] & E[\delta \mathbf{u} \boldsymbol{\zeta}^T \omega'(\phi)] & E[\delta \boldsymbol{\zeta}^T y \omega'(\phi)] \\ E[\delta \mathbf{z}_c^T \omega'(\phi)] & E[\delta \mathbf{u} \mathbf{z}_c^T \omega'(\phi)] & E[\delta \mathbf{z}_c^T y \omega'(\phi)] \\ E[\delta \mathbf{u}^T \omega'(\phi)] & E[\delta \mathbf{u} \mathbf{u}^T \omega'(\phi)] & E[\delta \mathbf{u}^T y \omega'(\phi)] \end{bmatrix}, \tag{6.19}$$

and Σ is the positive definite $(k+m+J) \times (k+m+J)$ matrix with $E[g_l(\phi, y, \mathbf{z}, \mathbf{u}, \delta) g_{l'}(\phi, y, \mathbf{z}, \mathbf{u}, \delta)]$ as its (l, l')th element. Thus, the asymptotic result in (6.18) requires that Γ in (6.19) is of full rank. Also, the asymptotic covariance matrix $(\Gamma^T \Sigma^{-1} \Gamma)^{-1}$ can be estimated by $(\hat{\Gamma}^T \hat{\Sigma}^{-1} \hat{\Gamma})^{-1}$, where $\hat{\Gamma}$ is the $(k+m+J) \times (k+2)$ matrix whose lth row is

$$\frac{1}{n} \sum_i \left. \frac{\partial g_l(\phi, y_i, \mathbf{z}_i, \mathbf{u}_i, \delta_i)}{\partial \phi} \right|_{\phi = \hat{\phi}}$$

and $\hat{\Sigma}$ is the $L \times L$ matrix whose (l,l')th element is

$$\frac{1}{n}\sum_i g_l(\hat{\phi}, y_i, \mathbf{z}_i, \mathbf{u}_i, \delta_i) g_{l'}(\hat{\phi}, y_i, \mathbf{z}_i, \mathbf{u}_i, \delta_i).$$

Once $\pi(\phi;\mathbf{u},y)$ is estimated, we can apply the approach in Section 6.2 to estimate parameters when the conditional density of y given $\mathbf{x}$ is parametric; we can also estimate some parameters in the conditional density of y given $\mathbf{x}$ nonparametrically. Details can be found in Wang et al. (2013).

Example 6.4. *Suppose that we are interested in estimating the parameters in the regression model*

$$y_i = \beta_0 + \beta_1 x_{1i} + \beta_2 x_{2i} + e_i, \tag{6.20}$$

where $E(e_i \mid \mathbf{x}_i) = 0$. Assume that y_i is subject to missingness and assume that

$$P(\delta_i = 1 \mid x_{1i}, x_{i2}, y_i) = \frac{\exp(\phi_0 + \phi_1 x_{1i} + \phi_2 y_i)}{1+\exp(\phi_0 + \phi_1 x_{1i} + \phi_2 y_i)}.$$

Thus, x_{2i} is the nonresponse instrument variable in this setup. A consistent estimator of ϕ can be obtained by solving

$$\hat{U}_2(\phi) \equiv \sum_{i=1}^n \left\{ \frac{\delta_i}{\pi(\phi;x_{1i},y_i)} - 1 \right\} (1, x_{1i}, x_{2i}) = (0,0,0). \tag{6.21}$$

Roughly speaking, the solution to (6.21) exists almost surely if $E\{\partial \hat{U}_2(\phi)/\partial \phi'\}$ is of full rank in the neighborhood of the true value of ϕ. If x_2 is a vector, then (6.21) is overidentified and the solution to (6.21) does not exist. In that case, the GMM algorithm can be used.

If x_{2i} is a categorical variable with category $\{1,\cdots,J\}$, then (6.21) can be written as

$$\hat{U}_2(\phi) \equiv \sum_{i=1}^n \left\{ \frac{\delta_i}{\pi(\phi;x_{1i},y_i)} - 1 \right\} (1, x_{1i}, \boldsymbol{\zeta}_i) = (0,0,0), \tag{6.22}$$

where $\boldsymbol{\zeta}_i$ is the J-dimensional row vector whose jth component is $I(x_{2i} = j)$.

Once the solution $\hat{\phi}$ to (6.21) or (6.22) is obtained, then a consistent estimator of $\beta = (\beta_0, \beta_1, \beta_2)$ can be obtained by solving

$$\hat{U}_1(\beta, \hat{\phi}) \equiv \sum_{i=1}^n \frac{\delta_i}{\hat{\pi}_i} \{y_i - \beta_0 - \beta_1 x_{1i} - \beta_2 x_{2i}\} (1, x_{1i}, x_{2i}) = (0,0,0) \tag{6.23}$$

for β. The asymptotic variance of $\hat{\beta} = \hat{\beta}(\hat{\phi})$, computed from (6.23), can be obtained by

$$V(\hat{\theta}) \cong \left(\Gamma_a' \Sigma_a^{-1} \Gamma_a\right)^{-1},$$

where

$$\begin{aligned}
\Gamma_a &= E\{\partial \hat{U}(\theta)/\partial \theta'\} \\
\Sigma_a &= V(\hat{U}) \\
\hat{U} &= (\hat{U}_1', \hat{U}_2')'
\end{aligned}$$

and $\theta = (\beta, \phi)$. The nonresponse instrument variable approach does not use a fully parametric model for $f(y \mid x)$ and so is less sensitive to the failure of the outcome model.

To solve a nonlinear equation $U(\phi) = 0$ such as (6.22), we can use the Newton method

$$\hat{\phi}^{(t+1)} = \hat{\phi}^{(t)} - \left\{ \dot{U}(\hat{\phi}^{(t)}) \right\}^{-1} U(\hat{\phi}^{(t)}), \tag{6.24}$$

where $\dot{U}(\phi) = \partial U(\phi)/\partial \phi'$. However, in (6.21) or (6.22), the partial derivative $\dot{U}(\phi)$ is not symmetric and the iterative computation in (6.24) can have numerical problems. To deal with the problem, we can use

$$\hat{\phi}^{(t+1)} = \hat{\phi}^{(t)} - \left\{ \dot{U}(\hat{\phi}^{(t)})' \dot{U}(\hat{\phi}^{(t)}) \right\}^{-1} \dot{U}(\hat{\phi}^{(t)})' U(\hat{\phi}^{(t)}), \tag{6.25}$$

which is essentially equivalent to finding $\hat{\phi}$ that minimizes $Q(\phi) = U(\phi)'U(\phi)$.

6.4 Pseudo likelihood approach

Now we switch to another type of conditional likelihood for parameter estimation with nonignorable missing data. Assume that $f(y|\mathbf{x};\theta)$ follows a parametric model and $g(\delta|\mathbf{x},y)$ is nonparametric. As discussed in Section 6.1, we still need a nonresponse instrument for the identifiability of unknown quantities. Tang et al. (2003) first studied this problem by assuming that the entire covariate vector $\mathbf{x}$ is a nonresponse instrument. Here, we assume a weaker assumption, in which $\mathbf{x} = (\mathbf{u},\mathbf{z})$ and $\mathbf{z}$ is a nonresponse instrument, i.e., (6.3)-(6.4) hold.

The main idea of this approach can be described as follows. Since δ and $\mathbf{z}$ are conditionally independent given $(y,\mathbf{u})$, the conditional probability density of $\mathbf{z}$ given $(\delta,y,\mathbf{u})$ satisfies

$$p(\mathbf{z} \mid y,\mathbf{u},\delta) = p(\mathbf{z} \mid y,\mathbf{u}).$$

Then, we can use the observed y_i's and $\mathbf{x}_i$'s to estimate $p(\mathbf{z}|y,\mathbf{u})$. If the parameter θ in $f(y|\mathbf{x};\theta)$ is of interest, then we can estimate it by maximizing

$$\prod_{i:\delta_i=1} p(\mathbf{z}_i \mid y_i,\mathbf{u}_i) = \prod_{i:\delta_i=1} \frac{f(y_i \mid \mathbf{u}_i,\mathbf{z}_i;\theta)p(\mathbf{z}_i|\mathbf{u}_i)}{\int f(y_i \mid \mathbf{u}_i,\mathbf{z};\theta)p(\mathbf{z}|\mathbf{u}_i)d\mathbf{z}}, \tag{6.26}$$

where the equality follows from the well-known Bayes' formula. Since $\mathbf{x}_i = (\mathbf{u}_i,\mathbf{z}_i)$, $i = 1,...,n$, are fully observed, the conditional probability density $p(\mathbf{z}|\mathbf{u})$ can be estimated using many well-established methods. Let $\hat{p}(\mathbf{z}|\mathbf{u})$ be an estimated conditional probability density of $\mathbf{z}$ given $\mathbf{u}$. Substituting this estimate into the likelihood in (6.26), we can obtain the following pseudo likelihood:

$$\prod_{i:\delta_i=1} \frac{f(y_i \mid \mathbf{u}_i,\mathbf{z}_i;\theta)\hat{p}(\mathbf{z}_i|\mathbf{u}_i)}{\int f(y_i \mid \mathbf{u}_i,\mathbf{z};\theta)\hat{p}(\mathbf{z}|\mathbf{u}_i)d\mathbf{z}}. \tag{6.27}$$

If $\hat{p}(\mathbf{z}|\mathbf{u})$ is obtained from a nonparametric method, the pseudo likelihood in (6.27) becomes semiparametric. For a parametric model $\hat{p}(\mathbf{z}|\mathbf{u}) = p(\mathbf{z}|\mathbf{u};\hat{\alpha})$, the pseudo likelihood in (6.27) becomes

$$\prod_{i:\delta_i=1} \frac{f(y_i \mid \mathbf{u}_i,\mathbf{z}_i;\theta)p(\mathbf{z}_i \mid \mathbf{u}_i;\hat{\alpha})}{\int f(y_i \mid \mathbf{u}_i,\mathbf{z};\theta)p(\mathbf{z} \mid \mathbf{u}_i;\hat{\alpha})d\mathbf{z}}. \tag{6.28}$$

The pseudo maximum likelihood estimator (PMLE) of θ, denoted by $\hat{\theta}_p$, can be obtained by solving

$$S_p(\theta;\hat{\alpha}) \equiv \sum_{\delta_i=1} [S(\theta;\mathbf{x}_i,y_i) - E\{S(\theta;\mathbf{u}_i,\mathbf{z},y_i) \mid y_i,\mathbf{u}_i;\theta,\hat{\alpha}\}] = 0$$

for θ, where $S(\theta;\mathbf{x},y) = S(\theta;\mathbf{u},\mathbf{z},y) = \partial \log f(y \mid \mathbf{x};\theta)/\partial\theta$ and

$$E\{S(\theta;\mathbf{u}_i,\mathbf{z},y_i) \mid y_i,\mathbf{u}_i;\theta,\hat{\alpha}\} = \frac{\int S(\theta;\mathbf{u}_i,\mathbf{z},y_i)f(y_i \mid \mathbf{u}_i,\mathbf{z};\theta)p(\mathbf{z} \mid \mathbf{u}_i;\hat{\alpha})d\mathbf{z}}{\int f(y_i \mid \mathbf{u}_i,\mathbf{z};\theta)p(\mathbf{z} \mid \mathbf{u}_i;\hat{\alpha})d\mathbf{z}}.$$

Since

$$\frac{\partial}{\partial\theta'}S_p(\theta;\hat{\alpha}) = \sum_{\delta_i=1}\left[\dot{S}(\theta;\mathbf{x}_i,y_i) - \frac{\partial}{\partial\theta'}E\{S(\theta;\mathbf{u}_i,\mathbf{z},y_i) \mid y_i,\mathbf{u}_i;\theta,\hat{\alpha}\}\right]$$

and

$$\begin{aligned}\frac{\partial}{\partial\theta'}E\{S(\theta;\mathbf{u}_i,\mathbf{z},y_i) \mid y_i,\mathbf{u}_i;\theta,\hat{\alpha}\} &= E\{\dot{S}(\theta;\mathbf{u}_i,\mathbf{z},y_i) \mid y_i,\mathbf{u}_i;\theta,\hat{\alpha}\} \\ &+ E\{S(\theta;\mathbf{u}_i,\mathbf{z},y_i)^{\otimes 2} \mid y_i,\mathbf{u}_i;\theta,\hat{\alpha}\} \\ &- E\{S(\theta;\mathbf{u}_i,\mathbf{z},y_i) \mid y_i,\mathbf{u}_i;\theta,\hat{\alpha}\}^{\otimes 2},\end{aligned}$$

where $\dot{S}(\theta;\mathbf{u}_i,\mathbf{z},y_i)=\partial S(\theta;\mathbf{u},\mathbf{z},y)/\partial\theta'$, the Fisher-scoring method for obtaining the PMLE is given by

$$\hat{\theta}_p^{(t+1)}=\hat{\theta}_p^{(t)}+\left\{\mathcal{I}_p\left(\hat{\theta}^{(t)},\hat{\alpha}\right)\right\}^{-1}S_p(\hat{\theta}^{(t)},\hat{\alpha}),$$

where

$$\mathcal{I}_p(\theta,\hat{\alpha})=\sum_{\delta_i=1}\left[E\{S(\theta;\mathbf{u}_i,\mathbf{z},y_i)^{\otimes 2}\mid y_i,\mathbf{u}_i;\theta,\hat{\alpha}\}-E\{S(\theta;\mathbf{u}_i,\mathbf{z},y_i)\mid y_i,\mathbf{u}_i;\theta,\hat{\alpha}\}^{\otimes 2}\right].$$

In particular, when $\mathbf{x}=\mathbf{z}$ ($\mathbf{u}=0$), we can estimate $p(\mathbf{x})$ by the empirical distribution putting mass n^{-1} on each observed $\mathbf{x}_i$, $i=1,\ldots,n$. The pseudo likelihood becomes

$$\prod_{i:\delta_i=1}\frac{f(y_i\mid\mathbf{x}_i;\theta)}{\sum_{l=1}^n f(y_i\mid\mathbf{x}_l;\theta)}.$$

The parameter θ can be estimated by maximizing the pseudo likelihood in (6.27) over θ. Under some regularity conditions, consistency and asymptotic normality of this estimator $\hat{\theta}$ is established in Tang et al. (2003), Jiang and Shao (2012), and Shao and Zhao (2012). That is,

$$\sqrt{n}(\hat{\theta}-\theta_0)\to_d N(0,\Sigma),$$

where θ_0 is the true value of the parameter θ and Σ is a covariance matrix. Due to the use of pseudo likelihood (the substitution of $p(\mathbf{z}|\mathbf{u})$ by its estimator), the form of Σ is very complicated. Shao and Zhao (2012) proposed an approach for estimating Σ. Alternatively, we can apply bootstrapping to estimate Σ; see Shao and Zhao (2012).

We now show an example from Jiang and Shao (2012), who considered a longitudinal y_i. The Health and Retirement Study (HRS) of about 22,000 Americans over the age of 50 and their spouses was conducted by the University of Michigan (see more details at the website http://hrsonline.isr.umich.edu/). The study is a biannual longitudinal household survey conducted from 1997 to 2006. We only consider 19,043 households and the univariate variable, household's income at year 1997. Missing values exist and the percentage of missing data is about 67.2%. This high missing rate may be, partly due to the fact that household income is regarded the total of several components (e.g., stocks, pensions, and annuities) and the total income is treated as a missing value if any one of these components is missing. We consider $y_i=\log(w_i+\sqrt{1+w_i^2})$, where w_i is the income of the ith household, and assume that

$$y_i=\beta_0+\beta_1x_i+\varepsilon_i,$$

where x_i is the number of years of education treated as a covariate that ranges from 0 to 17 with mean 12.74, ε_i is distributed as $N(0,\sigma^2)$ and ε_i's are independent.

For comparison, we applied two approaches for parameter estimation: (i) the method of using data from subjects without any missing value, i.e., ignoring subjects with incomplete data, and (ii) the approach of maximizing pseudo likelihood (6.27). Method (i) is justified when nonresponse is ignorable. For each approach, we applied a bootstrap method with $B=200$ bootstrap samples for estimating the standard errors of the estimates. In this example, it is of interest to estimate the mean household income, in addition to the parameters β and σ. We obtained estimates of the mean household income $E(w_i)$ based on the estimated parameters and the inverse of the transformation $y=\log(w+\sqrt{1+w^2})$. All parameter estimates and their estimated standard errors are given in Table 6.1.

It can be seen from Table 6.1 that the estimates obtained by ignoring subjects with incomplete data are very different from those obtained by applying a missing data method. In terms of the mean household income, ignoring subjects with incomplete data results in negatively biased estimates, which indicates that household income of a nonrespondent is typically higher than that of a respondent.

Table 6.1 Parameter estimates (Standard errors based on 200 bootstrap samples)

	Method	
	Complete Case	PMLE
β_0	8.845 (0.202)	9.568 (0.154)
β_1	0.176 (0.016)	0.136 (0.015)
σ	1.579 (0.146)	1.370 (0.116)
1997 mean household income	19218 (1012)	35338 (2361)

6.5 Exponential tilting (ET) model

Under the response model in (6.14), we can assume that δ_i are generated from a Bernoulli distribution with probability $\pi_i(\phi) = \pi(\mathbf{u}_i, y_i; \phi)$ for some ϕ. If y_i were observed throughout the sample, the likelihood function of ϕ would be

$$L(\phi) = \prod_{i=1}^{n} \{\pi_i(\phi)\}^{\delta_i} \{1 - \pi_i(\phi)\}^{(1-\delta_i)}, \tag{6.29}$$

and the maximum likelihood estimator (MLE) of ϕ could be obtained by solving the score equation $S(\phi) = \partial \log\{L(\phi)\}/\partial\phi = 0$. The score equation can be expressed as

$$\sum_{i=1}^{n} S_i(\phi) = \sum_{i=1}^{n} \{\delta_i - \pi_i(\phi)\}\mathbf{h}_i(\phi) = 0, \tag{6.30}$$

where $\mathbf{h}_i(\phi) = \partial \text{logit}\{\pi_i(\phi)\}/\partial\phi$, and $\text{logit}(\pi) = \log\{\pi/(1-\pi)\}$. However, as some y_i's are missing, the score equation (6.30) is not applicable. Instead, we can consider maximizing the observed likelihood function

$$L_{obs}(\phi) = \prod_{i=1}^{n} \{\pi_i(\phi)\}^{\delta_i} \left[\int \{1 - \pi_i(\phi)\} f(y|\mathbf{x}_i) dy\right]^{1-\delta_i},$$

where $f(y|\mathbf{x})$ is the true conditional distribution of y given $\mathbf{x}$. The MLE of ϕ can be obtained by solving the observed score function $S_{obs}(\phi) = \partial \text{log} L_{obs}(\phi)/\partial\phi = 0$. Finding the solution to the observed score equation can be computationally challenging because it involves integration with unknown parameters. An alternative way of finding the MLE of ϕ is to solve the mean score equation $\bar{S}(\phi) = 0$, where

$$\bar{S}(\phi) = \sum_{i=1}^{n} [\delta_i S_i(\phi) + (1-\delta_i) E\{S_i(\phi)|\mathbf{x}_i, \delta_i = 0\}], \tag{6.31}$$

where $S_i(\phi)$ is defined in (6.30).

To compute the conditional expectation in (6.31), we use the following relationship:

$$\begin{aligned} & Pr(y_i \in B \mid \mathbf{x}_i, \delta_i = 0) \\ = \; & Pr(y_i \in B \mid \mathbf{x}_i, \delta_i = 1) \times \frac{Pr(\delta_i = 0 \mid \mathbf{x}_i, y_i \in B) / Pr(\delta_i = 1 \mid \mathbf{x}_i, y_i \in B)}{Pr(\delta_i = 0 \mid \mathbf{x}_i) / Pr(\delta_i = 1 \mid \mathbf{x}_i)}. \end{aligned}$$

Thus, we can write the conditional distribution of the missing data given x as

$$f_0(y_i \mid \mathbf{x}_i) = f_1(y_i \mid \mathbf{x}_i) \times \frac{O(\mathbf{x}_i, y_i)}{E\{O(\mathbf{x}_i, Y_i) \mid \mathbf{x}_i, \delta_i = 1\}}, \tag{6.32}$$

where $f_\delta(y_i \mid \mathbf{x}_i) = f(y_i \mid \mathbf{x}_i, \delta_i = \delta)$ and

$$O(\mathbf{x}_i, y_i) = \frac{Pr(\delta_i = 0 \mid \mathbf{x}_i, y_i)}{Pr(\delta_i = 1 \mid \mathbf{x}_i, y_i)} \tag{6.33}$$

is the conditional odds of nonresponse. If the response probability in (6.14) follows from a logistic regression model

$$\pi(\mathbf{u}_i, y_i) \equiv Pr(\delta_i = 1 \mid \mathbf{u}_i, y_i) = \frac{\exp(\phi_0 + \phi_1 \mathbf{u}_i + \phi_2 y_i)}{1 + \exp(\phi_0 + \phi_1 \mathbf{u}_i + \phi_2 y_i,)} \tag{6.34}$$

the odds function (6.33) can be written as $O(x_i, y_i) = \exp\{-\phi_0 - \phi_1 \mathbf{u}_i - \phi_2 y_i\}$ and the expression (6.32) can be simplified to

$$f_0(y_i \mid \mathbf{x}_i) = f_1(y_i \mid \mathbf{x}_i) \times \frac{\exp(-\phi_2 y_i)}{E\{\exp(-\phi_2 Y) \mid \mathbf{x}_i, \delta_i = 1\}}, \tag{6.35}$$

where $f_1(y \mid \mathbf{x})$ is the conditional density of y given $\mathbf{x}$ and $\delta = 1$. Model (6.35) states that the density for the nonrespondents is an exponential tilting of the density for the respondents. If $\phi_2 = 0$, the the response mechanism is ignorable and $f_0(y|\mathbf{x}) = f_1(y|\mathbf{x})$. Kim and Yu (2011) also used an exponential tilting model to compute the conditional expectation $E_0(y|\mathbf{x})$ nonparametrically from the respondents. Equation (6.35) implies that we only need the response model (6.34) and the conditional distribution of study variable given the auxiliary variables for respondents $f_1(y|\mathbf{x})$, which is relatively easy to verify from the observed part of the sample.

To discuss how to solve the mean score equation in (6.31) using the exponential tilting model in (6.35), we assume the parametric model for $f_1(y|\mathbf{x})$ is correctly specified with parameter γ. Thus, we can write

$$f_1(y|\mathbf{x}) = f_1(y|\mathbf{x};\gamma), \tag{6.36}$$

for some γ. To estimate the response model parameter ϕ, we first need to obtain a consistent estimator of γ. For example, the MLE of γ can be computed by solving

$$S_1(\gamma) \equiv \sum_{i=1}^{n} \delta_i S_1(\gamma; \mathbf{x}_i, y_i) = 0, \tag{6.37}$$

where $S_1(\gamma; \mathbf{x}, y) = \delta_i \partial \log f_1(y \mid \mathbf{x}; \gamma)/\partial \gamma$. Using $\hat{\gamma}$ obtained from (6.37), the mean score equation can be written, by (6.31) and (6.35), as

$$\bar{S}(\phi, \hat{\gamma}) = \sum_{\delta_i=1} S(\phi; \delta_i, \mathbf{x}_i, y_i) + \sum_{\delta_i=0} \int S(\phi; \delta_i, \mathbf{x}_i, y) f_0(y \mid \mathbf{x}_i; \hat{\gamma}, \phi)\, dy = 0. \tag{6.38}$$

To solve (6.38) for ϕ, either a Newton method or the EM algorithm can be used.

To discuss the use of the EM algorithm to solve (6.38), we first consider the simple case when y is a categorical variable with M categories, taking values in $\{1, 2, \cdots, M\}$. Let $f_\delta(y \mid \mathbf{x}; \gamma)$ be the probability mass function of y conditional on $\mathbf{x}$ and δ. In this case, we can express (6.38) as

$$\bar{S}(\phi, \hat{\gamma}) = \sum_{\delta_i=1} S(\phi; \delta_i, \mathbf{x}_i, y_i) + \sum_{\delta_i=0} \left\{ \frac{\sum_{y=1}^{M} S(\phi; \delta_i, \mathbf{x}_i, y) \exp(-\phi_2 y) f_1(y \mid \mathbf{x}_i; \hat{\gamma}, \phi)}{\sum_{y=1}^{M} \exp(-\phi_2 y) f_1(y \mid \mathbf{x}_i; \hat{\gamma}, \phi)} \right\} = 0. \tag{6.39}$$

To solve (6.39) by the EM algorithm, we first compute the fractional weights

$$w_{iy}^{*(t)} = \frac{\exp(-\hat{\phi}_2^{(t)} y) f_1(y \mid \mathbf{x}_i; \hat{\gamma})}{\sum_{y=1}^{M} \exp(-\hat{\phi}_2^{(t)} y) f_1(y \mid \mathbf{x}_i; \hat{\gamma})}$$

using the current value $\hat{\phi}^{(t)}$ of ϕ. This is the E-step of the EM algorithm. In the M-step, the parameter estimate is updated by solving

$$\sum_{\delta_i=1} S(\phi; \delta_i, \mathbf{x}_i, y_i) + \sum_{\delta_i=0} \sum_{y=1}^{M} w_{iy}^{*(t)} S(\phi; \delta_i, \mathbf{x}_i, y) = 0.$$

If Y is continuous, an algorithm similar to the Monte Carlo EM algorithm can be implemented as follows:

(Step 1) Generate $y_{ij}^* \sim f(y|\mathbf{x}_i, \delta_i = 1, \hat{\gamma})$ for each nonrespondent i and $j = 1, 2, \cdots, m$, where $\hat{\gamma}$ is obtained from (6.37).

(Step 2) Using the current value of $\phi^{(t)}$ and the Monte Carlo sample from (Step 1), compute

$$\bar{S}_p(\phi|\phi^{(t)}) = \sum_{i=1}^{n} \left\{ \delta_i S(\phi; \delta_i, \mathbf{x}_i, y_i) + (1-\delta_i) \sum_{j=1}^{m} w_{ij}^{*(t)} S(\phi; \delta_i, \mathbf{x}_i, y_{ij}^*) \right\}, \quad (6.40)$$

where

$$w_{ij}^{*(t)} = \frac{O_{ij}^{*(t)}}{\sum_{k=1}^{m} O_{ik}^{*(t)}}, \quad (6.41)$$

$O_{ij}^{*(t)} = 1/\pi_{ij}^*(\phi^{(t)}) - 1$, and $\pi_{ij}^*(\phi) = \pi(\mathbf{u}_i, y_{ij}^*; \phi)$.

(Step 3) Find the solution $\phi^{(t+1)}$ to $\bar{S}_p(\phi|\phi^{(t)}) = 0$ where $\bar{S}_p(\phi|\phi^{(t)})$ is computed from (Step 2).

(Step 4) Update $t = t+1$ and repeat (Step 2)-(Step 3) until convergence.

In the above algorithm, (Step 1) and (Step 2) correspond to the E-step and (Step 3) corresponds to the M-step of the EM algorithm. Unlike the usual Monte Carlo EM algorithm, (Step1) is not repeated. The proposed method can be regarded as a special application of the parametric fractional imputation of Kim (2011) under nonignorable nonresponse. If the response model is of a logistic form in (6.34), the fractional weights in (6.41) can be simply expressed as

$$w_{ij}^{*(t)} = \frac{\exp(-\phi_2^{(t)} y_{ij}^*)}{\sum_{k=1}^{m} \exp(-\phi_2^{(t)} y_{ik}^*)}.$$

Instead of using the above Monte Carlo EM algorithm, one can estimate

$$\begin{aligned} E_0\{S(\phi; \mathbf{x}_i, Y) \mid \mathbf{x}_i\} &= \int S(\phi; \delta_i, \mathbf{x}_i, y) f_0(y \mid \mathbf{x}_i) dy \\ &\cong \frac{\int S(\phi; \delta_i, \mathbf{x}_i, y) f_1(y \mid \mathbf{x}_i; \hat{\gamma}) \exp(-\phi_2 y) dy}{\int f_1(y \mid \mathbf{x}_i; \hat{\gamma}) \exp(-\phi_2 y) dy} \end{aligned}$$

by

$$\bar{S}_0(\phi \mid \mathbf{x}_i; \hat{\gamma}, \phi) = \frac{\sum_{\delta_j=1} S(\phi; \delta_i, \mathbf{x}_i, y_j) f_1(y_j \mid \mathbf{x}_i; \hat{\gamma}) \exp(-\phi_2 y_j)/\hat{f}_1(y_j)}{\sum_{\delta_j=1} f_1(y_j \mid \mathbf{x}_i; \hat{\gamma}) \exp(-\phi_2 y_j)/\hat{f}_1(y_i)}, \quad (6.42)$$

where

$$\hat{f}_1(y) = n_R^{-1} \sum_{i=1}^{n} \delta_i f_1(y \mid \mathbf{x}_i; \hat{\gamma})$$

is a consistent estimator of $f_1(y) = \int f(y \mid \mathbf{x}, \delta = 1) f(\mathbf{x} \mid \delta = 1) d\mathbf{x}$. A fully efficient estimator of ϕ can be obtained by solving

$$S_2(\phi, \hat{\gamma}) \equiv \sum_{i=1}^{n} \{\delta_i S(\phi; \delta_i, \mathbf{x}_i, y_i) + (1-\delta_i) \bar{S}_0(\phi \mid \mathbf{x}_i; \hat{\gamma}, \phi)\} = 0. \quad (6.43)$$

Riddles and Kim (2013) provides some asymptotic properties of the PSA estimator using $\hat{\phi}$ obtained from (6.43).

Example 6.5. *We consider an example of a fully nonparametric approach for categorical data. Assume that both $\mathbf{x}_i = (z_i, u_i)$ and y_i are categorical with category $\{(i, j); i \in S_z \times S_u\}$ and S_y, respectively. Suppose that we are interested in estimating $\theta_k = Pr(Y = k)$, for $k \in S_y$. Under complete* *response, the parameter is estimated by* $\hat{\theta}_k = n^{-1}\sum_{i=1}^{n} I(y_i = k)$*. Now, we have nonresponse in* y *and let* δ_i *be the response indicator function for* y_i*. We assume that the response probability satisfies*

$$Pr(\delta = 1 \mid \mathbf{x}, y) = \pi(u, y; \phi).$$

To estimate ϕ*, we first compute the observed conditional probability of* y *among the respondents:*

$$\hat{p}_1(y \mid \mathbf{x}_i) = \frac{\sum_{\delta_j=1} I(\mathbf{x}_j = \mathbf{x}_i, y_j = y)}{\sum_{\delta_j=1} I(\mathbf{x}_j = \mathbf{x}_i)}.$$

The EM algorithm can be implemented by (6.43) with

$$\bar{S}_0(\phi \mid \mathbf{x}_i; \phi) = \frac{\sum_{\delta_j=1} S(\phi; \delta_i, u_i, y_j)\hat{p}_1(y_j \mid \mathbf{x}_i)O(\phi; u_i, y_j)/\hat{p}_1(y_j)}{\sum_{\delta_j=1} S(\phi; \delta_i, u_i, y_j)\hat{p}_1(y_j \mid \mathbf{x}_i)O(\phi; u_i, y_j)/\hat{p}_1(y_j)},$$

where $O(\phi; u, y) = \{1 - \pi(u, y; \phi)\}/\pi(u, y; \phi)$ *and*

$$\hat{p}_1(y) = n_R^{-1}\sum_{i=1}^{n} \delta_i \hat{p}_1(y \mid \mathbf{x}_i).$$

Alternatively, we can use

$$\bar{S}_0(\phi \mid \mathbf{x}_i; \phi) = \frac{\sum_{y \in S_y} S(\phi; \delta_i, u_i, y)\hat{p}_1(y \mid \mathbf{x}_i)O(\phi; u_i, y)}{\sum_{y \in S_y} \hat{p}_1(y \mid \mathbf{x}_i)O(\phi; u_i, y)}. \tag{6.44}$$

Once $\hat{\pi}(u, y) = \pi\left(u, y; \hat{\phi}\right)$ *is computed, we can use*

$$\hat{\theta}_{k,ET} = n^{-1}\left\{\sum_{\delta_i=1} I(y_i = k) + \sum_{\delta_i=0}\sum_{y \in S_y} w_{iy}^* I(y = k)\right\},$$

where w_{iy}^* *is the fractional weights computed by*

$$w_{iy}^* = \frac{\{\hat{\pi}(u_i, y)\}^{-1} - 1\}\hat{p}_1(y|x_i)}{\sum_{y \in S_y}\{\hat{\pi}(u_i, y)^{-1} - 1\}\hat{p}_1(y|x_i)}.$$

Example 6.6. *To investigate the performance of the estimators, a limited simulation study is considered. In the simulation, the samples are generated from*

$$\begin{pmatrix} x_{1i} \\ x_{2i} \end{pmatrix} \sim N\left[\begin{pmatrix} 1 \\ 2 \end{pmatrix}, \begin{pmatrix} 1 & 0 \\ 0 & 1 \end{pmatrix}\right]$$

and

$$y_i = -1 + x_{1i} + 0.5x_{2i} + e_i, \quad e_i \sim N(0, 1).$$

We assume that (x_{1i}, x_{2i}, z_i) *are observed throughout the sample, but we observe* y_i *only when* $\delta_i = 1$*, where* $\delta_i \sim Bernoulli(\pi_i)$ *with*

$$\pi_i = \frac{\exp(-0.5 + 0.5x_{1i} + 0.7y_i)}{1 + \exp(-0.5 + 0.5x_{1i} + 0.7y_i)}.$$

We use $n = 800$ *in the simulation with* $B = 2{,}000$ *Monte Carlo samples.*

From the sample, we consider four estimators:

1. *Simple mean from the complete data:* $\hat{\theta}_1 = n^{-1}\sum_{i=1}^{n} y_i$.

2. *The PSA estimator using the GMM method in §6.3. The estimator is*

$$\hat{\theta}_2 = \frac{\sum_{i=1}^n \delta_i y_i/\hat{\pi}_i}{\sum_{i=1}^n \delta_i/\hat{\pi}_i},$$

where

$$\hat{\pi}_i = \frac{\exp(\hat{\phi}_0 + \hat{\phi}_1 x_{1i} + \hat{\phi}_2 y_i)}{1 + \exp(\hat{\phi}_0 + \hat{\phi}_1 x_{1i} + \hat{\phi}_2 y_i)}$$

and $(\hat{\phi}_0, \hat{\phi}_1, \hat{\phi}_2)$ *is computed by solving (6.21).*

3. *The PSA estimator based on exponential tilting model (ET-PSA) is given by*

$$\hat{\theta}_{PSA} = \frac{\sum_{i=1}^n \delta_i y_i/\hat{\pi}_i}{\sum_{i=1}^n \delta_i/\hat{\pi}_i}, \tag{6.45}$$

where $\hat{\pi}_i = \pi(\mathbf{x}_i, y_i; \hat{\phi})$ *and* $\hat{\phi}$ *is computed by solving the mean score equation in (6.43).*

4. *The empirical likelihood estimator applied to the exponential tilting PSA estimator in 3. Note that the ET-PSA estimator in (6.45) does not necessarily satisfy the calibration constraints. That is, we may not have*

$$\sum_{i=1}^n \frac{\delta_i}{\hat{\pi}_i}\mathbf{x}_i = \sum_{i=1}^n \mathbf{x}_i.$$

To impose the calibration constraint, we can consider using

$$\hat{\theta}_\omega = \sum_{i=1}^n \frac{\delta_i}{\hat{\pi}_i}\omega_i y_i$$

where ω_i *are determined to maximize*

$$l(\omega) = \sum_{i=1}^n \log(\omega_i)$$

subject to $\sum_{i=1}^n \omega_i = 1$ *and*

$$\sum_{i=1}^n \frac{\delta_i}{\hat{\pi}_i}\omega_i(1, x_{1i}, x_{2i}) = \sum_{i=1}^n w_i(1, x_{1i}, x_{2i}).$$

This is a typical technique of empirical likelihood method for calibration. The solution can be written as

$$\hat{\omega}_i = \frac{1}{\hat{\lambda}_0 + (\delta_i/\hat{\pi}_i - 1)\hat{\lambda}_1'\mathbf{x}_i},$$

where $\mathbf{x}_i = (1, x_{1i}, x_{2i})'$ *and* $(\hat{\lambda}_0, \hat{\lambda}_1')$ *are constructed to satisfy the constraints.*

Table 6.2 **Monte Carlo mean and variance of the point estimators in Example 6.6**

Parameter		Complete	GMM-PSA	ET-PSA	ET-CPSA
θ	Mean	1.0004	0.9910	1.0028	1.0049
	Var	0.0028	0.0122	0.0111	0.0081

Table 6.2 shows that the estimators are all nearly unbiased. The ET PSA estimator is slightly more efficient than the PSA estimator computed by the GMM method in Section 6.3. The empirical likelihood calibration estimator is significantly more efficient because the linear regression model is true. Table 6.3 also shows that the mean score method based on the ET model is significantly more efficient than the GMM estimator.

Table 6.3 Estimates of parameters in the response model in Example 6.6

Parameter		GMM	ET
ϕ_0	Bias (Var)	0.04 (0.073)	0.02(0.025)
ϕ_1	Bias (Var)	-0.00 (0.143)	0.01(0.068)
ϕ_2	Bias (Var)	0.03 (0.100)	0.00(0.044)

6.6 Latent variable approach

Another approach of modeling nonignorable nonresponse is to assume a latent variable that is related to the survey variable and then assume that the study variable is observed if and only if the latent variable exceeds a threshold (say zero). The latent variable approach is very popular in econometrics in explaining self-selection bias (Heckman, 1979). O'Muircheartaigh and Moustaki (1999) also consider the latent variable approach to model item nonresponse in attitude scale.

To explain the idea, consider the following example considered in Little and Rubin (2002, Example 15.7).

Example 6.7. *Suppose that the original study variable y follows a normal-theory linear model given by*

$$y_i = \mathbf{x}_i'\beta + e_i \tag{6.46}$$

and $e_i \sim N(0, \sigma^2)$. Let δ_i be the response indicator function for y_i. The response model is

$$\delta_i = \begin{cases} 1 & \text{if } z_i > 0 \\ 0 & \text{if } z_i \le 0, \end{cases}$$

where z_i is the latent variable representing the level of survey participation and follows

$$z_i = \mathbf{x}_i'\gamma + u_i$$

and

$$\begin{pmatrix} e_i \\ u_i \end{pmatrix} \sim N\left[\begin{pmatrix} 0 \\ 0 \end{pmatrix}, \begin{pmatrix} \sigma^2 & \rho\sigma \\ \rho\sigma & 1 \end{pmatrix}\right]. \tag{6.47}$$

By the property of normal distribution, we can derive

$$\begin{aligned} P(\delta_i = 1 \mid \mathbf{x}_i, y_i) &= Pr(z_i \ge 0 \mid \mathbf{x}_i, y_i) \\ &= 1 - \Phi\left\{ -\frac{\mathbf{x}_i'\gamma + \rho\sigma^{-1}(y_i - \mathbf{x}_i'\beta)}{\sqrt{1-\rho^2}} \right\}. \end{aligned}$$

When $\rho \neq 0$, the response probability depends on y_i, which is subject to missingenss, and the response mechanism becomes nonignorable. If $\rho = 0$, then the response mechanism is ignorable.

In the above example, z_i is always missing but is useful in representing the response probability as a function of parameters in the latent variable model. The latent model is then viewed as a *hurdle model* since crossing a hurdle or threshold leads to participation. The classic early application of the latent model to nonignorable missing (or selection bias) was to labor supply, where z is the unobserved desire or propensity to work, while y is the actual hours worked. In Example 6.7, the observed likelihood is

$$L_{obs} = \prod_{i=1}^{n} \{P(z_i \le 0 \mid \mathbf{x}_i)\}^{1-\delta_i} \{f(y_i \mid \mathbf{x}_i, z_i > 0) P(z_i > 0 \mid \mathbf{x}_i)\}^{\delta_i}.$$

This likelihood function is applicable to general models, not just linear models with joint normal errors.

In the normal case,

$$E(y_i \mid \mathbf{x}_i, z_i > 0) = \mathbf{x}_i'\beta + \rho\sigma\lambda(\mathbf{x}_i'\gamma)$$

where $\lambda(z) = \phi(z)/\{1-\Phi(z)\}$, which is often called the *inverse Mills ratio* (Amemiya, 1985). To estimate the parameters, two-step estimation procedure, proposed by Heckman (1979), can be easily implemented as follows:

[Step 1] Estimate γ by applying the probit regression of δ_i on $\mathbf{x}_i$ since $P(\delta = 1 \mid \mathbf{x}) = \Phi(\mathbf{x}'\gamma)$.

[Step 2] Using only the cases with $\delta_i = 1$, fit a linear regression model

$$y_i = \mathbf{x}_i'\beta + \sigma_{12}q_i + v_i$$

where v_i is an error term and $q_i = \lambda(\mathbf{x}_i'\hat{\gamma})$ with $\hat{\gamma}$ obtained from [Step 1].

Instead of the above two-step method, an EM algorithm can also be used. Note that, as $\lambda(z) \cong a + bz$, we may write

$$E(y_i \mid \mathbf{x}_i, z_i > 0) \cong a + \mathbf{x}_i'\beta + b\mathbf{x}_i'\gamma,$$

which leads to obvious multicollinearity problems (Nawata and Nagase, 1996). To avoid this non-identifiability problem, one regressor in estimating γ may be excluded from the model in estimating β, which might limit the applicability of the latent variable approach in practice. For more details of the latent variable approach to handle selection bias, see Chapter 16 of Cameron and Trivedi (2005).

6.7 Callbacks

We now assume a nonignorable response mechanism of the form

$$Pr(\delta_i = 1 \mid \mathbf{x}_i, y_i) = \pi(\phi; \mathbf{x}_i, y_i) = \frac{\exp(\phi_0 + \mathbf{x}_i\phi_1 + y_i\phi_2)}{1 + \exp(\phi_0 + \mathbf{x}_i\phi_1 + y_i\phi_2)} \tag{6.48}$$

and discuss how to obtain a consistent estimator of the response probability under the existence of missing data. Clearly, the score equation

$$\sum_{i=1}^{n} \{\delta_i - \pi(\phi; \mathbf{x}_i, y_i)\}(\mathbf{x}_i, y_i) = 0$$

cannot be solved because y_i are not observed when $\delta_i = 0$.

To estimate the parameters in (6.48), we consider the special case when there are some callbacks among nonrespondents. That is, among the elements with $\delta_i = 0$, further efforts are made to obtain the observation of y_i. Let $\delta_{2i} = 1$ if the element i is selected for a callback or $\delta_i = 1$ and $\delta_{2i} = 0$ otherwise. We assume that the selection mechanism for the callback depends only on δ_i. That is,

$$Pr(\delta_2 = 1 \mid \mathbf{x}, y, \delta) = \begin{cases} 1 & \text{if } \delta = 1 \\ \nu & \text{if } \delta = 0 \end{cases} \tag{6.49}$$

for some $\nu \in (0, 1]$. The following lemma shows that the response probability can be estimated from the original sample and the callback sample.

Lemma 6.2. *Assume that the response mechanism satisfies (6.48) and the followup sample is randomly selected among nonrespondents with probability ν. Then, the response probability among the set with $\delta_{i2} = 1$ can be expressed as*

$$Pr(\delta_i = 1 \mid \mathbf{x}_i, y_i, \delta_{2i} = 1) = \frac{\exp(\phi_0^* + \mathbf{x}_i\phi_1^* + y_i\phi_2^*)}{1 + \exp(\phi_0^* + \mathbf{x}_i\phi_1^* + y_i\phi_2^*)}, \tag{6.50}$$

where $\phi_0^* = \phi_0 - \ln(\nu)$, $(\phi_1^*, \phi_2^*) = (\phi_1, \phi_2)$, and (ϕ_0, ϕ_1, ϕ_2) is defined in (6.48).

Proof. By Bayes' formula,

$$\frac{Pr(\delta=1\mid \mathbf{x},y,\delta_2=1)}{Pr(\delta=0\mid \mathbf{x},y,\delta_2=1)}=\frac{Pr(\delta_2=1\mid \mathbf{x},y,\delta=1)}{Pr(\delta_2=1\mid \mathbf{x},y,\delta=0)}\times\frac{Pr(\delta=1\mid \mathbf{x},y)}{Pr(\delta=0\mid \mathbf{x},y)}.$$

By (6.49), the above formula reduces to

$$\frac{Pr(\delta=1\mid \mathbf{x},y,\delta_2=1)}{Pr(\delta=0\mid \mathbf{x},y,\delta_2=1)}=\frac{1}{v}\times\frac{Pr(\delta=1\mid \mathbf{x},y)}{Pr(\delta=0\mid \mathbf{x},y)}.$$

Taking the logarithm of the above equality, we have

$$\phi_0^*+\phi_1^*\mathbf{x}+\phi_2^*y=\phi_0-\ln(v)+\phi_1\mathbf{x}+\phi_2 y.$$

Because the above relationship holds for all $\mathbf{x}$ and y, we have $\phi_0^*=\phi_0-\ln(v)$ and $(\phi_1^*,\phi_2^*)=(\phi_1,\phi_2)$. □

By Lemma 6.2, the MLE of ϕ^* can be obtained by maximizing the conditional likelihood. That is, we solve

$$\sum_{i=1}^{n}\delta_{2i}\left\{\delta_i-\pi(\phi^*;\mathbf{x}_i,y_i)\right\}(\mathbf{x}_i,y_i)=0 \tag{6.51}$$

and then apply the transformation in Lemma 6.2. In particular, the MLE for the slope (ϕ_1,ϕ_2) in (6.48) can be directly computed by solving (6.51). Asymptotic variance of $(\hat{\phi}_1,\hat{\phi}_2)$ is directly obtained by the asymptotic variance of $(\hat{\phi}_1^*,\hat{\phi}_2^*)$. For more theoretical details, see Scott and Wild (1997).

In practice, the sample from the callback is also subject to missingness and, in this case, the score equation (6.51) is not directly applicable. Now, assume that there are several followups to increase the number of respondents. Let A_1 be the set of respondents who provided answers to the surveys at the initial contact. Suppose that there are $T-1$ followups made to those who remain nonrespondents in the survey. Let $A_2(\subset A)$ be the set of respondents who provided answers to the surveys at the time of the second contact. By definition, A_2 contains those already provided answers in the first contact. Thus, $A_1\subset A_2$. Similarly, we can define A_3 be the set of respondents who provided answers at the time of the third contact, or the second followup. Continuing the process, we can define $A_1,\cdots,A_T$ such that

$$A_1\subset\cdots\subset A_T.$$

Suppose that there are T attempts (or $T-1$ followups) to obtain the survey response y_i and let δ_{it} be the response indicator function for y_i at the t-th attempt. If $\delta_{iT}=0$, then the unit never responds and it is called *hard-core nonrespondent* (Drew and Fuller, 1980). Using the definition of A_t, we can write $\delta_{it}=1$ if $i\in A_t$ and $\delta_{it}=0$ otherwise.

When the study variable y is categorical with K categories, Drew and Fuller (1980) proposed using a multinomial distribution with $T\times K+1$ cells where the cell probabilities are defined by

$$\begin{aligned}\pi_{tk} &= \gamma(1-p_k)^{t-1}p_k f_k\\ \pi_0 &= (1-\gamma)+\gamma\sum_{k=1}^{K}(1-p_k)^T f_k\end{aligned}$$

where p_k is the response probability for category K, f_k is the population proportion such that $\sum_{k=1}^{K}f_k=1$ and $1-\gamma$ is a proportion of hard-core nonrespondents. Thus, π_{tk} means the response probability that an individual in category k will respond at time t and π_0 is the probability that an individual will not have responded after T trials. Under simple random sampling, the maximum likelihood estimator of the parameter can be easily obtained by maximizing the log-likelihood

$$\log L = \sum_{t=1}^{T}\sum_{k=1}^{K}n_{tk}\log\pi_{tk}+n_0\log\pi_0,$$

where n_{tk} is the number of elements in the k-th category responding on the t-th contact and n_0 is the number of individual who did not respond up to the T-th contact.

Alho (1990) considered the same problem with a continuous y variable under simple random sampling. Let p_{it} be the conditional probability of $\delta_{it} = 1$, conditional on y_i and $\delta_{i,t-1} = 0$, and assume the logistic regression model

$$p_{it} = P(\delta_{it} = 1 | \delta_{i,t-1} = 0, y_i) = \frac{\exp(\alpha_t + x_i\phi_1 + y_i\phi_2)}{1 + \exp(\alpha_t + x_i\phi_1 + y_i\phi_2)}, \quad t = 1, 2, \cdots, T, \tag{6.52}$$

for the conditional response probability of δ_{it}. Here, we assume $\delta_{i0} \equiv 0$.

To estimate the parameters in (6.52), Alho (1990) also assumed that $(\delta_{i1}, \delta_{i2} - \delta_{i1}, \cdots, \delta_{iT} - \delta_{i,T-1}, 1 - \delta_{iT})$ follows a multinomial distribution with parameter vector $(\pi_{i1}, \pi_{i2}, \cdots, \pi_{iT}, 1 - \sum_{t=1}^{T} \pi_{it})$, where

$$\pi_{it} = Pr(\delta_{i,t-1} = 0, \delta_{it} = 1 \mid y_i). \tag{6.53}$$

Thus, we can write $\pi_{it} = p_{it} \prod_{k=1}^{t-1}(1 - p_{ik})$. Under this setup, Alho (1990) considered maximizing the following conditional likelihood.

$$\begin{aligned} L(\phi) &= \prod_{\delta_{iT}=1} Pr(\delta_{i1} = 1 \mid y_i, \delta_{iT} = 1)^{\delta_{i1}} \prod_{t=2}^{T} \{Pr(\delta_{it} = 1 \mid y_i, \delta_{i,t-1} = 0, \delta_{iT} = 1)\}^{\delta_{it}} \\ &= \prod_{R_i=1} \left(\frac{\pi_{i1}}{1 - \pi_{i,T+1}}\right)^{\delta_{i1}} \prod_{t=2}^{T} \left(\frac{\pi_{it}}{1 - \pi_{i,T+1}}\right)^{\delta_{it} - \delta_{i,t-1}}, \end{aligned} \tag{6.54}$$

where $\pi_{i,T+1} = 1 - \sum_{t=1}^{T} \pi_{it}$. To avoid the nonidentifiability problem, Alho (1990) imposed

$$\sum_{i \in A_{t-1}} \delta_{it} \exp(-\alpha_t - \phi y_i) = n - (n_1 + \cdots + n_t), \tag{6.55}$$

for $t = 1, 2, \cdots, T$. Note that (6.55) computes α_t given ϕ. To incorporate the observed auxiliary information outside A_t, one can add the following constraints

$$\sum_{i=1}^{n} (1 - \delta_{i,t-1}) \frac{\delta_{iT}}{\hat{p}_{it}} = \sum_{i=1}^{n} (1 - \delta_{i,t-1}), \quad t = 1, 2, \cdots, T \tag{6.56}$$

$$\sum_{i=1}^{n} \frac{\delta_{iT}}{(1 - \hat{\pi}_{i,T+1})} x_i = \sum_{i=1}^{n} x_i. \tag{6.57}$$

A constrained optimization algorithm can be used to find the constrained maximum likelihood estimators.

Instead of the maximum likelihood method from the conditional likelihood, a calibration approach can also be used. Kim and Im (2012) proposed solving

$$\sum_{i=1}^{n} \delta_{i,t-1}(x_i, y_i) + \sum_{i=1}^{n} (1 - \delta_{i,t-1}) \frac{\delta_{it}}{p_{it}} (x_i, y_i) = \sum_{i=1}^{n} \delta_{i,T-1}(x_i, y_i) + \sum_{i=1}^{n} (1 - \delta_{i,T-1}) \frac{\delta_{iT}}{p_{iT}} (x_i, y_i), \tag{6.58}$$

for $t = 1, 2, \cdots, T-1$, and

$$\sum_{i=1}^{n} (1 - \delta_{i,t-1}) \frac{\delta_{it}}{p_{it}} = \sum_{i=1}^{n} (1 - \delta_{i,t-1}), \quad t = 1, 2, \cdots, T. \tag{6.59}$$

Note that both terms in (6.58) estimates $\sum_{i=1}^{n} y_i$ unbiasedly under the conditional response model.

Now, under the conditional response model in (6.52), we have, by (6.59),

$$\sum_{i=1}^{n} (1 - \delta_{i,t-1}) \delta_{it} \{1 + \exp(-\alpha_t - \phi_1 x_i - \phi_2 y_i)\} = \sum_{i=1}^{n} (1 - \delta_{i,t-1}) \tag{6.60}$$

and the solution $\hat{\alpha}_t$ to (6.60) can be written

$$\exp(-\hat{\alpha}_t) = \frac{\sum_{i=1}^n (1-\delta_{i,t-1})(1-\delta_{it})}{\sum_{i=1}^n (1-\delta_{i,t-1})\delta_{it}\exp(-\phi_1 x_i - \phi_2 y_i)}. \tag{6.61}$$

Inserting (6.61) into (6.58) and (6.59), we have

$$\begin{aligned} &(\hat{X}_{R(t)}, \hat{Y}_{R(t)}) + \hat{N}_{M(t)}\left(\frac{\sum_{i=1}^n w_{i,t}(\phi)(x_i, y_i)}{\sum_{i=1}^n w_{i,t}(\phi)}\right) \\ &= (\hat{X}_{R(T)}, \hat{Y}_{R(T)}) + \hat{N}_{M(T)}\left(\frac{\sum_{i=1}^n w_{i,T}(\phi)(x_i, y_i)}{\sum_{i=1}^n w_{i,T}(\phi)}\right), \end{aligned}$$

for $t = 1, 2, \cdots, T-1$, where $(\hat{X}_{R(t)}, \hat{Y}_{R(t)}) = \sum_{i=1}^n \delta_{it}(x_i, y_i)$, $\hat{N}_{M(t)} = \sum_{i=1}^n (1-\delta_{it})$, and $w_{it}(\phi) = (1-\delta_{i,t-1})\delta_{it}\exp(-\phi_1 x_i - \phi_2 y_i)$. Thus, we have $p+q$ parameters ($p = \dim(x)$ and $q = \dim(y)$) with $(p+q)(T-1)$ equations. When $T > 2$, we have more equations than parameters and so we can apply the generalized method of moment (GMM) technique to compute the estimates.

6.8 Capture–recapture (CR) experiment

In this section, we consider the case of making two independent attempts to obtain a response for (x, y), where y is subject to missingness and x is always observed. The classical capture–recapture (CR) sampling setup can be applied to estimate the response probability. Capture–recapture (CR) sampling is very popular in estimating the population size of wildlife animals. Amstrup et al. (2005) provided a comprehensive summary of the existing methods for CR analysis. Huggins and Hwang (2011) reviewed the conditional likelihood approach in CR experiments.

To apply the conditional likelihood approach, we assume that the two response indicators, δ_{1i} and δ_{2i}, are assumed to be independently generated from Bernoulli distributions with probabilities

$$\pi_{1i}(\phi) = \Pr(\delta_{1i} = 1 | x_i, y_i) = \frac{\exp(\phi_0 + \phi_1 x_i + \phi_2 y_i)}{1 + \exp(\phi_0 + \phi_1 x_i + \phi_2 y_i)}$$

and

$$\pi_{2i}(\phi^*) = \Pr(\delta_{2i} = 1 | x_i, y_i) = \frac{\exp(\phi_0^* + \phi_1^* x_i + \phi_2^* y_i)}{1 + \exp(\phi_0^* + \phi_1^* x_i + \phi_2^* y_i)},$$

respectively, where $\phi = (\phi_0, \phi_1, \phi_2)$ and $\phi^* = (\phi_0^*, \phi_1^*, \phi_2^*)$. Write $\Phi = (\phi, \phi^*)$. An efficient estimator of Φ can be obtained by maximizing the conditional likelihood

$$L_C(\Phi) = \prod_{i \in A_1/A_2} \frac{\pi_{1i}(\phi)\{1-\pi_{2i}(\phi^*)\}}{p_i(\phi, \phi^*)} \prod_{i \in A_1 \cap A_2} \frac{\pi_{1i}(\phi)\pi_{2i}(\phi^*)}{p_i(\phi, \phi^*)} \prod_{i \in A_2/A_1} \frac{\{1-\pi_{1i}(\phi)\}\pi_{2i}(\phi^*)}{p_i(\phi, \phi^*)},$$

where $A_2/A_1 = A_2 \cap A_1^c$ and A_1 is the set of sample elements with $\delta_{1i} = 1$, A_2 is the set of sample elements with $\delta_{2i} = 1$, and $p_i(\phi, \phi^*) = 1 - \{1-\pi_{1i}(\phi)\}\{1-\pi_{2i}(\phi^*)\}$. The conditional likelihood is obtained by considering the conditional distribution of $(\delta_{1i}, \delta_{i2})$ given that unit i is selected in either of the two samples. The log-likelihood of the conditional distribution is

$$l_C(\Phi) = \sum_{i \in A_1} \log(\pi_{1i}) + \sum_{i \in A_2} \log(\pi_{2i}) + \sum_{i \in A_1/A_2} \log(1-\pi_{2i}) + \sum_{i \in A_2/A_1} \log(1-\pi_{1i}) - \sum_{i \in A_1 \cup A_2} \log(p_i).$$

The conditional maximum likelihood estimator (CMLE) that maximizes the conditional likelihood can be obtained by solving $S_C(\Phi) = 0$ where $S_C(\Phi) = \partial l_C(\Phi)/\partial \Phi = (S_{C1}'(\Phi), S_{C2}'(\Phi))'$ with

$$S_{C1}(\Phi) \triangleq \sum_{i \in A_1} (1, \mathbf{x}_i', y_i)' - \sum_{i \in A_1 \cup A_2} \frac{\pi_{1i}(\phi)}{p_i(\phi, \phi^*)}(1, \mathbf{x}_i', y_i)'$$

and

$$S_{C2}(\Phi) \triangleq \sum_{i \in A_2} (1, \mathbf{x}_i', y_i)' - \sum_{i \in A_1 \cup A_2} \frac{\pi_{2i}(\phi)}{p_i(\phi, \phi^*)} (1, \mathbf{x}_i', y_i)'.$$

Once the CMLE of Φ, denoted by $\hat{\Phi}$, is obtained, we can construct the following propensity score estimator of $\theta = E(Y)$ based on $A_1 \cup A_2$ by

$$\hat{\theta} = \frac{\sum_{i \in A_1 \cup A_2} p_i^{-1}(\hat{\phi}, \hat{\phi}^*) y_i}{\sum_{i \in A_1 \cup A_2} p_i^{-1}(\hat{\phi}, \hat{\phi}^*)}.$$

Asymptotic properties of the above PS estimator can be obtained by a standard linearization argument.

Exercises

1. Consider a bivariate categorical variable (x, y) where x takes values among $\{1, \cdots, K\}$ and y is either 0 or 1. We assume a fully nonparametric model on (x, y) such that $Pr(X = i, Y = j) = p_{ij}$ with $\sum_{i=1}^{K} \sum_{j=1}^{2} p_{ij} = 1$. Assume that x_i is fully observed and y_i is subject to missingness with probability

$$P(\delta_i = 1 \mid x_i, y_i) = \frac{\exp(\phi_0 + \phi_1 y_i)}{1 + \exp(\phi_0 + \phi_1 y_i)},$$

   for some (ϕ_0, ϕ_1). Answer the following questions:

   (a) Show that the model for observed data is identifiable if $K \geq 3$.

   (b) Discuss how to construct a GMM estimator of (ϕ_0, ϕ_1).

   (c) Discuss how to construct a pseudo likelihood estimator of (ϕ_0, ϕ_1).

2. Under the setup of Example 2.2, answer the following questions:

   (a) Show that

$$E(y_i \mid x_i, y_i > 0; \theta) = x_i'\beta + \sigma \lambda \left(-\frac{x_i'\beta}{\sigma} \right)$$

   where $\lambda(z) = \phi(z) / \{1 - \Phi(z)\}$.

   (b) Discuss how to implement an EM algorithm for estimating β when σ is known.

   (c) Discuss how to implement an EM algorithm for estimating $\theta = (\beta, \sigma)$.

3. Under the setup of Example 6.7, discuss how to implement an EM algorithm for estimating γ and β.

4. Suppose that $f(y \mid x, \delta = 1)$ is a normal distribution with mean $\beta_0 + \beta_1 x$ and variance σ^2. Assume that

$$Pr(\delta = 1 \mid x, y) = \frac{\exp(\phi_0 + \phi_1 x + \phi_2 y)}{1 + \exp(\phi_0 + \phi_1 x + \phi_2 y)}.$$

   Using (6.35), show that the conditional distribution $f(y \mid x, \delta = 0)$ also follows from a normal distribution with mean $\beta_0 - \phi_2 \sigma^2 + \beta_1 x$ and variance σ^2.

5. Suppose that we are interested in estimating $\theta = E(Y)$. Consider two response probability models, the conditional model using y only and the conditional model using x and y. That is, we have

$$\hat{\theta}_1 = \frac{1}{n} \sum_{i=1}^{n} \delta_i y_i / \pi_{1i}$$

   and

$$\hat{\theta}_2 = \frac{1}{n} \sum_{i=1}^{n} \delta_i y_i / \pi_{2i}$$

where $\pi_{1i} = E(\delta_i \mid y_i)$ and $\pi_{i2} = E(\delta_i \mid x_i, y_i)$. Prove that

$$V(\hat{\theta}_1) \leq V(\hat{\theta}_2).$$

6. Assume that $y_i \mid \delta_i$ follows a normal distribution with mean $\mu_1 \delta_i + \mu_0(1-\delta_i)$ and variance σ^2. Assume $\pi = Pr(\delta = 1)$ is known. Prove that the response probability can be written as

$$Pr(\delta = 1 \mid y) = \frac{\exp(\phi_0 + \phi_1 y)}{1 + \exp(\phi_0 + \phi_1 y)} \tag{6.62}$$

for some (ϕ_0, ϕ_1). Express ϕ_0 and ϕ_1 in terms of μ_0, μ_1, and π.

7. Consider the problem of estimating $\theta = E(Y)$ under the existence of missing data in y. Assume that the response model satisfies (6.62) with known ϕ_1. Answer the following questions.
   (a) Show that, using (6.35) or other formulas,

$$\hat{\mu}_0 = \frac{\sum_{i=1}^n \delta_i \exp(-\phi y_i) y_i}{\sum_{i=1}^n \delta_i \exp(-\phi y_i)} \tag{6.63}$$

   is asymptotically unbiased for $\mu_0 = E(Y \mid \delta = 0)$.
   (b) The prediction estimator

$$\hat{\theta}_p = \frac{1}{n}\sum_{i=1}^n \{\delta_i y_i + (1-\delta_i)\hat{\mu}_0\},$$

   where $\hat{\mu}_0$ is defined in (6.63), is algebraically equivalent to the PSA estimator

$$\hat{\theta}_{PSA} = \frac{1}{n}\sum_{i=1}^n \delta_i \frac{1}{\hat{\pi}_i} y_i,$$

   where $\hat{\pi}_i$ is the estimated response probability computed by

$$\sum_{i=1}^n \left(\frac{\delta_i}{\hat{\pi}_i} - 1\right) = 0.$$

8. Derive (6.54).
9. Suppose that we are interested in estimating the current employment status of a certain population and we have four followups in the survey. Suppose that we have obtained the results in Table 6.4.

Table 6.4 Realized responses in a survey of employment status

status	T=1	T=2	T=3	T=4	No reponse
Employment	81,685	46,926	28,124	15,992	
Unemployment	1,509	948	597	352	32350
Not in labor force	57882	32308	19086	10790	

   (a) Compute the full sample likelihood function under the assumption that there is no nonresponse after four followups.
   (b) Compute the conditional likelihood function and obtain the parameter values that maximize the conditional likelihood.
   (c) Use the Drew and Fuller (1980) method to compute the maximum likelihood estimator.
   (d) Discuss how to compute the standard errors of the MLE in (c).

10. Assume that two voluntary samples, A_1 and A_2, are obtained with a nested structure. That is, $A_2 \subset A_1$. Let δ_{1i} and δ_{2i} be the response indicator functions of the first sample A_1 and the second sample A_2, respectively. Assume that A_1 has the probability of survey participation given by

$$Pr(\delta_{1i} = 1 \mid y_i) = \frac{\exp(\phi_0 + \phi_1 y_i)}{1 + \exp(\phi_0 + \phi_1 y_i)}$$

for some (ϕ_0, ϕ_1). To estimate ϕ_1, we obtain a second voluntary sample A_2, by asking the same questions again, where the probability of the second survey participation

$$Pr(\delta_{2i} = 1 \mid y_i, \delta_{1i} = 1) = \frac{\exp(\phi_0^* + \phi_1 y_i)}{1 + \exp(\phi_0^* + \phi_1 y_i)}$$

for some ϕ_0^*. Thus, we assume that the two response probabilities are the same up to an intercept term. Answer the following questions:

(a) Show that a consistent estimator of (ϕ_0^*, ϕ_1) is obtained by solving

$$\sum_{i \in A_1} \delta_{2i} \{1 + \exp(-\phi_0^* - \phi_1 y_i)\} (1, y_i) = \sum_{i \in A_1} (1, y_i).$$

(b) Assuming that the marginal probability $\pi = Pr(\delta_{1i} = 1)$ is known, discuss how to construct a PSA estimator of $\theta = E(Y)$.

(c) Derive the asymptotic variance of the PSA estimator in (b).

Chapter 7

Longitudinal and clustered data

In a longitudinal study, we collect data from every sampled subject (or unit) at multiple time points. Under cluster sampling, we obtain data from units within each sampled cluster. Longitudinal or clustered data are often encountered in medical studies, population health, social studies, and economics. Related statistical analyses typically estimate or make an inference on the mean of the study response variable or the relationship between the response and some covariates. Longitudinal or clustered data look similar with multivariate data, but the main difference is the former studies one response variable measured at different time points or units, whereas the latter concerns several different variables. There are two major approaches for longitudinal or clustered data analysed under complete data. One is based on modeling the marginal distribution (or mean and variance) of the responses without requiring a correct specification of the correlation structure of longitudinal data; for example, the generalized estimation equation (GEE) approach. The linear model approach is a special case of GEE. The other approach is based on a mixed-effect model, which applies to the conditional distribution (or mean and variance) of the responses given some random effects.

Missing data in the study variable is a serious impediment to performing a valid statistical analysis, because the response probability usually (directly or indirectly) depends on the value of the response and missing mechanism is often nonignorable. In this chapter we introduce some methods of handling longitudinal or cluster data with missing values. These methods are introduced in each section that makes a particular assumption about missing data mechanism.

7.1 Ignorable missing data

Let y_{it} be the response at time point t for subject i, $\mathbf{y}_i = (y_{i1}, ..., y_{iT})$, δ_{it} be the indicator of whether y_{it} is observed, $\boldsymbol{\delta}_i = (\delta_{i1}, ..., \delta_{iT})$, and $\mathbf{x}_{it}$ be a covariate vector whose values are always observed, $\mathbf{x}_i = (\mathbf{x}_{i1}, ..., \mathbf{x}_{iT})$. The covariate $\mathbf{x}_i$ may be cross-sectional or longitudinal.

We assume that $(\mathbf{y}_i, \boldsymbol{\delta}_i, \mathbf{x}_i)$, $i = 1, ..., n$, are independent. Then the general definition of ignorable missing data reduces to

$$q(\boldsymbol{\delta}_i|\mathbf{y}_i, \mathbf{x}_i) = q(\boldsymbol{\delta}_i|\mathbf{y}_{i,\text{obs}}, \mathbf{x}_i) \tag{7.1}$$

where $\mathbf{y}_{i,\text{obs}}$ contains observed components of $\mathbf{y}_i$. An assumption stronger than (7.1) is that the response mechanism is covariate-dependent, i.e.,

$$q(\boldsymbol{\delta}_i|\mathbf{y}_i, \mathbf{x}_i) = q(\boldsymbol{\delta}_i|\mathbf{x}_i). \tag{7.2}$$

If (7.2) does not hold, then (7.1) appears unnatural when $\mathbf{y}_i$ is a cluster of data; for example, if y_{it}'s are responses from a sampled household, then it is often not true that the probability of a unit not responding depends on the respondents in the same household. In the situation where components of $\mathbf{y}_i$ are sampled at T ordered time points, (7.1) is unnatural unless missing pattern is monotone in the sense that if y_{it} is missing, so is y_{is} for any $s > t$, as it is hard to imagine that the probability of observing y_{it} depends on an observed y_{is} at a future time point $s > t$.

Thus, in this section we focus on monotone missing data under assumption (7.1) or general type missing data under assumption (7.2). If we adopt a parametric approach, then methods for multivariate responses in previous chapters can be applied here. If we assume a covariate-dependent response mechanism, then we do not need a parametric assumption on $f(\mathbf{y}_i)$ or $f(\mathbf{y}_i|\mathbf{x}_i)$. Methods described in Chapter 5 can be applied too.

In the rest of this section we consider an imputation method that assumes monotone ignorable missingness but does not require a parametric model on $\mathbf{y}_i$ (Paik, 1997). To use this method, we need to assume y_{i1} observed or use one benchmarking covariate as y_{i1} (such as the baseline observations in a clinical study).

For a missing y_{jt} with the last observed value at time point $r < t$, y_{jt} is imputed by $\hat{\phi}_{t,r}(\mathbf{y}_{jr})$, where $\mathbf{y}_{jr} = (y_{j1},...,y_{jr})$ and $\hat{\phi}_{t,r}$ is an estimated conditional expectation

$$\phi_{t,r}(\mathbf{y}_{ir}) = E(y_{it}|\mathbf{y}_{ir},\mathbf{x}_i,\delta_{i(r+1)} = 0,\delta_{ir} = 1) = E(y_{it}|\mathbf{y}_{ir},\mathbf{x}_i,\delta_{i(r+1)} = 1), \tag{7.3}$$

using data from all units with $\delta_{i(r+1)} = 1$, $r = 1,...,t-1$, $t = 2,...,T$. When $\mathbf{y}_i$ is multivariate normal, the conditional expectation in (7.3) is linear in $\mathbf{y}_{ir}$. Thus, $\hat{\phi}_{t,r}$ is the linear regression function fitted using y_{it} as the response and $y_{i1},...,y_{ir}$ as predictors. Data for this regression fitting are from units with $\delta_{i(r+1)} = 1$; the observed $\mathbf{y}_{ir}$ is used as a predictor and either observed y_{it} or previously imputed values of missing y_{it} are used as responses.

Note that the previously imputed values can be used as responses, but not as predictors in the regression fitting. For each fixed t, imputation can be done sequentially for $r = t-1, t-2,...,2$ so that previously imputed responses can be used.

When $\mathbf{y}_i$ is not normal, however, the conditional expectation in (7.3) is not linear in $\mathbf{y}_{ir}$, even if $\phi_{t,r}(\mathbf{y}_{ir})$ is linear in the case of no missing data. Hence, some nonparametric or semiparametric regression methods need to be applied to obtain $\hat{\phi}_{t,r}$. Since nonparametric or semiparametric regression will be presented later in more complicated situations, we omit the discussion here.

After missing values are imputed, the mean of y_{it} for each t can be estimated by the sample mean at t by treating imputed values as observed. If a regression between y_{it} and $\mathbf{x}_i$ needs to be fitted, we can also use standard methods by treating imputed values as observed.

To assess the variances of point estimators, however, we cannot treat imputed values as observed data and apply standard variance estimation methods. Adjustments have to be made or a bootstrap method that consists of a re-imputation component can be applied.

7.2 Nonignorable monotone missing data

In this section we consider nonignorable missing data with a monotone missing pattern, i.e., if y_{it} is missing at a time point t, then y_{is} is also missing at any $s > t$. Monotone missingness is also referred to as *dropout*. In this section, we introduce methods under different parametric-nonparametric assumptions on the propensity $q(\boldsymbol{\delta}|\mathbf{x},\mathbf{y})$ and the conditional probability density $p(\mathbf{y}|\mathbf{x})$.

7.2.1 Parametric models

We first consider the full parametric case, i.e., both $p(\mathbf{y}|\mathbf{x})$ and $q(\boldsymbol{\delta}|\mathbf{x},\mathbf{y})$ are parametric, say $p(\mathbf{y}|\mathbf{x}) = f(\mathbf{y}|\mathbf{x};\theta)$ and $q(\boldsymbol{\delta}|\mathbf{x},\mathbf{y}) = g(\boldsymbol{\delta}|\mathbf{x},\mathbf{y};\phi)$, where f and g are known functions and θ and ϕ are unknown parameter vectors. As we pointed out in Chapter 6, in general the parameters θ and ϕ are not identifiable, i.e., two different sets of (θ,ϕ) may produce the same data.

The concept of the nonresponse instrument in handling nonignorable missing data has been introduced in Chapter 6. The same idea can be applied here. Let $\mathbf{z}$ be a nonresponse instrument, i.e., $\mathbf{x} = (\mathbf{u},\mathbf{z})$ and

$$q(\boldsymbol{\delta}|\mathbf{x},\mathbf{y}) = q(\boldsymbol{\delta}|\mathbf{u},\mathbf{y}) \quad \text{and} \quad p(\mathbf{y}|\mathbf{u},\mathbf{z}) \neq p(\mathbf{y}|\mathbf{u}), \tag{7.4}$$

then θ and ϕ are identifiable and can be estimated by maximizing the parametric likelihood

$$\prod_{i:\delta_i=1} f(\mathbf{y}_i|\mathbf{x}_i;\theta)g(\boldsymbol{\delta}_i|\mathbf{u}_i,\mathbf{y}_i;\phi) \prod_{i:\delta_i=0} \int f(\mathbf{y}|\mathbf{x}_i;\theta)g(\boldsymbol{\delta}_i|\mathbf{u}_i,\mathbf{y};\phi)d\mathbf{y}.$$

The integral may not have an explicit form and numerical methods are needed.

Parametric methods can be sensitive to model violations. We introduce two semiparametric methods next.

7.2.2 *Nonparametric $p(\mathbf{y}|\mathbf{x})$*

We consider nonparametric $p(\mathbf{y}|\mathbf{x})$ and parametric propensity $q(\boldsymbol{\delta}|\mathbf{x},\mathbf{y}) = q(\boldsymbol{\delta}|\mathbf{u},\mathbf{y}) = g(\boldsymbol{\delta}|\mathbf{u},\mathbf{y};\phi)$, where $\mathbf{x} = (\mathbf{u},\mathbf{z})$ and $\mathbf{z}$ is a nonresponse instrument satisfying (7.4). In addition, for longitudinal data, it makes sense to assume that the dropout at time point t is statistically unrelated to future values $y_{t+1},...,y_T$. Thus,

$$P(\delta_t = 1|\delta_{t-1} = 1, y_1,...,y_T,\mathbf{x}) = P(\delta_t = 1|\delta_{t-1} = 1, y_1,...,y_t,\mathbf{u}). \tag{7.5}$$

Consider now the situation where $\mathbf{u}$ has a continuous component $\mathbf{u}_c$ and a discrete component u_d taking values $1,...,R$. Assume that

$$P(\delta_t = 1|\delta_{t-1} = 1, y_1,...,y_t,\mathbf{u}) = \psi(\alpha_{tu_d} + \beta_{tu_d} y_t + \mathbf{w}_t \gamma_{tu_d}), \qquad t = 1,...,T, \tag{7.6}$$

where $\mathbf{w}_t = (y_1,...,y_{t-1},\mathbf{u}_c)$, ψ is an increasing function on (0,1], α_{tu_d}, β_{tu_d}, and components of γ_{tu_d} are unknown parameters possibly depending on u_d.

In applications, we may consider some special cases of (7.6). For example,

$$P(\delta_t = 1|\delta_{t-1} = 1, y_1,...,y_t,\mathbf{u}) = \psi(\alpha_t + \beta_t y_t), \qquad t = 1,...,T,$$

or

$$P(\delta_t = 1|\delta_{t-1} = 1, y_1,...,y_t,\mathbf{u}) = \psi(\alpha_t + \beta_t y_t + \gamma_t y_{t-1}), \qquad t = 1,...,T,$$

where γ_t is an unknown parameter.

We adopt the GMM method described in Chapter 6 to estimate the unknown parameters in the propensity. The key is to construct a set of L estimation functions under condition (7.6).

First, we consider the case where $\mathbf{x} = \mathbf{z}$ is a q-dimensional continuous covariate and $\mathbf{u} = 0$ in dropout propensity model (7.6), which means $\mathbf{w}_t = (y_1,...,y_{t-1})$. For each t, there are $t+1$ parameters in model (7.6). Therefore, the total number of parameters in the propensity for all time points is $T(T+3)/2$. To apply the GMM, we need at least $T(T+3)/2$ functions. When T is not small, optimization over $T(T+3)/2$ parameters simultaneously has two crucial issues. First, there could be a large computational rounding error which results in large standard errors of the GMM estimators. Second, the computation speed is slow since the optimization is done in a $T(T+3)/2$ dimensional space. Therefore, we consider estimating the $(t+1)$-dimensional parameter $\phi_t = (\alpha_t, \beta_t, \gamma_t^{\mathrm{T}})$ for each separate t. For $t = 1,...,T$, consider the following $L = t+q$ functions for the GMM:

$$\mathbf{g}_t(\vartheta,\mathbf{y},\mathbf{z},\boldsymbol{\delta}) = \begin{pmatrix} \delta_{t-1}[\delta_t\omega(\vartheta) - 1] \\ \mathbf{z}^{\mathrm{T}}\delta_{t-1}[\delta_t\omega(\vartheta) - 1] \\ \mathbf{w}_t^{\mathrm{T}}\delta_{t-1}[\delta_t\omega(\vartheta) - 1] \end{pmatrix}, \tag{7.7}$$

where $\vartheta = (\vartheta_1, \vartheta_2, \vartheta_3^{\mathrm{T}})$, ϑ_3 is a $(t-1)$-dimensional column vector, $\omega(\vartheta) = [\psi(\vartheta_1 + \vartheta_2 y_t + \mathbf{w}_t\vartheta_3)]^{-1}$, and $\mathbf{w}_t$ is defined in (7.6). Since ϕ_t is $(t+1)$-dimensional, the minimum requirement for q is $q = 1$. The GMM estimator $\hat{\phi}_t = (\hat{\alpha}_t, \hat{\beta}_t, \hat{\gamma}_t^{\mathrm{T}})$ can be obtained using the two-step algorithm described in Chapter 6 with g_l being the l-th component function of $\mathbf{g}_t$ in (7.7).

The following theorem establishes the consistency and asymptotic normality of the GMM estimator of ϕ_t for every t. It also derives a consistent estimator of the asymptotic covariance matrix of the GMM estimator.

Theorem 7.1. *Suppose that the parameter space Θ_t containing the true value ϕ_t is an open subset of $\mathcal{R}^{t+1}$ and model (7.6) holds. Assume further the following conditions.*

(C1) *$E(\|\mathbf{z}\|^2) < \infty$ and there exists a neighborhood $\mathbf{N}$ of ϕ_t such that*

$$E\left[\delta_t \sup_{\vartheta \in \mathbf{N}} \{(1+\|\mathbf{z}\|^2)\omega^2(\vartheta) + |\boldsymbol{\xi}_t||\boldsymbol{\xi}_{t+1}||\omega'(\vartheta)| + \|\boldsymbol{\xi}_{t+1}\|^2|\omega''(\vartheta)|\}\right] < \infty,$$

where $\boldsymbol{\xi}_t = (1, \mathbf{z}, y_1, ..., y_{t-1})^{\mathrm{T}}$, $\omega'(\vartheta) = \omega'(\vartheta_1 + \vartheta_2 y_t + \mathbf{w}_t\vartheta_3)$, $\omega''(\vartheta) = \omega''(\vartheta_1 + \vartheta_2 y_t + \mathbf{w}_t\vartheta_3)$, $\omega(s) = [\psi(s)]^{-1}$, $|\mathbf{a}|$ is the L_1 norm of a vector $\mathbf{a}$ and $\|\mathbf{a}\| = \sqrt{\mathrm{trace}(\mathbf{a}^{\mathrm{T}}\mathbf{a})}$ is the L_2 norm for a vector or matrix $\mathbf{a}$.

(C2) *The $(t+q)\times(t+1)$ matrix*

$$\boldsymbol{\Gamma}_t = \left[E\{\boldsymbol{\xi}_t\delta_t\omega'(\phi_t)\}, E\{\boldsymbol{\xi}_t y_t\delta_t\omega'(\phi_t)\}, E\{\boldsymbol{\xi}_t\mathbf{w}_t\delta_t\omega'(\phi_t)\}\right]$$

is of full rank.

Then, we have the following conclusions as $n \to \infty$.

(i) *There exists $\{\hat{\phi}_t\}$ such that $P(\mathbf{s}(\hat{\phi}_t) = 0) \to 1$, $\mathbf{G}(\hat{\phi}_t) \to_p 0$, and $\hat{\phi}_t \to_p \phi_t$, where $\mathbf{s}(\vartheta) = -\partial\mathbf{G}^{\mathrm{T}}(\vartheta)\hat{\mathbf{W}}\mathbf{G}(\vartheta)/\partial\vartheta$ and $\to_p$ denotes convergence in probability.*

(ii) *For any sequence $\{\tilde{\phi}\}$ satisfying $s(\tilde{\phi}) = 0$ and $\tilde{\phi} \to_p \phi_t$,*

$$\sqrt{n}(\tilde{\phi} - \phi_t) \to_d N\left(0, (\boldsymbol{\Gamma}^{\mathrm{T}}\boldsymbol{\Sigma}^{-1}\boldsymbol{\Gamma})^{-1}\right),$$

where $\to_d$ is convergence in distribution and $\boldsymbol{\Sigma}$ is the positive definite $(t+q)\times(t+q)$ matrix whose (l,l')th element is $E[g_l(\phi_t, \mathbf{y}, \mathbf{z}, \boldsymbol{\delta})g_{l'}(\phi_t, \mathbf{y}, \mathbf{z}, \boldsymbol{\delta})]$.

(iii) *Let $\hat{\boldsymbol{\Gamma}}$ be the $(t+q)\times(t+1)$ matrix whose l-th row is*

$$\frac{1}{n}\sum \frac{\partial g_l(\vartheta, \mathbf{y}_i, \mathbf{z}_i, \boldsymbol{\delta}_i)}{\partial\vartheta}\bigg|_{\vartheta=\hat{\phi}_t}$$

and $\hat{\boldsymbol{\Sigma}}$ be the $(t+q)\times(t+q)$ matrix whose (l,l')-th element is

$$\frac{1}{n}\sum g_l(\hat{\phi}_t, \mathbf{y}_i, \mathbf{z}_i, \boldsymbol{\delta}_i)g_{l'}(\hat{\phi}_t, \mathbf{y}_i, \mathbf{z}_i, \boldsymbol{\delta}_i).$$

Then $\hat{\boldsymbol{\Gamma}}^{\mathrm{T}}\hat{\boldsymbol{\Sigma}}^{-1}\hat{\boldsymbol{\Gamma}} \to_p \boldsymbol{\Gamma}^{\mathrm{T}}\boldsymbol{\Sigma}^{-1}\boldsymbol{\Gamma}$.

Consider now the general case where $\mathbf{x} = (\mathbf{u}, \mathbf{z})$, $\mathbf{u} = (u_d, \mathbf{u}_c)$, and $\mathbf{z} = (z_d, \mathbf{z}_c)$, where $\mathbf{u}_c$ and $\mathbf{z}_c$ are continuous r- and q-dimensional covariate vectors, respectively, and u_d and z_d are discrete covariates taking values $1, ..., K$ and $1, ..., M$, respectively. Assume that the dropout propensity follows model (7.6). To apply GMM to estimate the parameter vector $\phi_{tk} = (\alpha_{tk}, \beta_{tk}, \gamma_{tk}^{\mathrm{T}})$ in the category of $u_d = k$, we construct the following $L = r+t+q+M$ functions:

$$\mathbf{g}_t(\vartheta, \mathbf{y}, \mathbf{x}, \boldsymbol{\delta}) = I(u_d = k)\begin{pmatrix} \boldsymbol{\zeta}^{\mathrm{T}}\delta_{t-1}[\delta_t\omega(\vartheta) - 1] \\ \mathbf{z}_c^{\mathrm{T}}\delta_{t-1}[\delta_t\omega(\vartheta) - 1] \\ \mathbf{w}_t^{\mathrm{T}}\delta_{t-1}[\delta_t\omega(\vartheta) - 1] \end{pmatrix}, \tag{7.8}$$

where $\boldsymbol{\zeta}$ is the M-dimensional row vector whose l-th component is $I(z_d = l)$, $I(A)$ is the indicator function of A, $\mathbf{w}_t = (y_1, ..., y_{t-1}, \mathbf{u}_c)$, $\omega(\vartheta) = [\psi(\vartheta_1 + \vartheta_2 y_t + \mathbf{w}_t\vartheta_3)]^{-1}$, $\vartheta = (\vartheta_1, \vartheta_2, \vartheta_3^{\mathrm{T}})$, and ϑ_3 is a $(r+t-1)$-dimensional column vector. Because ϕ_{tk} is $(r+t+1)$-dimensional, we require that $M + q \geq 2$.

The consistency and asymptotic normality of the GMM estimators can be established under similar conditions in Theorem 7.1 by replacing $\boldsymbol{\xi}_t$ with $(\boldsymbol{\zeta}, \mathbf{z}_c, y_1, ..., y_{t-1}, \mathbf{u}_c)^{\mathrm{T}}$ and the general $\omega(\vartheta)$ with the specific $\omega(\vartheta)$ in (7.8).

Once parameters in the dropout propensity are estimated, we can obtain nonparametric estimators of some parameters in the marginal distribution of $\mathbf{y}$ or the joint distribution of $\mathbf{y}$ and $\mathbf{x}$. For example, the estimation of the marginal mean of $\mathbf{y}$ is often the main focus in areas such as clinical studies and sample surveys. We consider the general case where $\mathbf{x} = (\mathbf{u}, \mathbf{z})$ and both $\mathbf{u}$ and $\mathbf{z}$ may have continuous and discrete components.

First, consider the situation where $\mathbf{u}$ is continuous. For any $t = 1, ..., T$, let $\hat{\phi}_t = (\hat{\alpha}_t, \hat{\beta}_t, \hat{\gamma}_t^{\mathrm{T}})$ be the GMM estimator of the unknown parameter under model (7.6), and y_{ti}'s, $\mathbf{w}_{ti}$'s, and δ_{ti}'s be the realized values of y_t, $\mathbf{w}_t = (y_1, ..., y_{t-1}, \mathbf{u})$, and δ_t from the sampled unit $i = 1, ..., n$, respectively. Since

$$P(\delta_t = 1|\mathbf{x}, \mathbf{y}) = \prod_{s=1}^{t} P(\delta_s = 1|\mathbf{x}, \mathbf{y}, \delta_{s-1} = 1) = \prod_{s=1}^{t} \psi(\alpha_s + \beta_s y_s + \mathbf{w}_s \gamma_s),$$

which can be estimated by

$$\hat{\pi}_{ti} = \prod_{s=1}^{t} \psi(\hat{\alpha}_s + \hat{\beta}_s y_{si} + \mathbf{w}_{si} \hat{\gamma}_s), \tag{7.9}$$

the marginal distribution of y_t can be estimated by the empirical distribution putting mass p_{ti} to each observed y_{ti}, where p_{ti} is proportional to $\delta_{ti}/\hat{\pi}_{ti}$ for a fixed t. The marginal mean of y_t, $\mu_t = E(y_t)$, can be estimated by a Horvitz–Thompson (HT) type estimator

$$\tilde{\mu}_t = \frac{1}{n} \sum_{i=1}^{n} \frac{\delta_{ti} y_{ti}}{\hat{\pi}_{ti}} \qquad \text{or} \qquad \hat{\mu}_t = \sum_{i=1}^{n} \frac{\delta_{ti} y_{ti}}{\hat{\pi}_{ti}} \Big/ \sum_{i=1}^{n} \frac{\delta_{ti}}{\hat{\pi}_{ti}}. \tag{7.10}$$

Similarly, we can estimate $E(y_t y_s)$ for any $s \le t$ (and hence the covariance or the correlation between y_t and y_s) by using (7.10) with y_{ti} replaced by $y_{ti} y_{si}$. Replacing y_s by $\mathbf{x}$, we can also obtain estimators of covariances between y_t and any covariate.

For every t, we establish the following general theorem that can be applied to show the asymptotic normality of various estimators and derive asymptotic covariance estimators, which allows us to carry out a large sample inference such as setting confidence intervals.

Theorem 7.2. *Assume the conditions in Theorem 7.1 with $\boldsymbol{\xi}_t = (\boldsymbol{\zeta}, \mathbf{z}_c, \mathbf{w}_t)^{\mathrm{T}}$ for $t = 1, ..., T$. Let $\boldsymbol{f}(\vartheta, \mathbf{d})$ be an m-dimensional function with $E[\boldsymbol{f}(\phi, \mathbf{d})] = \varphi$, where $\phi = (\phi_1, ..., \phi_t)$ is the parameter vector in the dropout propensity and let $\mathbf{g}(\vartheta) = (\mathbf{g}_1(\vartheta)^{\mathrm{T}}, ..., \mathbf{g}_t(\vartheta)^{\mathrm{T}})^{\mathrm{T}}$ where $\vartheta = (\vartheta_1, ..., \vartheta_t)$ and $\mathbf{g}_s(\vartheta) = \mathbf{g}_s(\vartheta_s)$ is the estimation functions for estimating ϕ_s. Let*

$$\hat{\varphi} = \frac{1}{n} \sum_{i=1}^{n} \boldsymbol{f}(\hat{\phi}, \mathbf{d}_i), \tag{7.11}$$

where $\hat{\phi} = (\hat{\phi}_1, ..., \hat{\phi}_t)$ with $\hat{\phi}_s$ being the GMM estimator of ϕ_s in Theorem 7.1. Assume further the following condition:

(C3) $\|E[\boldsymbol{f}(\phi, \mathbf{d}) \boldsymbol{f}^{\mathrm{T}}(\phi, \mathbf{d})]\| < \infty$, $\|E[\nabla \boldsymbol{f}(\phi)]\| < \infty$, where $\nabla \boldsymbol{\eta}(\phi) = \partial \boldsymbol{\eta}(\vartheta)/\partial \vartheta|_{\vartheta = \phi}$ for a function $\boldsymbol{\eta}(\vartheta)$, and there exists a neighborhood $\boldsymbol{N}$ of ϕ such that

$$E\left[\sup_{\vartheta \in \boldsymbol{N}} \left\| \frac{\partial^2 \boldsymbol{f}(\vartheta, \mathbf{d})}{\partial \vartheta \partial \vartheta^{\mathrm{T}}} \right\| \right] < \infty.$$

Then we have the following conclusions as $n \to \infty$.

(i) $\sqrt{n}(\hat{\varphi} - \varphi) \to_d N(\mathbf{0}, \boldsymbol{\Omega})$. Here $\boldsymbol{\Omega} = \mathbf{K}^{\mathrm{T}} \boldsymbol{\Lambda} \mathbf{K}$, where $\boldsymbol{\Lambda} = E[\mathbf{h}(\phi, \varphi, \mathbf{d}) \mathbf{h}^{\mathrm{T}}(\phi, \varphi, \mathbf{d})]$, $\mathbf{H} = (\boldsymbol{\Gamma}_1^{\mathrm{T}} \boldsymbol{W} \boldsymbol{\Gamma}_1)^{-1} \boldsymbol{\Gamma}_1^{\mathrm{T}} \boldsymbol{W}$, $\mathbf{K} = [-\boldsymbol{\Gamma}_2 \mathbf{H}, \mathbf{I}_{m \times m}]^{\mathrm{T}}$, $\boldsymbol{\Gamma}_2 = E[\nabla \boldsymbol{f}(\phi, \mathbf{d})]$, $\mathbf{h}(\vartheta, \boldsymbol{\psi}, \mathbf{d}) = [\mathbf{g}^{\mathrm{T}}(\vartheta, \mathbf{d}), (\boldsymbol{f}(\vartheta, , \mathbf{d}) - \boldsymbol{\psi})^{\mathrm{T}}]^{\mathrm{T}}$ and

$$\boldsymbol{W} = \begin{pmatrix} \boldsymbol{\Sigma}_1^{-1} & & \\ & \ddots & \\ & & \boldsymbol{\Sigma}_t^{-1} \end{pmatrix}$$

with $\boldsymbol{\Sigma}_s = [E(\mathbf{g}_s(\phi) \mathbf{g}_s(\phi)^{\mathrm{T}})]^{-1}$.

(ii) Let $\hat{\boldsymbol{\Omega}} = \hat{\mathbf{K}}^{\mathrm{T}}\hat{\boldsymbol{\Lambda}}\hat{\mathbf{K}}$*, where*

$$\hat{\boldsymbol{\Lambda}} = \frac{1}{n}\sum_{i=1}^{n}\mathbf{h}(\hat{\phi},\hat{\varphi},\mathbf{d}_i)\mathbf{h}^{\mathrm{T}}(\hat{\phi},\hat{\varphi},\mathbf{d}_i), \quad \hat{\mathbf{H}} = (\hat{\boldsymbol{\Gamma}}_1^{\mathrm{T}}\hat{\boldsymbol{W}}\hat{\boldsymbol{\Gamma}}_1)^{-1}\hat{\boldsymbol{\Gamma}}_1^{\mathrm{T}}\hat{\boldsymbol{W}},$$

$$\hat{\boldsymbol{\Gamma}}_1 = \frac{1}{n}\sum_{i=1}^{n}\nabla\mathbf{g}(\hat{\phi},\mathbf{d}_i), \quad \hat{\mathbf{K}} = \left[-\hat{\boldsymbol{\Gamma}}_2\hat{\mathbf{H}},\mathbf{I}_{m\times m}\right]^{\mathrm{T}}, \quad \hat{\boldsymbol{\Gamma}}_2 = \frac{1}{n}\sum_{i=1}^{n}\left[\nabla\boldsymbol{f}(\hat{\phi},\mathbf{d}_i)\right]$$

$$\text{and} \qquad \hat{\boldsymbol{W}} = \begin{pmatrix} \hat{\boldsymbol{\Sigma}}_1^{-1} & & \\ & \ddots & \\ & & \hat{\boldsymbol{\Sigma}}_t^{-1} \end{pmatrix},$$

with $\hat{\boldsymbol{\Sigma}}_s = 1/n\sum_i \mathbf{g}_s(\hat{\phi},\mathbf{d}_i)\mathbf{g}_s(\hat{\phi},\mathbf{d}_i)^{\mathrm{T}}$*. Then* $\hat{\boldsymbol{\Omega}} \to_p \boldsymbol{\Omega}$*.*

If we take $\boldsymbol{f}(\vartheta,\mathbf{d}_i) = y_{ti}\delta_{ti}/\pi_{ti}(\vartheta)$, then $\hat{\varphi}$ in (7.11) is $\tilde{\mu}_t$ in (7.10). If

$$\boldsymbol{f}(\vartheta,\mathbf{d}_i) = (y_{ti}\delta_{ti}/\pi_{ti}(\vartheta),\ \delta_{ti}/\pi_{ti}(\vartheta))^{\mathrm{T}},$$

then Theorem 7.2 and the Delta method imply that

$$\sqrt{n}(\hat{\mu}_t - \mu_t) \to_d N(0,\mathbf{a}^{\mathrm{T}}\tilde{\boldsymbol{\Sigma}}\mathbf{a}),$$

where $\tilde{\boldsymbol{\Sigma}} = \mathbf{K}^{\mathrm{T}}\boldsymbol{\Lambda}\mathbf{K}$ with $\mathbf{K}$ and $\boldsymbol{\Lambda}$ given in Theorem 7.2 and $\mathbf{a} = (1,-\mu_t)^{\mathrm{T}}$. Furthermore, $\hat{\mathbf{a}}^{\mathrm{T}}\hat{\mathbf{K}}^{\mathrm{T}}\hat{\boldsymbol{\Lambda}}\hat{\mathbf{K}}\hat{\mathbf{a}}$ is a consistent estimator of $\mathbf{a}^{\mathrm{T}}\tilde{\boldsymbol{\Sigma}}\mathbf{a}$, where $\hat{\mathbf{K}}$ and $\hat{\boldsymbol{\Lambda}}$ are given in Theorem 7.2 and $\hat{\mathbf{a}} = (1,-\hat{\mu}_t)^{\mathrm{T}}$.

For estimating $E(y_t y_s)$, $s \le t$, or $E(y_t\mathbf{x})$, we can obtain the asymptotic results by replacing y_{ti} in the previous $\boldsymbol{f}(\vartheta,\mathbf{d}_i)$ by $y_{ti}y_{si}$ or $y_{ti}\mathbf{x}_i$. The details are omitted.

Next, consider the case where $\mathbf{u}$ has a discrete component u_d taking values $k = 1,...,K$, and (7.6) holds. For every k, we can apply the previous results using data with $u_d = k$ to obtain estimators of parameters in the conditional distribution of $\mathbf{y}$ or $(\mathbf{y},\mathbf{x})$ given $u_d = k$. Then, the parameters in the unconditional distribution of $\mathbf{y}$ or $(\mathbf{y},\mathbf{x})$ can be obtained by taking averages. For example, an estimator of $\mu_{t,k} = E(y_t|u_d = k)$ is

$$\hat{\mu}_{t,k} = \sum_{i=1}^{n} I(u_d = k)\frac{\delta_{ti}y_{ti}}{\hat{\pi}_{ti}} \Bigg/ \sum_{i=1}^{n} I(u_d = k)\frac{\delta_{ti}}{\hat{\pi}_{ti}}.$$

Asymptotic results for these estimators similar to those in Theorem 7.2 can be established.

7.2.3 Nonparametric propensity

Now we consider nonparametric propensity and the parametric model

$$p(\mathbf{y}|\mathbf{x}) = \prod_{t=1}^{T} f_t(y_t|\mathbf{v}_{t-1},\theta_t), \tag{7.12}$$

where $f_t(y_t|\mathbf{v}_{t-1},\theta_t)$ is the probability density of y_t given $\mathbf{v}_{t-1} = (y_1,...,y_{t-1},\mathbf{x})$, f_t's are known functions, and θ_t's are distinct unknown parameter vectors. The parameter of interest is $\theta = (\theta_1,...,\theta_T)$. We still assume that there is a nonresponse instrument $\mathbf{z}$ satisfying (7.4).

Consider first the case of $\mathbf{x} = \mathbf{z}$ ($\mathbf{u} = 0$). When $t = 1$, under the assumed conditions,

$$p(\mathbf{x}|y_1,\delta_1 = 1) = \frac{p(y_1|\mathbf{x})p(\mathbf{x})}{\int p(y_1|\mathbf{x})p(\mathbf{x})d\mathbf{x}}.$$

Hence, using the theory in Section 6.4, we consider the likelihood

$$\prod_{i:\delta_{i1}=1} \frac{f_1(y_{i1}|\mathbf{x}_i;\theta_1)p(\mathbf{x}_i)}{\int f_1(y_{i1}|\mathbf{x};\theta_1)p(\mathbf{x})d\mathbf{x}}.$$

Substituting $p(\mathbf{x})$ by the nonparametric empirical distribution of $\mathbf{x}$ putting mass n^{-1} to each $\mathbf{x}_i$, we obtain an estimator $\hat{\theta}_1$ by maximizing the pseudo likelihood

$$\prod_{i:\delta_{i1}=1} \frac{f_1(y_{i1}|\mathbf{x}_i;\theta_1)}{\sum_{j=1}^n f_1(y_{i1}|\mathbf{x}_j;\theta_1)}.$$

For $t = 2,...,T$, suppose that $\hat{\theta}_1,...,\hat{\theta}_{t-1}$ have been obtained. Consider the likelihood

$$\prod_{i:\delta_{it}=1} p(\mathbf{x}_i|y_{i1},...,y_{it},\delta_{it}=1) = \prod_{i:\delta_{it}=1} \frac{p(y_{i1},...,y_{it}|\mathbf{x}_i)p(\mathbf{x}_i)}{\int p(y_{i1},...,y_{it}|\mathbf{x})p(\mathbf{x})d\mathbf{x}}.$$

Under (7.12),

$$p(y_{i1},...,y_{it}|\mathbf{x}_i) = f_t(y_{it}|\mathbf{v}_{i(t-1)},\theta_t)\prod_{s=1}^{t-1} f_s(y_{is}|\mathbf{v}_{i(s-1)},\theta_s),$$

where $\mathbf{v}_{is} = (y_{i1},...,y_{is},\mathbf{x}_i)$. Replacing each θ_s by the previously obtained $\hat{\theta}_s$ and $p(\mathbf{x}_i)$ by the nonparametric empirical distribution of $\mathbf{x}$, we estimate θ_t by maximizing the pseudo likelihood

$$\prod_{i:\delta_{it}=1} \frac{f_t(y_{it}|\mathbf{v}_{i(t-1)},\theta_t)\prod_{s=1}^{t-1} f_s(y_{is}|\mathbf{v}_{i(s-1)},\hat{\theta}_s)}{\sum_{j=1}^{n}\left\{f_t(y_{it}|\mathbf{x}_j,y_{i1},...,y_{i(t-1)},\theta_t)\prod_{s=1}^{t-1} f_s(y_{is}|\mathbf{x}_j,y_{i1},...,y_{i(s-1)},\hat{\theta}_s)\right\}}. \tag{7.13}$$

Note that all observed values up to time t are included in this likelihood.

If we do not substitute $\theta_1,...,\theta_{t-1}$ by their estimates, in theory we can estimate $(\theta_1,...,\theta_t)$ by maximizing (7.13) with $\hat{\theta}_s$ replaced by θ_s, $s = 1,...,t-1$. However, the computation may not be feasible because the dimension of $(\theta_1,...,\theta_t)$ is much higher than the dimension of θ_t.

Consider now the general case where $\mathbf{x} = (\mathbf{u},\mathbf{z})$. Note that

$$\begin{aligned} p(\mathbf{z}|y_1,...,y_t,\mathbf{u},\delta_t=1) &= p(\mathbf{z}|y_1,...,y_t,\mathbf{u}) \\ &= \frac{p(y_1,...,y_t,\mathbf{u}|\mathbf{z})p(\mathbf{z})}{\int p(y_1,...,y_t,\mathbf{u}|\mathbf{z})p(\mathbf{z})d\mathbf{z}} \\ &= \frac{p(y_1,...,y_t|\mathbf{u},\mathbf{z})p(\mathbf{u}|\mathbf{z})p(\mathbf{z})}{\int p(y_1,...,y_t|\mathbf{u},\mathbf{z})p(\mathbf{u}|\mathbf{z})p(\mathbf{z})d\mathbf{z}} \\ &= \frac{p(y_1,...,y_t|\mathbf{u},\mathbf{z})p(\mathbf{z}|\mathbf{u})}{\int p(y_1,...,y_t|\mathbf{u},\mathbf{z})p(\mathbf{z}|\mathbf{u})d\mathbf{z}} \end{aligned}$$

First, if $\mathbf{u} = u$ is a discrete covariate, then we can substitute $p(\mathbf{z}|u)$ by the empirical distribution of $\mathbf{z}$ conditioned on u, which results in the following likelihood for the estimation of θ_t:

$$\prod_u \prod_{i:\delta_{it}=1,u_i=u} \frac{f_t(y_{it}|\mathbf{v}_{i(t-1)},\theta_t)\prod_{s=1}^{t-1} f_s(y_{is}|\mathbf{v}_{i(s-1)},\hat{\theta}_s)}{\sum_{u_j=u}\left\{f_t(y_{it}|\mathbf{x}_j,y_{i1},...,y_{i(t-1)},\theta_t)\prod_{s=1}^{t-1} f_s(y_{is}|\mathbf{x}_j,y_{i1},...,y_{i(s-1)},\hat{\theta}_s)\right\}},$$

where $\hat{\theta}_1,...,\hat{\theta}_{t-1}$ are estimators from the previous steps. Next, consider the case where $\mathbf{u}$ is continuous and a parametric model on $p(\mathbf{z}|\mathbf{u}) = g(\mathbf{z}|\mathbf{u};\xi)$ is assumed, where ξ is an unknown parameter vector. Since $\mathbf{u}$ and $\mathbf{z}$ have no missing data, ξ can be estimated by $\hat{\xi}$ using the likelihood based on $\mathbf{x}_1,...,\mathbf{x}_n$, which leads to the following likelihood for the estimation of θ_t:

$$\prod_{i:\delta_{it}=1} \frac{f_t(y_{it}|\mathbf{v}_{i(t-1)},\theta_t)\prod_{s=1}^{t-1} f_s(y_{is}|\mathbf{v}_{i(s-1)},\hat{\theta}_s)g(\mathbf{z}_i|\mathbf{u}_i;\hat{\xi})}{\int f_t(y_{it}|\mathbf{u}_i,\mathbf{z},y_{i1},...,y_{i(t-1)},\theta_t)\prod_{s=1}^{t-1} f_s(y_{is}|\mathbf{u}_i,\mathbf{z},y_{i1},...,y_{i(s-1)},\hat{\theta}_s)g(\mathbf{z}|\mathbf{u}_i;\hat{\xi})d\mathbf{z}}.$$

Finally, consider the case where $\mathbf{u}$ is continuous, a parametric model on $p(\mathbf{u}|\mathbf{z}) = h(\mathbf{u}|\mathbf{z};\zeta)$ is assumed, where ζ is an unknown parameter vector, and ζ is estimated by $\hat{\zeta}$ using the likelihood based on $\mathbf{x}_1,...,\mathbf{x}_n$. Then, the following likelihood can be used for the estimation of θ_t:

$$\prod_{i:\delta_{it}=1} \frac{f_t(y_{it}|\mathbf{v}_{i(t-1)},\theta_t)\prod_{s=1}^{t-1} f_s(y_{is}|\mathbf{v}_{i(s-1)},\hat{\theta}_s)h(\mathbf{u}_i|\mathbf{z}_i;\hat{\zeta})}{\sum_{j=1}^{n}\left\{f_t(y_{it}|\mathbf{u}_i,\mathbf{z}_j,y_{i1},...,y_{i(t-1)},\theta_t)\prod_{s=1}^{t-1} f_s(y_{is}|\mathbf{u}_i,\mathbf{z}_j,y_{i1},...,y_{i(s-1)},\hat{\theta}_s)h(\mathbf{u}_i|\mathbf{z}_j)\right\}}.$$

In any case it is assumed that $f_t(y_t|\mathbf{v}_{t-1},\theta_t)$ depends on $\mathbf{z}$, i.e., $\mathbf{z}$ is a useful covariate, although $f_t(y_t|\mathbf{v}_{t-1},\theta_t)$ may not depend on $\mathbf{u}$.

To consider asymptotic properties, we focus on the situation where $\mathbf{x} = \mathbf{z}$. The following two additional conditions are needed:

$$\pi_t = P(\delta_t = 1) > 0, \quad t = 1,...,T, \tag{7.14}$$

and, for any θ_t in the parameter space that is not the same as the true parameter value θ_t^0 and any function ψ of $(y_1,...,y_t,\theta_t)$,

$$P\left((y_1,...,y_t): \frac{f_t(y_t|\mathbf{v}_{t-1},\theta_t)}{f_t(y_t|\mathbf{v}_{t-1},\theta_t^0)} = \psi(y_1,...,y_t,\theta_t) \text{ for any } \mathbf{x}\right) < 1. \tag{7.15}$$

Consistency and asymptotic normality of $\hat{\theta}_t$ can be established using a standard argument.

We now derive an asymptotic representation of $\sqrt{n}(\hat{\theta}_t - \theta_t^0)$, which allows us to obtain an easy-to-compute consistent estimator of the asymptotic covariance matrix of $\sqrt{n}(\hat{\theta}_t - \theta_t^0)$ without knowing its actual form. The asymptotic covariance matrix of $\sqrt{n}(\hat{\theta}_t - \theta_t^0)$ is very complicated because of the fact that $\hat{\theta}_t$ is defined in terms of previous estimators $\hat{\theta}_1,...,\hat{\theta}_{t-1}$ and the empirical distribution of $\mathbf{x}$.

Theorem 7.3. *Assume (7.6), (7.12), (7.14), (7.15), and the following two conditions.*

1. *The functions f_t's in (7.12) are continuously twice differentiable with respect to θ_t and $E\left[\frac{\partial^2 H_t(\varphi_t^0)}{\partial\theta_t\partial\theta_t'}\right]$ is positive definite, where $H_t(\varphi_t) = \delta_t \log G_t(\varphi_t)$ and*

$$G_t(\varphi_t) = \frac{f_t(y_t|\mathbf{v}_{t-1},\theta_t)\prod_{s=1}^{t-1} f_s(y_s|\mathbf{v}_{s-1},\theta_s)p(\mathbf{x})}{\int f_t(y_t|\mathbf{x},y_1,...,y_{t-1},\theta_t)\prod_{s=1}^{t-1} f_s(y_s|\mathbf{x},y_1,...,y_{s-1},\theta_s)p(\mathbf{x})d\mathbf{x}}.$$

2. *There exists an open subset Ω_t containing θ_t^0 such that*

$$\sup_{\theta_t\in\Omega_t}\left\|\frac{\partial^2 H_t(\theta_t,\varphi_{t-1}^0)}{\partial\theta_t\partial\theta_j'}\right\| < M_{tj}, \qquad j = 1,...,t,$$

where M_{tj} are integrable functions and $\|A\|^2 = \text{trace}(A^TA)$ for a matrix A.

Then, as $n\to\infty$,

$$\sqrt{n}(\hat{\theta}_t - \theta_t^0) = \frac{1}{\sqrt{n}}\sum_{i=1}^{n}\psi_t(W_t^{(i)},A_t,\varphi_t^0) + o_p(1) \to_d N(0,\Sigma_t), \tag{7.16}$$

where $\to_d$ denotes convergence in distribution, $o_p(1)$ denotes a quantity converging to 0 in probability, Σ_t is the covariance matrix of $\psi_t(W_{it},A_t,\varphi_t^0)$, $W_{it} = (\mathbf{v}_{it},\delta_{it})$, $i = 1,...,n$, $A_1 = A_{11}$, $A_t = (A_{t-1},A_{t1},...,A_{tt}), t\geq 2$,

$$A_{tj} = E\left[\frac{\partial^2 H_t(\varphi_t^0)}{\partial\theta_t\partial\theta_j'}\right], \quad j = 1,...,t,$$

$$\psi_t(W_{it},A_t,\varphi_t^0) = -A_{tt}^{-1}\left\{\frac{\partial H_{it}(\varphi_t^0)}{\partial\theta_t} + 2h_{1t}(\mathbf{x}_i,\varphi_t^0) + \sum_{j=1}^{t-1} A_{tj}\psi_j(W_{ij},A_j,\varphi_j^0)\right\}, \quad (7.17)$$

$$\psi_1(W_{i1},A_1,\varphi_1^0) = -A_{11}^{-1}\left\{\frac{\partial H_{i1}(\theta_1^0,F)}{\partial\theta_1} + 2h_{11}(\mathbf{x}_i,\varphi_1^0)\right\}, \quad (7.18)$$

and F is the distribution function of $\mathbf{x}_i$.

The functions ψ_t, $t = 1,...,T$, are defined iteratively according to (7.17)-(7.18) and, hence, their covariance matrices are very complicated. One may apply a bootstrap method to obtain estimators of Σ_t's, but in each bootstrap replication, maximizing a bootstrap analog of (7.13) is required, which results in a very large amount of computation. Instead, we propose the following estimator of Σ_t, utilizing the representation in (7.16). Let $D_{it} = \psi_t(W_{it},A_t,\varphi_t^0)$. Since $\Sigma_t = \text{Var}(D_{it})$, the sample covariance matrix based on $D_{1t},...,D_{nt}$ is a consistent estimator of Σ_t. However, D_{it} contains the unknown φ_t^0 and A_t. Substituting D_{it} by $\hat{D}_{it} = \psi_t(W_{it},\hat{A}_t,\hat{\varphi}_t)$, $i = 1,...,n$, where $\hat{A}_t = (\hat{A}_{t-1},\hat{A}_{t1},...,\hat{A}_{tt})$ and

$$\hat{A}_{tj} = \frac{1}{n}\sum_{i=1}^{n}\frac{\partial^2 H_t^{(i)}(\varphi_t)}{\partial\theta_t\partial\theta_j'}\bigg|_{\varphi_t=\hat{\varphi}_t}, \quad j = 1,...,t,$$

we define the sample covariance matrix based on $\hat{D}_{1t},...,\hat{D}_{nt}$ as our estimator $\hat{\Sigma}_t$. This estimator is easy to compute, using (7.17)-(7.18). Under the conditions listed in Theorem 7.4, $\hat{\Sigma}_t$ is consistent.

Theorem 7.4. *Assume that the conditions in Theorem 7.3 hold and that*

1. $\sup_{\|w\|\le c}\|\psi_t(w,\hat{A}_t,\hat{\varphi}_t) - \psi_t(w,A_t,\varphi_t^0)\| = o_p(1)$ *for any* $c > 0$.
2. *There exist a constant* $c_0 > 0$ *and a function* $h(w) \ge 0$ *such that* $E[h(W_t^{(1)})] < \infty$ *and* $P(\|\psi_t(w,\hat{A}_t,\hat{\varphi}_t)\|^2 \le h(w)$ *for all* $\|w\| \ge c_0) \to 1$.

Then, as $n \to \infty$, $\|\hat{\Sigma}_t - \Sigma_t\| = o_p(1)$.

The proofs of Theorems 7.3 - 7.4 can be found in Shao and Zhao (2012).

7.3 Past-value-dependent missing data

Longitudinal data with nonmonotone missing responses are typically nonignorable and hard to handle. Some assumptions are needed. For example, if we assume that

$$q(\boldsymbol{\delta}_i|\mathbf{y}_i,\mathbf{x}_i) = q(\boldsymbol{\delta}_i|y_{it})$$

without necessarily assuming a parametric form for q, then the method described in §7.2.3 can be applied. We omit the details that can be found in Tang et al. (2003) and Jiang and Shao (2012).

In this section, we consider longitudinal data with nonmonotone missing responses under the following past-value-dependent missingness:

$$q(\delta_{it}|\mathbf{y}_i,\mathbf{x}_i,\delta_{is},s\ne t) = q(\delta_{it}|\mathbf{y}_{i(t-1)},\mathbf{x}_i), \quad t = 2,...,T, \quad (7.19)$$

where $\mathbf{y}_{i(t-1)} = (y_{i1},...,y_{i(t-1)})$. In other words, at time point t, the missingness propensity depends on covariates and all past y-values (whether or not they are observed), but does not depend on the current and future y-values. This propensity is ignorable if we add the monotone missingness condition, but it is nonignorable if missing is nonmonotone, since some of $y_{i1},...,y_{i(t-1)}$ may be missing. We still assumed that at $t = 1$, there is no missing value.

7.3.1 *Three different approaches*

The first approach is parametric modeling under (7.19). With parametric models assumed for $q(\delta_{it}|\mathbf{y}_{i(t-1)},x_i)$ and $p(\mathbf{y}_i|\mathbf{x}_i)$, the parametric approach estimates model parameters using the maximum likelihood or some Bayesian methods. Zhou and Kim (2012) provides a unified approach of the PSA estimation under parametric models for the propensity scores with monotone missing data.

The second approach is to artificially create a dataset with monotone missing data by using only observed y-values from a sampled subject up till its first missing y-value. After that, the missing and artificially discarded data are "missing" at random. We can then apply methods appropriate for monotone missing data under ignorable missing (e.g., Section 7.1) to the reduced dataset. We call this method *censoring at the first missing* or *censoring* for short. Although the censoring approach produces consistent estimators, it is not efficient when T is not small, since many observed data can be discarded.

The third approach uses the same idea in the imputation method for monotone missing data described in Section 7.1. However, because missing is nonignorable, the imputation procedure is much more complicated. Furthermore, in most cases we have to use nonparametric or at least semi-parametric regression in the imputation process.

7.3.2 *Imputation models under past-value-dependent nonmonotone missing*

It can be shown that, under missing mechanism (7.19),

$$\begin{aligned} & E(y_{it}|\mathbf{y}_{i(t-1)},\mathbf{x}_i,\delta_{i1}=\cdots=\delta_{i(t-1)}=1,\delta_{it}=0) \\ = \; & E(y_{it}|\mathbf{y}_{i(t-1)},\mathbf{x}_i,\delta_{i1}=\cdots=\delta_{i(t-1)}=1,\delta_{it}=1) \end{aligned} \qquad t=2,\ldots,T. \tag{7.20}$$

This means that the conditional expectation of a missing y_{it}, given that y_{it} is the first missing value and given observed values $y_{i1},\ldots,y_{i(t-1)}$ and $\mathbf{x}_i$, is the same as the conditional expectation of an observed y_{it} given that $y_{i1},\ldots,y_{i(t-1)}$ are observed and given observed values $y_{i1},\ldots,y_{i(t-1)}$ and $\mathbf{x}_i$. We can make use of this to carry out imputation.

Also, it can be shown that, for a missing y_{it} with $r+1$ as the first time point of having a missing value ($r=1,\ldots,t-2$),

$$\begin{aligned} & E(y_{it}|\mathbf{y}_{ir},\mathbf{x}_i,\delta_{i1}=\cdots=\delta_{ir}=1,\delta_{i(r+1)}=0,\delta_{it}=0) \\ = \; & E(y_{it}|\mathbf{y}_{ir},\mathbf{x}_i,\delta_{i1}=\cdots=\delta_{ir}=1,\delta_{i(r+1)}=1,\delta_{it}=0) \end{aligned} \qquad r=1,\ldots,t-2, \quad t=3,\ldots,T. \tag{7.21}$$

This means that the conditional expectation of a missing y_{it}, given that $y_{i(r+1)}$ is the first missing value and given observed values $y_{i1},\ldots,y_{ir}$ and $\mathbf{x}_i$, is the same as the conditional expectation of a missing y_{it}, given that $y_{i1},\ldots,y_{i(r+1)}$ are observed and given observed values $y_{i1},\ldots,y_{ir}$ and $\mathbf{x}_i$.

We use (7.20)-(7.21) as imputation models. Like the monotone missing case, the number of imputation models is $T(T-1)/2$. Note that models in (7.20) are the same as those in (7.3) with $r=t-1$, but models in (7.21) are different from those in (7.3).

We now explain how to use (7.20)-(7.21) for imputation. Let t be a fixed time point > 1 (it does not matter which t we start). First, consider subjects whose first missing occurs at time point t, i.e., $r=t-1$. Denote the first line of (7.20) by $\phi_{t,t-1}(\mathbf{y}_{i(t-1)},\mathbf{x}_i)$. If the function $\phi_{t,t-1}$ is known, then a natural imputed value for a missing y_{it} is $\phi_{t,t-1}(\mathbf{y}_{i(t-1)},\mathbf{x}_i)$. Since $\phi_{t,t-1}$ is usually unknown, we have to estimate it. Since $\phi_{t,t-1}$ cannot be estimated by regressing missing y_{it} on $(\mathbf{y}_{i(t-1)},\mathbf{x}_i)$ based on data from subjects with missing y_{it} values, we need to use (7.20), i.e., the fact that $\phi_{t,t-1}$ is the same as the quantity on the second line of (7.20), which can be estimated by regressing y_{it} on $(\mathbf{y}_{i(t-1)},\mathbf{x}_i)$, using data from all subjects having observed y_{it} and observed $\mathbf{y}_{i(t-1)}$. Denote the resulting estimate by $\hat{\phi}_{t,t-1}$. (The form of $\hat{\phi}_{t,t-1}$ will be given in Section 7.3.3.) Then, the missing y_{it} of subject i whose first missing is at time point t can be imputed by $\hat{\phi}_{t,t-1}(\mathbf{y}_{i(t-1)},\mathbf{x}_i)$. Model (7.20) allows us to use data from subjects without any missing values in estimating the regression function $\phi_{t,t-1}$.

The case of $r<t-1$ is more complicated. For a subject whose first missing is at time point $r+1$ with $r<t-1$, a missing y_{it} can be imputed by $\phi_{t,r}(\mathbf{y}_{ir},\mathbf{x}_i)$, which denotes the quantity on the first line of (7.21), if $\phi_{t,r}$ is known. Since $\phi_{t,r}$ is unknown, we need to estimate it. To estimate $\phi_{t,r}$ by regression we need some values of y_{it} as responses. Unlike the case of $r=t-1$, the conditional expectation on the second line of (7.21) is also conditional on a missing y_{it} ($\delta_{it}=0$), although $y_{i1},\ldots,y_{ir}$ and $\mathbf{x}_i$ are observed. Suppose that imputation is carried out sequentially for $r=t-1,t-2,\ldots,1$. Then,

for a given $r < t-1$, the missing y_{it} values from subjects whose first missing is at time point $r+2$ have already been imputed. (For $r = t-2$, imputed values are obtained in the previous discussion for subjects whose first missing is at $t = r+2$.) We can then fit a regression between imputed y_{it} and observed $(\mathbf{y}_{ir}, \mathbf{x}_i)$, using data from all subjects with already imputed y_{it} (as responses) and observed $y_{i1}, \ldots, y_{ir}$ and $\mathbf{x}_i$ (as predictors) and $\delta_{i(r+1)} = 1$. Denote the resulting estimate by $\hat{\phi}_{t,r}$. (The form of $\hat{\phi}_{t,r}$ will be given in Section 7.3.3.) Then the missing y_{it} of subject i whose first missing is at time point $r+1$ can be imputed by $\hat{\phi}_{t,r}(\mathbf{y}_{ir}, \mathbf{x}_i)$. Model (7.21) allows us to use previously imputed values of y_{it} in the regression estimation of $\phi_{t,r}$.

We illustrate the proposed imputation process in the case of $T = 4$ (Table 7.1). The horizontal direction in Table 7.1 corresponds to time points and the vertical direction corresponds to 8 different missing patterns, where each pattern is represented by a 4-dimensional vector of 0's and 1's with 0 indicating a missing value and 1 indicating an observed value. It does not matter at which $t = 2, \ldots, T$ the imputation starts, but within each t, imputation is sequential.

- [Step A] Consider first the imputation at $t = 3$. There are two steps (the block in Table 7.1 under title $t = 3$). At step 1, we impute the missing data at $t = 3$ with the first missing at time 3 ($r = 2$), i.e., patterns 2 and 6. According to imputation model (7.20), we fit a regression using the data in patterns 3 and 8 indicated by $+$ (used as predictors) and $\times$ (used as responses). Then, imputed values (indicated by $\bigcirc$) are obtained from the fitted regression using the data indicated by $*$ as predictors. At step 2, we impute the missing data at $t = 3$ with the first missing at time 2 ($r = 1$), i.e., patterns 1 and 5. According to imputation model (7.21), we fit a regression using data in patterns 2 and 6 indicated by $+$ (as predictors) and $\otimes$ (previously imputed values used as responses). Then, imputed values (indicated by $\bigcirc$) are obtained from the fitted regression using the data indicated by $*$ as predictors.
- [Step B] Consider next the imputation at $t = 2$ (the block in Table 7.1 under title $t = 2$). This is the simplest case: the missing data at $t = 2$ are in patterns 1, 4, 5, and 7; we fit a regression using the data in patterns 2, 3, 6, and 8 indicated by $+$ (as predictors) and $\times$ (as responses). Then, imputed values (indicated by $\bigcirc$) are obtained from the fitted regression using the data indicated by $*$ as predictors.
- [Step C] Finally, consider the imputation at $t = 4$ (the block in Table 7.1 under title $t = 4$). At step 1, we impute the missing data at time 4 with the first missing at time 4 (pattern 3). According to imputation model (7.20), we fit a regression using the data in pattern 8 indicated by $+$ (as predictors) and $\times$ (as responses). Then, imputed values (indicated by $\bigcirc$) are obtained from the fitted regression using the data indicated by $*$ as predictors. At step 2, the missing values at $t = 4$ with the first missing at time 3 are in pattern 2. According to imputation model (7.21), we fit a regression using the data in pattern 3 indicated by $+$ (as predictors) and $\otimes$ (previously imputed values used as responses). Then, imputed values (indicated by $\bigcirc$) are obtained from the fitted regression using the data indicated by $*$ as predictors. At step 3, the missing values at $t = 4$ with the first missing at time 2 are in patterns 1 and 4. According to imputation model (7.21), we fit a regression using the data in patterns 2 and 3 indicated by $+$ (as predictors) and $\otimes$ (previously imputed values used as responses). Then, imputed values (indicated by $\bigcirc$) are obtained from the fitted regression using the data indicated by $*$ as predictors.

One may wonder why we don't use observed y_{it} values in the estimation of $\phi_{t,r}$ when $r < t-1$. For a subject with missing y_{it} and the first missing at time point $r+1 < t$, in general,

$$E(y_{it} | \mathbf{y}_{ir}, \mathbf{x}_i, \delta_{i1} = \cdots = \delta_{ir} = 1, \delta_{it} = 0) \neq E(y_{it} | \mathbf{y}_{ir}, \mathbf{x}_i, \delta_{i1} = \cdots = \delta_{ir} = 1, \delta_{it} = 1)$$

unless the missing is ignorable. Therefore, we cannot use observed y_{it} values in the estimation of $\phi_{t,r}$ when $r < t-1$.

After missing values are imputed, the mean of y_{it} for each t can be estimated by the sample mean at t by treating imputed values as observed. If a regression model between y_{it} and $\mathbf{x}_i$ needs to be fitted, we can also use standard methods by treating imputed values as observed.

Table 7.1 Illustration of imputation process when $T = 4$

	$t=3$								$t=2$			
	Step 1: $r=2$				Step 2: $r=1$				$r=1$			
	Time				Time				Time			
Pattern	1	2	3	4	1	2	3	4	1	2	3	4
(1,0,0,0)					*		○		*	○		
(1,1,0,0)	*	*	○		+		⊗		+	×		
(1,1,1,0)	+	+	×						+	×		
(1,0,1,0)									*	○		
(1,0,0,1)					*		○		*	○		
(1,1,0,1)	*	*	○		+		⊗		+	×		
(1,0,1,1)									*	○		
(1,1,1,1)	+	+	×						+	×		
	$t=4$											
	Step 1: $r=3$				Step 2: $r=2$				Step 3: $r=1$			
	Time				Time				Time			
Pattern	1	2	3	4	1	2	3	4	1	2	3	4
(1,0,0,0)									*			○
(1,1,0,0)					*	*		○	+			⊗
(1,1,1,0)	*	*	*	○	+	+		⊗	+			⊗
(1,0,1,0)									*			○
(1,0,0,1)												
(1,1,0,1)												
(1,0,1,1)												
(1,1,1,1)	+	+	+	×								

+: observed data used in regression fitting as predictors
×: observed data used in regression fitting as responses
⊗: imputed data used in regression fitting as responses
*: observed data used as predictors in imputation
○: imputed values

7.3.3 Nonparametric regression imputation

The imputation procedure described in Section 7.3.2 requires that we regress observed or already imputed y_{it} on $(\mathbf{y}_{ir}, \mathbf{x}_i)$, using subjects with some particular missing patterns. In this section we specify the regression method. Because missing is nonignorable, conditional expectations in (7.20)-(7.21) depend not only on the distribution of $\mathbf{y}_i$, but also on the propensity. Thus, parametric regression requires parametric models on both $p(\mathbf{y}_i|\mathbf{x}_i)$ and the propensity. Furthermore, even if $p(\mathbf{y}_i|\mathbf{x}_i)$ is normal, conditional expectations in (7.20)-(7.21) are not linear because of the nonignorable missing mechanism.

Nonparametric regression model is robust because it avoids specifying a parametric model on the propensity that cannot be verified using data. We now describe a Kernel nonparametric regression method to estimate $\phi_{t,r}$.

Let $Z_{ir} = (\mathbf{y}_{ir}, \mathbf{x}_i)$ and

$$\phi_{t,t-1}(u) = E(y_{it}|Z_{ir} = u, \delta_{i1} = \cdots = \delta_{it} = 1).$$

Under the condition given in (7.20), the Kernel regression estimator of $\phi_{t,t-1}(u)$ is

$$\hat{\phi}_{t,t-1}(u) = \sum_{i=1}^{n} \kappa_{t,t-1}\left(\frac{u - Z_{i(t-1)}}{h}\right) I_{t,t-1,i} y_{it} \Big/ \sum_{i=1}^{n} \kappa_{t,t-1}\left(\frac{u - Z_{i(t-1)}}{h}\right) I_{t,t-1,i},$$

where $\kappa_{t,t-1}$ is a probability density function, $h > 0$ is a bandwidth, and

$$I_{t,t-1,i} = \begin{cases} 1 & \delta_{i1} = \cdots = \delta_{it} = 1 \\ 0 & \text{otherwise.} \end{cases}$$

For $t = 2, ..., T$, a missing y_{it} with observed $Z_{i(t-1)}$ is imputed by

$$\tilde{y}_{it} = \hat{\phi}_{t,t-1}(Z_{i(t-1)}).$$

For $r = 1, ..., t-2$, let

$$\phi_{t,r}(u) = E(y_{it}|Z_{ir} = u, \delta_{i1} = \cdots = \delta_{i(r+1)} = 1, \delta_{it} = 0),$$

the conditional expectation in (7.21). Its Kernel estimator is

$$\hat{\phi}_{t,r}(u) = \sum_{i=1}^{n} \kappa_{t,r}\left(\frac{u - Z_{ir}}{h}\right) I_{t,r,i}\tilde{y}_{it} \Big/ \sum_{i=1}^{n} \kappa_{t,r}\left(\frac{u - Z_{ir}}{h}\right) I_{t,r,i},$$

where $\tilde{y}_{it}$ is a previously imputed value and

$$I_{t,r,i} = \begin{cases} 1 & \delta_{it} = 0, \delta_{i1} = \cdots = \delta_{i(r+1)} = 1 \\ 0 & \text{otherwise.} \end{cases}$$

A missing y_{it} with the first missing at $r+1$ is imputed by

$$\tilde{y}_{it} = \hat{\phi}_{t,r}(Z_{ir}).$$

7.3.4 *Dimension reduction*

The dimension of the regressor Z_{ir} increases with T and the dimension of the covariate vector. As the dimension of regressor increases, the number of observations needed for Kernel regression escalates exponentially. Unless we have a very large sample under each imputation model, Kernel regression imputation in Section 7.3.3 may break down because of the sparseness of relevant data points. This is the so-called *curse of dimensionality* well known in nonparametric regression.

Nonparametric regression imposes no condition on $p(\mathbf{y}_i|\mathbf{x}_i)$, and no assumption on the propensity other than the past-data-dependent missing assumption (7.19). To deal with the curse of dimensionality, we consider the following semiparametric model:

$$q(\delta_{it}|\mathbf{y}_i, \mathbf{x}_i, \delta_{is}, s \neq t) = q(\delta_{it}|Z'_{i(t-1)}\beta_{t-1}, \Delta_{i(t-1)}), \quad t = 2, ..., T, \tag{7.22}$$

where $\Delta_{i(t-1)} = (\delta_{i1}, ..., \delta_{i(t-1)})$. That is, the dependence of the propensity on the high-dimensional $Z_{i(t-1)}$ is through a one-dimensional projection $Z'_{i(t-1)}\beta_{t-1}$. It follows from (7.22) that

$$\begin{aligned} & E(y_{it}|Z'_{i(t-1)}\beta_{t,t-1}, \delta_{i1} = \cdots = \delta_{i(t-1)} = 1, \delta_{it} = 0) \\ = \; & E(y_{it}|Z'_{i(t-1)}\beta_{t,t-1}\delta_{i1} = \cdots = \delta_{i(t-1)} = 1, \delta_{it} = 1) \end{aligned} \qquad t = 2, ..., T, \tag{7.23}$$

and, when $r+1$ is the first time point of having a missing value,

$$\begin{aligned} & E(y_{it}|Z'_{ir}\beta_{t,r}\delta_{i1} = \cdots = \delta_{ir} = 1, \delta_{i(r+1)} = 0, \delta_{it} = 0) \\ = \; & E(y_{it}|Z'_{ir}\beta_{t,r}\delta_{i1} = \cdots = \delta_{ir} = 1, \delta_{i(r+1)} = 1, \delta_{it} = 0) \end{aligned} \qquad r = 1, ..., t-2, \quad t = 3, ..., T, \tag{7.24}$$

where $\beta_{t,r}$ is $\beta_{t-1}(A_{ir})$ with all components of A_{ir} equal to 1, $r = 1, ..., t-1$. Under (7.22), (7.23) and (7.24) replace (7.20) and (7.21), respectively, as our imputation models.

If $\beta_{t,r}$'s are known, then we can apply the imputation method in Section 7.3.2 using a one-dimensional Kernel regression with Z_{ir} replaced by $Z'_{ir}\beta_{t,r}$. In general, $\beta_{t,r}$'s are unknown. We first apply the sliced inverse regression (Li, 1991) under model (7.22) to obtain consistent estimators of $\beta_{t,r}$'s. We then apply the one-dimensional Kernel regression based on (7.23)-(7.24) with $\beta_{t,r}$'s replaced by their estimators.

The sliced inverse regression

For each t and r, we obtain an estimator of $\beta_{t,r}$ using the sliced inverse regression based on model (7.22) and the observed data on $\delta_{i(r+1)}$ and Z_{ir} from subjects with $\delta_{i1} = \cdots = \delta_{ir} = 1$. The detailed procedure is given as follows. Let $t = 2,...,T$ be fixed and $r+1$ be the first missing time point.

1. Compute $D = [(\bar{Z}_{r1} - \bar{Z}_r)(\bar{Z}_{r1} - \bar{Z}_r)' + (\bar{Z}_{r0} - \bar{Z}_r)(\bar{Z}_{r0} - \bar{Z}_r)']/2$, where $\bar{Z}_{r1}$ is the sample mean of Z_{ir}'s from subjects with $\delta_{i1} = \cdots = \delta_{i(r+1)} = 1$, $\bar{Z}_{r0}$ is the sample mean of Z_{ir}'s from subjects with $\delta_{i1} = \cdots = \delta_{ir} = 1$ and $\delta_{i(r+1)} = 0$, and $\bar{Z}_r$ is the sample mean of Z_{ir}'s from subjects with $\delta_{i1} = \cdots = \delta_{ir} = 1$.
2. Compute S, the sample covariance matrix of Z_{ir}'s from subjects with $\delta_{i1} = \cdots = \delta_{ir} = 1$.
3. Compute $\hat{\beta}_{t,r}$, our estimator of $\beta_{t,r}$, which is the eigenvector corresponding to the largest eigenvalue of the matrix $D^{-1}S$.

One-dimensional Kernel regression imputation

The regression functions

$$\phi_{t,t-1}(u'\beta_{t,t-1}) = E(y_{it}|Z'_{i(t-1)}\beta_{t,t-1} = u'\beta_{t,t-1}, \delta_{i1} = \cdots = \delta_{i(t-1)} = 1, \delta_{it} = 1)$$

and

$$\phi_{t,r}(u'\beta_{t,r}) = E(y_{it}|Z'_{ir}\beta_{t,r} = u'\beta_{t,r}, \delta_{i1} = \cdots = \delta_{ir} = 1, \delta_{i(r+1)} = 1, \delta_{it} = 0)$$

are both one-dimensional functions. Once we have estimators $\hat{\beta}_{t,r}$, they can be estimated by

$$\hat{\phi}_{t,r}(u'\hat{\beta}_{t,r}) = \sum_{i=1}^{n} \kappa\left(\frac{u'\hat{\beta}_{t,r} - Z'_{ir}\hat{\beta}_{t,r}}{h}\right) I_{t,r,i}\tilde{y}_{it} \Big/ \sum_{i=1}^{n} \kappa\left(\frac{u'\hat{\beta}_{t,r} - Z'_{ir}\hat{\beta}_{t,r}}{h}\right) I_{t,r,i},$$

$r = 1,...,t-1$, $t = 2,...,T$, where $\tilde{y}_{it}$ is y_{it} when $r = t-1$ and is imputed from $\hat{\phi}_{t,r}(Z'_{ir}\hat{\beta}_{t,r})$ if $r < t-1$.

One of the conditions for the sliced inverse regression is that $\mathbf{y}_i$ has an elliptically symmetric distribution such as multivariate normal. If this condition does not hold, then $\hat{\beta}_{t,r}$ may be inconsistent and the sample means based on imputed data may be biased.

Last-value-dependent missing

Another dimension reduction method is based on the assumption of last-value-dependent missing mechanism

$$q(\delta_{it}|\mathbf{y}_i, \mathbf{x}_i, \delta_{is}, s \neq t) = q(\delta_{it}|y_{i(t-1)}), \quad t = 2,...,T. \tag{7.25}$$

Under (7.25),

$$E(y_{it}|y_{i(t-1)}, \delta_{it} = 0, \delta_{i(t-1)} = 1) = E(y_{it}|y_{i(t-1)}, \delta_{it} = 1, \delta_{i(t-1)} = 1), \quad t = 2,...,T, \tag{7.26}$$

and

$$E(y_{it}|y_{ir}, \delta_{it} = \cdots = \delta_{i(r+1)} = 0, \delta_{ir} = 1) = E(y_{it}|y_{ir}, \delta_{it} = \cdots = \delta_{i(r+2)} = 0, \delta_{i(r+1)} = \delta_{ir} = 1)$$
$$r = 1,...,t-2, t = 2,...,T, \tag{7.27}$$

where, for each missing y_{it}, y_{ir} is the last observed component from the same unit. Equations (7.26) and (7.27) replace equations (7.20) and (7.21), respectively. The imputation procedure is similar to that in Section 7.3.2. Since the conditional expectations in (7.26)-(7.27) involve only one y-value (the last observed y_{ir}), the Kernel regression in Section 7.3.3 can be applied with the one-dimensional "covariate" y_{ir}. Thus, we achieve the dimensional reduction using assumption (7.25).

7.3.5 *Simulation study*

We present here results from a simulation study to evaluate the performance of the sample mean as an estimator of $E(y_{it})$ based on several methods, when the sample size $n = 1,000$, the total number of time points $T = 4$, and there are no covariates.

For comparison, we consider five estimators: the sample mean of the complete data, which is used as the gold standard; the sample mean of subjects without any missing value, which ignores missing data; the sample mean based on censoring and linear regression imputation (see Section 7.3.1), which first discards all observations of a subject after the first missing time point in order to create a dataset with "monotone missing" and then apply linear regression imputation to the created monotone missing dataset as described in Paik (1997); the sample mean based on the imputation method introduced in Sections 7.3.2 and 7.3.3, that we call nonparametric regression imputation; and the sample mean based on the imputation method introduced in Section 7.3.2 and the sliced inverse regression in Section 7.3.4, that we call *semiparametric regression imputation*. Censoring and linear regression imputation and semiparametric regression imputation produce consistent estimators if $E(y_{it}|Z_{ir})$ is linear for all t and r (e.g., y_i is multivariate normal), otherwise they produce biased estimators.

We simulated data in two situations, a normal case and a log-normal case. In the normal case, $\mathbf{y}_i$'s were independently generated from a multivariate normal distribution with mean vector $(1.33, 1.94, 2.73, 3.67)$ and the covariate matrix having an AR(1) structure with correlation coefficient 0.7; all data at $t = 1$ were observed; missing data at $t = 2, 3, 4$ were generated according to

$$P(\delta_{it} = 0|Z_{i(t-1)}, A_{i(t-1)}) = \Phi(0.6 - 0.6Z'_{i(t-1)}\beta_{t-1}), \quad (7.28)$$

where Φ is the standard normal distribution function, β_{t-1} is a $(t-1)$-vector whose j-th component is

$$\frac{j + (1-\delta_j)j}{\sum_{k=1}^{T}\{k + (1-\delta_k)k\}}, \quad j = 1, ..., t-1.$$

The probabilities of missing patterns under model (7.28) are given in Table 7.2. In the log-normal case, the log of the components of $\mathbf{y}_i$'s were independently generated from the multivariate normal distribution with mean vector $(1.33, 1.77, 2.25, 2.76)$ and the same covariance matrix as in the normal case. The missing mechanism remains the same except that the right-hand side of (7.28) is changed to $\Phi(2 - 0.5Z'_{i(t-1)}\beta_{t-1})$. The probabilities of missing patterns are also given in Table 7.2.

Table 7.2 Probabilities of missing patterns in the simulation study ($T = 4$)

	Missing pattern	Probability of missing pattern: Normal case		Log-normal case	
Monotone	(1,0,0,0)	0.051	total = 0.195	0.111	total = 0.187
	(1,1,0,0)	0.054		0.043	
	(1,1,1,0)	0.090		0.033	
Intermittent	(1,0,0,1)	0.089	total = 0.432	0.105	total = 0.336
	(1,0,1,0)	0.055		0.031	
	(1,0,1,1)	0.139		0.118	
	(1,1,0,1)	0.149		0.082	
Complete	(1,1,1,1)	0.373		0.477	

Table 7.3 reports (based on 1,000 simulation runs) the relative bias and variance of mean estimators, the mean of bootstrap variance (BV) estimators (based on 200 bootstrap replications), the coverage probability of approximate 95% confidence intervals (CI) obtained using point estimator $\pm 1.96 \times \sqrt{\text{bootstrap variance}}$, and the length of CI. The following is a summary of the results in Table 7.3.

Table 7.3 Simulation results for mean estimation

Method	Quantity	Normal case			Log-normal case		
		$t=2$	$t=3$	$t=4$	$t=2$	$t=3$	$t=4$
I	relative bias	0.0%	0.0%	0.0%	0.0%	0.0%	0.0%
	variance $\times 10^3$	0.981	0.954	0.917	0.154	0.430	1.193
	bootstrap variance $\times 10^3$	0.999	0.995	0.992	0.157	0.416	1.164
	CI coverage rate	94.9%	95.9%	95.6%	94.5%	94.5%	95.0%
	CI length	0.124	0.123	0.123	1.539	2.497	4.181
II	relative bias	10.2%	6.8%	3.5%	31.3%	28.4%	17.1%
	variance $\times 10^3$	1.334	1.448	1.212	0.322	0.868	1.789
	bootstrap variance $\times 10^3$	1.393	1.424	1.269	0.334	0.835	1.755
	CI coverage rate	0.0%	0.1%	4.3%	0.0%	0.0%	4.4%
	CI length	0.146	0.148	0.139	2.242	3.529	5.128
III	relative bias	0.0%	0.1%	0.1%	8.5%	14.6%	15.8%
	variance $\times 10^3$	1.281	1.984	2.859	0.261	1.176	3.726
	bootstrap variance $\times 10^3$	1.328	2.094	3.066	0.244	1.018	3.323
	CI coverage rate	95.3%	95.1%	95.1%	62.0%	33.9%	34.2%
	CI length	0.143	0.179	0.216	1.913	3.875	7.025
IV	relative bias	0.1%	0.1%	-0.3%	0.1%	0.9%	0.8%
	variance $\times 10^3$	1.349	2.986	3.884	0.182	0.848	2.365
	bootstrap variance $\times 10^3$	1.399	3.047	4.369	0.182	0.824	2.441
	CI coverage rate	95.6%	96.0%	96.5%	93.8%	94.2%	95.1%
	CI length	0.146	0.215	0.257	1.658	3.459	5.912
V	relative bias	0.1%	0.4%	0.2%	0.1%	5.9%	6.2%
	variance $\times 10^3$	1.349	1.643	1.764	0.182	1.167	2.212
	bootstrap variance $\times 10^3$	1.375	1.736	1.906	0.182	1.054	2.812
	CI coverage rate	95.6%	94.0%	94.8%	93.8%	88.8%	90.0%
	CI length	0.146	0.163	0.170	1.658	3.960	6.049

Method I: complete data
Method II: ignoring all missing data
Method III: censoring and linear regression imputation
Method IV: the nonparametric regression imputation in Sections 7.3.2 and 7.3.3
Method V: the semiparametric regression imputation in Sections 7.3.2 and 7.3.4

1. Bias. The nonparametric regression imputation method produces estimators with negligible biases at all time points in both normal and log-normal cases. The sample mean based on ignoring all missing data is clearly biased. Although in some cases the bias is small, the corresponding CI has very low coverage probability, because the variance of the sample mean is also very small. The sample means based on censoring and linear regression imputation are well estimated in the normal case with negligible biases, but significantly biased in the log-normal case because a wrong linear regression is used in imputation. The semiparametric regression imputation method produces negligible biases in the normal case but large biases in the log-normal case because the elliptically symmetric distribution condition does not hold.
2. Variance. In the normal case where all three imputation methods are correct, semiparametric regression imputation is the most efficient method, and nonparametric regression imputation is the least efficient. This is because semiparametric regression uses more information. The censoring and linear regression imputation method is more efficient than nonparametric regression imputation when linear regression is correct (in the normal case), but is less efficient than semiparametric regression because it discards data. In the log-normal case, however, nonparametric

regression imputation is the only correct method and has a smaller variance than the other two methods, which shows the robustness of nonparametric regression.

3. Bootstrap and CI. The bootstrap variance estimator performs well in all cases, even when the mean estimator is biased. The related CI has a coverage probability close to the nominal level of 95% when the mean estimator has little bias.

7.3.6 Wisconsin Diabetes Registry Study

The following analysis of data from the Wisconsin Diabetes Registry Study (WDRS) is given in Xu (2007). The WDRS is a geographically defined population-based incident cohort study. All individuals with newly diagnosed type I diabetes between May 1987 and April 1992 in southern and central Wisconsin were invited to enroll in the study. Subjects were asked to measure glycosylated haemoglobin (GHb) at each ordinary visit to a local physician, or every four months by submitting a blood specimen using prestamped mailing kits.

One of the main interests in this study is how GHb changes with duration defined as the number of years after diagnosis at year 0. The average GHb values within each year is the response from each subject. However, compliance was not constant, and some subjects could go one or two years without submitting blood samples. In some cases, they resumed mailing blood samples. At the beginning of the study, the number of subjects was 521. Table 7.4 shows the number of subjects with responses from the baseline year to the 5th year.

Table 7.4 Realized missing percentages in a WDRS survey

Duration (years)	0	1	2	3	4	5
Number of observed data	521	450	417	441	357	321
Missing percentage	0%	14%	20%	15%	31%	38%

To estimate the mean of GHb from duration 0 (baseline) to duration 5, four methods are applied: the sample mean of subjects without any missing value (the naive method), the sample mean based on censoring and linear regression imputation, the sample mean based on nonparametric regression imputation, and the sample mean based on semiparametric regression imputation. The estimated means are plotted against durations in Figure 7.1. The estimated mean of GHb based on nonparametric and semiparametric regression imputation increases initially and levels off after duration 3 (3 years after diagnosis), whereas the estimated means of GHb based on the other two methods display continued increase after duration 3. They have a larger increase than the naive method of ignoring missing data.

Clinical experience indicates that GHb values do not increase several years after the diagnosis of diabetes, which supports the theory that nonparametric regression imputation provides the most plausible result from the medical or epidemiological point of view.

7.4 Random-effect-dependent missing data

The random-effect-dependent propensity model assumes that there exists a subject-level unobserved random effect $\mathbf{b}_i$ such that

$$q(\delta_i|\mathbf{y}_i,\mathbf{x}_i,\mathbf{b}_i) = q(\delta_i|\mathbf{x}_i,\mathbf{b}_i), \tag{7.29}$$

which is often called the *shared parameter model* (Follmann and Wu, 1995). Since $\mathbf{b}_i$ is unobserved, this propensity is nonignorable, which is common when a mixed-effect model is assumed for complete data or when cluster sampling is applied in surveys. As a specific example for (7.29), imagine a cluster sampling case, where $\mathbf{y}_i$ is from a particular household (cluster) and a single person completes survey forms for all persons in the household. It is likely that the missing probability

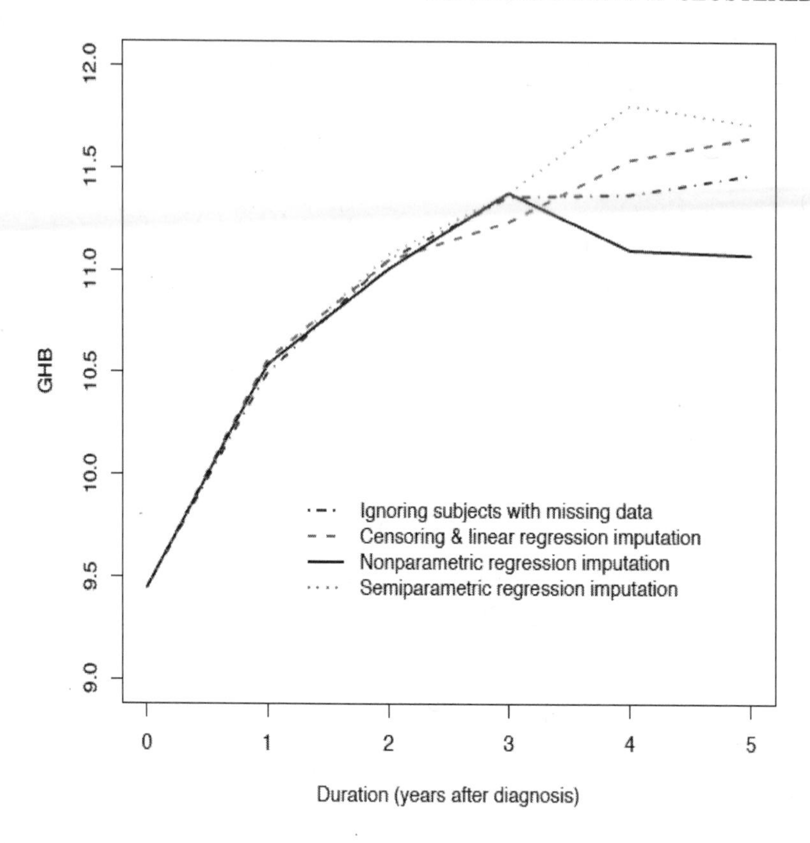

Figure 7.1 Estimated means of GHb by duration.

depends on a household-level variable (the person who completes survey forms), not on any within-household variable. We assume that, in addition to (7.29), $\mathbf{y}_i$ has at least one observed component.

When there are no missing data, we assume a linear mixed-effect model

$$\mathbf{y}_i = \mathbf{x}_i\beta + \mathbf{z}_i\mathbf{b}_i + \varepsilon_i, \tag{7.30}$$

where $\mathbf{b}_i$ is a subject-level random-effect vector, $E(\mathbf{b}_i) = 0$, $\mathbf{b}_i$ is independent of $\mathbf{x}_i$, $\mathbf{z}_i$ is a submatrix of $\mathbf{x}_i$, ε_i is a within-subject error vector, $\mathbf{b}_i$ and ε_i are independent and unobserved, $E(\varepsilon_i) = 0$, and $V(\varepsilon_i) = V(\mathbf{x}_i)$, a covariance matrix possibly depending on $\mathbf{x}_i$. Under (7.30), assumption (7.29) holds if and only if the missing probability of a component y_{it} depends on $(\mathbf{b}_i, \mathbf{x}_i)$ but not on ε_i.

7.4.1 Three existing approaches

Under missing mechanism (7.29), there are three existing approaches.

The first one is the parametric likelihood approach. Under (7.29),

$$p(\mathbf{y}_i, \delta_i, \mathbf{b}_i|\mathbf{x}_i) = p(\mathbf{y}_i|\mathbf{b}_i, \mathbf{x}_i)p(\delta_i|\mathbf{y}_i, \mathbf{b}_i, \mathbf{x}_i)p(\mathbf{b}_i|\mathbf{x}_i) = p(\mathbf{y}_i|\mathbf{b}_i, \mathbf{x}_i)p(\delta_i|\mathbf{b}_i, \mathbf{x}_i)p(\mathbf{b}_i|\mathbf{x}_i),$$

where $p(\cdot|\cdot)$ is the generic notation for conditional likelihood. Let $\mathbf{y}_{i,\text{mis}}$ be the missing components of $\mathbf{y}_i$. Assuming parametric models on $p(\mathbf{y}_i|\mathbf{b}_i, \mathbf{x}_i)$, $p(\delta_i|\mathbf{b}_i, \mathbf{x}_i)$, and $p(\mathbf{b}_i|\mathbf{x}_i)$, we obtain the following parametric likelihood

$$\prod_i \int \left(\int p(\mathbf{y}_i|\mathbf{b}_i, \mathbf{x}_i) d\mathbf{y}_{i,\text{mis}} \right) p(\delta_i|\mathbf{b}_i, \mathbf{x}_i) p(\mathbf{b}_i|\mathbf{x}_i) d\mathbf{b}_i. \tag{7.31}$$

The integration is necessary since $\mathbf{y}_{i,\text{mis}}$'s and $\mathbf{b}_i$'s are not observed. The parameters in $p(\mathbf{y}_i|\mathbf{b}_i, \mathbf{x}_i)$ can be estimated as long as they can be identified under some conditions. The likelihood in (7.31) involves intractable integrals except for some very special cases, hence, the difficulty of computing maximum likelihood. To avoid the difficulty, parametric fractional imputation of Kim (2011) can be applied. Yang et al. (2013) provides a detailed description of the FI method for the shared parameter model in (7.29). In general, shared parameter models require untestable assumptions and the parametric approach is sensitive to parametric assumptions on various conditional probability densities.

The second approach is a semiparametric method specifying only the first- and second-order conditional moments. Assumptions (7.29)-(7.30) imply that

$$E(\mathbf{y}_i|\mathbf{x}_i, \mathbf{b}_i, \delta_i) = \mathbf{x}_i\beta + \mathbf{z}_i\mathbf{b}_i \qquad \text{and} \qquad V(\mathbf{y}_i|\mathbf{x}_i, \mathbf{b}_i, \delta_i) = V(\mathbf{x}_i).$$

Then, we have a conditional model

$$E(\mathbf{y}_i|\mathbf{x}_i, \delta_i) = E[E(\mathbf{y}_i|\mathbf{x}_i, \mathbf{b}_i, \delta_i)|\mathbf{x}_i, \delta_i] = \mathbf{x}_i\beta + \mathbf{z}_i E(\mathbf{b}_i|\mathbf{x}_i, \delta_i). \tag{7.32}$$

If $E(\mathbf{b}_i|\mathbf{x}_i, \delta_i)$ can be approximated by a simple form, then β may be estimated using this approximate conditional model (ACM). This approach is known as the ACM approach. However, one must deal with the following two issues.

- How to find a reasonable approximation $E(\mathbf{b}_i|\mathbf{x}_i, \delta_i)$? Follmann and Wu (1995) showed that, if $p(\delta_i|\mathbf{b}_i)$ is in an exponential family and if δ_{it}'s are conditionally i.i.d. given $\mathbf{b}_i$, then $E(\mathbf{b}_i|\delta_i)$ is a monotone function of a summary statistic S_i. As a result, they suggested that $E(\mathbf{b}_i|\delta_i)$ can be approximated by a linear or polynomial function of S_i. This approximation, however, may not be good enough. Furthermore, the exponential family assumption is somewhat restrictive.
- The parameter β may not be identifiable after $E(\mathbf{b}_i|\mathbf{x}_i, \delta_i)$ is approximated through some simple function, i.e., β may be confounded with some parameters in the ACM (Albert and Follmann, 2000). For example, suppose that the covariate and $\mathbf{b}_i = b_i$ are both univariate so that $E(y_{it}|x_{it}, b_i) = \beta_0 + \beta_1 x_{it} + b_i x_{it}$. If S_i is univariate and $E(b_i|\mathbf{x}_i, \delta_i)$ is approximated by $\gamma_0 + \gamma_1 S_i$, then the ACM is $E(y_{it}|x_{it}, b_i) \approx \beta_0 + \beta_1 x_{it} + \gamma_0 x_{it} + \gamma_1 S_i x_{it}$ and β_1 is confounded with γ_0. Non-identifiability seriously limits the scope of the application of the ACM approach, although in some situations β_1 can be estimated with additional work.

The third approach is the method of grouping, which can be applied according to the following steps.

1. Find a summary statistic S_i such that

$$E(\mathbf{b}_i|\mathbf{x}_i, \delta_i) = E(\mathbf{b}_i|S_i) \qquad \text{and} \qquad E(\mathbf{b}_i\mathbf{b}_i'|\mathbf{x}_i, \delta_i) = E(\mathbf{b}_i\mathbf{b}_i'|S_i). \tag{7.33}$$

2. Assume that S_i is discrete and takes values $s_1, ..., s_L$. Then we divide the sample into L groups according the value of S_i.
3. Obtain estimate $\hat{\beta}_l$ using data in the l-th group and the following model:

$$\begin{aligned} E(\mathbf{y}_{i,\text{obs}}|\mathbf{x}_{i,\text{obs}}, S_i = s_l) &= \mathbf{x}_{i,\text{obs}}\beta + \mathbf{z}_{i,\text{obs}} E(\mathbf{b}_i|S_i = s_l) \\ V(\mathbf{y}_{i,\text{obs}}|\mathbf{x}_{i,\text{obs}}, S_i = s_l) &= \mathbf{z}_{io} V(\mathbf{b}_i|S_i = s_l)\mathbf{z}_{i,\text{obs}}' + E(V(\mathbf{x}_i)|\mathbf{x}_{i,\text{obs}}, S_i = s_l), \end{aligned} \tag{7.34}$$

where $\mathbf{y}_{i,\text{obs}}$ is the vector of observed $\mathbf{y}_i$-values and $\mathbf{x}_{i,\text{obs}}$ and $\mathbf{z}_{i,\text{obs}}$ are submatrices of $\mathbf{x}_i$ and $\mathbf{z}_i$

corresponding to $\mathbf{y}_{i,\text{obs}}$. In (7.34), $E(\mathbf{b}_i|S_i = s_l)$ is viewed as an unobserved random effect. We can use the weighted least squares estimator

$$\hat{\beta}_l = \left(\sum_{i \in \mathcal{G}_l} \mathbf{x}'_{i,\text{obs}} \hat{V}^{-1}_{i,\text{obs}} \mathbf{x}_{i,\text{obs}} \right)^{-1} \sum_{i \in \mathcal{G}_l} \mathbf{x}'_{i,\text{obs}} \hat{V}^{-1}_{i,\text{obs}} \mathbf{y}_{i,\text{obs}},$$

where $\mathcal{G}_l$ is the l-th group, $\hat{V}_{i,\text{obs}}$ is an estimator of $V_{i,\text{obs}} = V(\mathbf{y}_{i,\text{obs}}|\mathbf{x}_{i,\text{obs}}, S_i = s_l)$ as specified in (7.34). Many statistical packages can be used to obtain $\hat{\beta}_l$. Since we view $E(\mathbf{b}_i|S_i = s_l)$ as an unobserved random effect, we do not need to estimate or approximate it.

4. Because $E(\mathbf{b}_i|S_i)$ is not necessarily equal to 0, $\hat{\beta}_l$ with a fixed l is not approximately unbiased for β. Let $p_l = P(S_i = s_l)$. Since

$$\sum_l p_l E(\mathbf{b}_i|S_i = s_l) = E(\mathbf{b}_i) = 0,$$

$\sum_l p_l \hat{\beta}_l$ is approximately unbiased for β. After replacing the unknown p_l by its estimator n_l/n, where n_l is the number of subjects in $\mathcal{G}_l$, we obtain the following approximately unbiased estimator of β:

$$\hat{\beta} = \sum_l \frac{n_l}{n} \hat{\beta}_l.$$

The key to this grouping method is that, although each $\hat{\beta}_l$ is biased for β ($E(\mathbf{b}_i|S_i = l) \neq 0$), the linear combination $\hat{\beta}$ is approximately unbiased for β because the average of the $E(\mathbf{b}_i|S_i = s_l)$ is 0. This method avoids the estimation of $E(\mathbf{b}_i|S_i)$ (which is what the original ACM does) and, hence, it does not have the problem of parameter confounding.

If the summary statistic S_i is continuous (or discrete but takes many values), then we need to replace S_i by a function of S_i taking discrete values. Details are given in Section 7.4.2.

7.4.2 *Summary statistics*

As we discussed previously, the starting point for the ACM or the group method is to find a summary statistic. We define a summary statistic S_i to be a function of $(\mathbf{x}_i, \delta_i)$ such that (7.33) holds. The second condition in (7.33) is not needed if we use ordinary least squares instead of weighted least squares in model fitting. A trivial summary statistic is $(\mathbf{x}_i, \delta_i)$ itself, but a simple S_i with low dimension is desired.

The following lemma is useful for finding an S_i satisfying (7.33).

Lemma 7.1. *Assume (7.29)-(7.30). A sufficient condition for (7.33) is that there exists a measurable function g such that*

$$p(\delta_i|\mathbf{x}_i, \mathbf{b}_i) = g(\mathbf{b}_i, S_i).$$

We now derive S_i under some nonparametric or semiparametric models on the missing mechanism.

Conditionally i.i.d. model

We start with the simplest case where $p(\delta_i|\mathbf{x}_i, \mathbf{b}_i) = p(\delta_i|\mathbf{b}_i)$ (i.e., the propensity does not depend on covariates) and components of δ_i are conditionally i.i.d., given $\mathbf{b}_i$. That is,

$$p(\delta_i|\mathbf{b}_i) = \prod_{t=1}^{T} P(\delta_{it} = 1|\mathbf{b}_i)^{\delta_{it}} P(\delta_{it} = 0|\mathbf{b}_i)^{1-\delta_{it}} = P(\delta_{it} = 1|\mathbf{b}_i)^{R_i} P(\delta_{it} = 0|\mathbf{b}_i)^{T-R_i},$$

which is a function of $\mathbf{b}_i$ and $R_i = \sum_{t=1}^{T} \delta_{it}$ = the number of observed components of $\mathbf{y}_i$. According to Lemma 7.1, $S_i = R_i$ is a discrete summary statistic.

Conditionally independent model with time trend

Wu and Follmann (1999) considered a propensity model with time trend where the components of δ_i are independent given $\mathbf{b}_i$ and satisfy

$$P(\delta_{it}=1|\mathbf{x}_i,\mathbf{b}_i)=\frac{\exp\{\phi_1(\mathbf{b}_i)+\phi_2(\mathbf{b}_i)s_t\}}{1+\exp\{\phi_1(\mathbf{b}_i)+\phi_2(\mathbf{b}_i)s_t\}},\quad t=1,\ldots,T, \tag{7.35}$$

where $s_1,\ldots,s_T$ are time-related values (fixed and identical for all sampled subjects) and ϕ_1 and ϕ_2 are unknown nonparametric functions of $\mathbf{b}_i$. Then,

$$\begin{aligned} p(\delta_i|\mathbf{x}_i,\mathbf{b}_i) &= \prod_{t=1}^{T}\left[\frac{P(\delta_{it}=1|\mathbf{x}_i,\mathbf{b}_i)}{P(\delta_{it}=0|\mathbf{x}_i,\mathbf{b}_i)}\right]^{\delta_{it}} P(\delta_{it}=0|\mathbf{x}_i,\mathbf{b}_i) \\ &= \frac{\exp\{\phi_1(\mathbf{b}_i)R_i+\phi_2(\mathbf{b}_i)R_{si}\}}{\prod_{t=1}^{T}[1+\exp\{\phi_1(\mathbf{b}_i)+\phi_2(\mathbf{b}_i)s_t\}]}, \end{aligned}$$

where $R_i=\sum_{t=1}^{T}\delta_{it}$ and $R_{si}=\sum_{t=1}^{T}\delta_{it}s_t$. According to Lemma 7.1, the summary statistic is $S_i=(R_i,R_{si})$. The component R_{si} accounts for the time trend in the propensity model.

An extension to model (7.35) is to replace s_t by a covariate x_{it}. The resulting summary statistic is $S_i=(R_i,R_{xi})$ with $R_i=\sum_{t=1}^{T}\delta_{it}$ and $R_{xi}=\sum_{t=1}^{T}\delta_{it}x_{it}$.

Markov dependency model

In previous cases the components of δ_i are assumed to be conditionally independent given $(\mathbf{x}_i,\mathbf{b}_i)$. To consider a conditionally dependent model, we assume $p(\delta_i|\mathbf{x}_i,\mathbf{b}_i)=p(\delta_i|\mathbf{b}_i)$ and that δ_{it}'s follow a stationary Markov chain model, given $\mathbf{b}_i$. Let $\phi_{u,v}=P(\delta_{it}=u|\delta_{i(t-1)}=v,\mathbf{b}_i)$. Then

$$\begin{aligned} p(\delta_i|\mathbf{b}_i) &= P(\delta_{i1}=1|\mathbf{b}_i)^{\delta_{i1}}P(\delta_{i1}=0|\mathbf{b}_i)^{1-\delta_{i1}} \\ &\quad\times\prod_{t=2}^{T}\left[\phi_{0,0}^{(1-\delta_{it})(1-\delta_{i(t-1)})}\phi_{1,0}^{\delta_{it}(1-\delta_{i(t-1)})}\phi_{0,1}^{(1-\delta_{it})\delta_{i(t-1)}}\phi_{1,1}^{\delta_{it}\delta_{i(t-1)}}\right] \\ &= P(\delta_{i1}=1|\mathbf{b}_i)^{\delta_{i1}}P(\delta_{i1}=0|\mathbf{b}_i)^{1-\delta_{i1}}\phi_{0,0}^{T-\delta_{iT}+G_i}\phi_{1,0}^{R_i-\delta_{iT}-G_i}\phi_{0,1}^{R_i-\delta_{i1}-G_i}\phi_{1,1}^{G_i}, \end{aligned}$$

where $R_i=\sum_{t=1}^{T}\delta_{it}$ and $G_i=\sum_{t=2}^{T}\delta_{it}\delta_{i(t-1)}$. It follows from Lemma 5.1 that a summary statistic is $S_i=(\delta_{i1},\delta_{iT},R_i,G_i)$. Compared with the summary statistic R_i in the i.i.d. case, we find that three statistics, δ_{i1}, δ_{iT}, and G_i, are needed to account for Markov dependency.

If $p(\delta_i|\mathbf{x}_i,\mathbf{b}_i)$ depends on $\mathbf{x}_i$, similar but more complicated summary statistics can be derived.

Monotone missing data

Since monotone missing has fewer missing patterns, simpler summary statistics can often be obtained. For example, when missing is monotone, $\sum_{t=1}^{R_i}\delta_{it}s_t=\sum_{t=1}^{R_i}s_t$; the summary statistic $(\delta_{i1},\delta_{iT},R_i,G_i)$ under the Markov dependency model reduces to R_i, because $(\delta_{i1},\delta_{iT},G_i)$ is a function of R_i.

Approximate summary statistics

Some summary statistics, such as R_i and R_{si}, are discrete. But some are continuous or discrete with many categories and the grouping method cannot be directly applied.

Let $\tilde{S}_i$ be a function of S_i such that (1) $\tilde{S}_i$ is discrete with a reasonable number of categories and (2) in each group defined by a category of $\tilde{S}_i$, values of $E(b_i|S_i)$ are similar. We can then call $\tilde{S}_i$ an approximate summary statistic and apply the grouping method using $\tilde{S}_i$ to create groups. If $E(b_i|S_i)$ is observed, then any method in classification or clustering may be applied to find a good approximate summary statistic $\tilde{S}_i$. Some examples can be found in Xu and Shao (2009).

7.4.3 *Simulation study*

Some simulations were conducted to study the performance of the grouping method. We used the following response model

$$y_{it} = \beta_0 + \beta_1 x_{it} + b_{i0} + b_{i1} x_{it} + e_{it}, \tag{7.36}$$

where $i = 1, ..., 1,000$ is the index for subjects, each subject has $t = 1, ..., 5$ repeated measurements, $\beta_0 = 1$ and $\beta_1 = 1$ are parameters, x_{it} is a one-dimensional covariate to be specified later, b_{i0} and b_{i1} are random effect variables following a bivariate normal distribution with mean 0, $V(b_{i0}) = 1$, $V(b_{i1}) = 2$, and correlation coefficient 0.3, and e_{it}'s are i.i.d. and follows a standard normal distribution and are independent of $b_i = (b_{i0}, b_{i1})$.

After data were generated according to (7.36), missing data were generated according to

$$\text{logit}\{P(\delta_{it} = 0 | b_i, x_{it})\} = \gamma_0 + \gamma_1 b_{i1} + \lambda_0 x_{it} + \lambda_1 b_{i1} x_{it}, \tag{7.37}$$

where δ_{it}'s are conditionally independent given b_i and x_i. The overall missingness proportions are between 35% and 50%.

The following methods for the estimation of β_1 were compared in the simulation: (1) the naive maximum likelihood estimator ignoring missing data; (2) ACM-R, the ACM method using $R_i =$ the number of observed components in $\mathbf{y}_i$ as the summary statistic; (3) ACM-S, the ACM method using a correctly derived S_i from model (7.37) as the summary statistic; (4) GRP-R, the grouping method using R_i as the summary statistic; (5) GRP-S, the grouping method using a correctly derived S_i from model (7.37) as the summary statistic. Note that R_i is a correct summary statistic if $\lambda_0 = \lambda_1 = 0$ in (7.37).

For the grouping method, if S_i is continuous, then an approximate summary statistic is used. For the ACM, S_i is included as a covariate in the following linear model

$$y_{it} = \beta_0^* + \beta_1^* x_{it} + \beta_2^* S_i + \beta_3^* S_i x_{it} + b_{i0}^* + b_{i1}^* x_{it} + \varepsilon_{ij}^*.$$

The estimator $\hat{\beta}_1^*$ is not an estimator of β_1. Instead, we use $\hat{\beta}_1^* + \hat{\beta}_3^* \bar{S}$ as an estimator of β_1, where $\bar{S}$ is the sample mean of S_i's.

From the 500 replications of the simulation, we computed the empirical bias and mean squared error (MSE) of the estimator of β_1 and the coverage probability (CP) of the related 95% confidence interval using bootstrapping for variance estimation. The results are given in Table 7.5. Four different situations (Cases I-IV) with different missing mechanisms and covariate types were considered. The following is a description of these four cases and a summary of the results in Table 7.5.

Case I. The parameters in model (7.37) are $\gamma_0 = 0.5$, $\gamma_1 = 1$, and $\lambda_0 = \lambda_1 = 0$, which corresponds to the conditional i.i.d. model in Section 7.4.2 and a correct summary statistic is R_i. Thus, ACM-S and GRP-S are the same as ACM-R and GRP-R, respectively.

The naive method has bias about 10% of the true value ($\beta_1 = 1$) and the CP of the 95% confidence interval is only 16.4%. The ACM and grouping methods are almost the same. They are much better than the naive method and have nearly 95% CP.

Case II. To account for time-dependent missingness, we used a propensity model that depends on both random effects and a time covariate $x_{it} = s_t$, where $s_t = t$. The parameters in model (7.37) are $\gamma_0 = \gamma_1 = 0$, $\lambda_0 = -0.2$, and $\lambda_1 = 0.3$. Under this time-dependent missing mechanism, the proportion of missing data increases as t increases. According to the discussion in Section 7.4.2 and by the fact that $\gamma_0 = \gamma_1 = 0$, a correct summary statistic is $S_i = \sum_{t=1}^{5} a_{it} s_t$ instead of R_i.

The naive method is heavily biased and has CP = 0. Since R_i is not the right summary statistic, both ACM-R and GRP-R are biased with low CP values. ACM-S and GRP-S use the correct summary statistic S_i. However, ACM-S still has a 4.6% relative bias, which results in a low CP of 89.2%. This may be caused by the fact that $E(\mathbf{b}_i | S_i)$ is not linear in S_i. GRP-S is much better than GRP-R, indicating the importance of having a correct summary statistic. It is also better than ACM-S, since no linear approximation to $E(\mathbf{b}_i | S_i)$ is required in the grouping method.

Case III. We considered a discrete covariate $x_{it} = x_i$ that does not depend on t and takes only two values, 0.5 and -1. The parameters in (7.37) are $\gamma_0 = 0.1$, $\gamma_1 = 0.2$, $\lambda_0 = 0.3$ and $\lambda_1 = 1$. Subjects with different x_i values have opposite signs of random slopes in the missing mechanism. The result in Section 7.4.2 indicates that a correct summary statistic is the 2-dimensional statistic $S_i = (R_i, x_i)$.

The naive method fares well in terms of bias and its CP is 92.8%. This may be because the dataset has two subsets according to the value of x_i and, within each subset, the naive method is biased (e.g., the results in Case I) but the biases are canceled in the overall estimation. The ACM-R does poorly compared to the naive method. We did not use ACM-S in this case, because how to apply ACM using a 2-dimensional summary statistic was not addressed previously. GRP-R uses an incorrect summary statistic, but its performance is acceptable. GRP-S uses a correct summary statistic and performs the best among all methods under consideration.

Case IV. We considered a continuous covariate $x_{it} \sim N(\log t, 0.5)$. The parameters in (7.37) are $\gamma_0 = \gamma_1 = 0$, $\lambda_0 = 0.1$ and $\lambda_1 = \sqrt{2}/10$. According to the result in Section 7.4.2, a correct summary statistic is $S_i = \sum_{t=1}^{5} a_{it}x_{it}$, which is a continuous statistic. Thus, for the grouping approach, we apply the GUIDE (Loh, 2002) to find an approximate summary statistic for grouping.

The naive method has a 4.7% bias and a low CP of 81.8%. ACM-R and ACM-S are about the same, although the former uses the wrong summary statistic R_i and the latter uses the correct summary statistic S_i and is less biased. GRP-R uses the wrong summary statistic R_i and is slightly worse than ACM-S. GRP-S uses the correct summary statistic and has the smallest bias and a CP closest to 95% among all methods under consideration.

Table 7.5 Simulation results based on 500 runs

	Case I			Case II		
Method	Bias	MSE	CP(%)	Bias	MSE	CP(%)
Naive	0.0985	0.0120	16.4	0.3695	0.1405	0
ACM-R	−0.0026	0.0026	94.8	0.0657	0.0097	78.2
ACM-S				0.0455	0.0074	89.2
GRP-R	−0.0013	0.0026	95.4	0.1413	0.0286	46.4
GRP-S				0.0210	0.0093	94.6
	Case III			Case IV		
Method	Bias	MSE	CP(%)	Bias	MSE	CP(%)
Naive	−0.0091	0.0049	92.8	0.0467	0.0068	81.8
ACM-R	−0.0126	0.0052	91.8	0.0090	0.0049	91.0
ACM-S				0.0052	0.0049	91.0
GRP-R	−0.0058	0.0078	94.2	0.0085	0.0063	91.5
GRP-S	−0.0028	0.0051	95.0	0.0028	0.0063	94.0

Naive: maximum likelihood estimator ignoring missing data
ACM-R: ACM with summary statistic R
ACM-S: ACM with summary statistic S
GRP-R: grouping with summary statistic R
GRP-S: grouping with summary statistic S
R: number of observed components
S: derived summary statistic

7.4.4 *Modification of diet in renal disease*

We now present a real data example in Xu and Shao (2009). The modification of diet in renal disease (MDRD) study was a randomized clinical trial of patients with progressive renal disease.

The intervention examined here is the two levels of dietary intake of protein and phosphorous. The primary interest of the trial is to compare two treatments in terms of the rate reduction of their renal disease. The primary outcome measure is the decline in glomerular filtration rate (GFR), a continuous measure of how rapidly the kidneys filter blood. GFR is measured serially every 4 months. The primary period of interest is from month 16 to month 36.

A total of 520 patients were randomized to one of the two treatment levels: low protein diet (Diet L) and very low protein diet (Diet VL). Relevant summaries of the two treatment groups are given in Table 7.6.

Table 7.6 MDRD study with 2 treatments

	Treatment	
	Diet VL	Diet L
Number of patients	259	261
Proportion of missing data	44%	45%
Median number of GFR per patient	3	3
Number (%) of patients with complete data	41 (15.8%)	39 (14.9%)
Number (%) of patients with monotone missingness	191 (73.7%)	197 (75.5%)
Number (%) of patients with intermittent missingness	27 (10.5%)	25 (9.6%)

We consider the following response model when there is no missing:

$$\mathrm{GFR}_{it} = \beta_0 + \beta_1 s_t + \beta_2 x_i + \beta_3 x_i s_t + b_{i0} + b_{i1} s_t + \varepsilon_{it},$$

where x_i is the treatment indicator ($x_i = 0$ for Diet VL and $x_i = 1$ for Diet L), $s_t = 12 + 4t$ is the duration of the study that does not depend on i, $t = 1, ..., 6$, β_j's are unknown parameters, and b_{i0} and b_{i1} are random subject effects. Note that β_1 is the slope over time for GFR with Diet VL treatment and $\beta_1 + \beta_3$ is the slope over time for GFR with Diet L treatment. Thus, β_3 is the difference in slope for GFR between two treatments and is the main focus of our analysis. A positive β_3 leads to the conclusion that Diet L is better whereas a negative β_3 leads to the conclusion that Diet VL is better.

The overall proportion of missing data is about 45%. One issue related to the missing data is that when a patient's GFR drops to some level the kidney function would be impaired and no outcome could be obtained as a result. We calculate the ordinary least squares estimates of GFR slopes for patients with at least two observations. The individual slopes from the patients with missing data are significantly more negative than those from the patients with complete data (p-value = 0.004922, Wilcoxon rank sum test). Another characteristic of the missing data in this example is the notable time trend of missingness proportions. We analyze this dataset under the following propensity model:

$$\mathrm{logit}\{(\delta_{it} = 0 | b_i, x_{it})\} = \phi_2(b_i) s_t.$$

Here, the summary statistic is $S_i = \sum_{t=1}^{6} \delta_{it} s_t$. Because the majority of subjects have monotone missingness (Table 7.6), in this example S_i is close to R_i = the number of observed components of $\mathbf{y}_i$.

We apply the five methods in the simulation study in Section 7.4.3 to the MDRD data: (1) The naive method ignoring missing data; (2) ACM-R, the ACM using R_i as the summary statistic; (3) ACM-S, the ACM using S_i as the summary statistic; (4) GRP-R, the grouping method using R_i as the summary statistic; (5) GRP-S, the grouping method using S_i as the summary statistic. When we apply ACM-S, we fit the model

$$\mathrm{GFR}_{it} = \beta_0^* + \beta_1^* s_t + \beta_2^* x_i + \beta_3^* S_i + \beta_4^* x_i s_t + \beta_5^* x_i S_i + \beta_6^* s_t S_i + \beta_7^* x_i s_t S_i + b_{i0}^* + b_{i1}^* s_t + \varepsilon_{it}.$$

We consider $\hat{\beta}_4^* + \hat{\beta}_6^* (\bar{S}_1 - \bar{S}_0) + \hat{\beta}_7^* \bar{S}_1$, instead of $\hat{\beta}_4^*$, as an estimator of β_3, where $\bar{S}$ is the sample

Table 7.7 Estimates and 95% confidence upper and lower limits of β_3 in the MDRD study

Method	Estimate	Lower limit	Upper limit
Naive	0.06	−0.04	0.17
ACM-R	0.06	−0.11	0.21
ACM-S	0.01	−0.16	0.18
GRP-R	−0.03	−0.19	0.13
GRP-S	−0.04	−0.16	0.08

mean of S_i's and $\bar{S}_k$ is the sample mean of S_i under treatment k $(= 0, 1)$. For ACM-R, we can simply substitute S_i by R_i.

Table 7.7 shows the estimates and 95% confidence upper and lower limits of β_3 using the 5 methods. The estimates from the naive method, ACM-R, and ACM-S are all positive, whereas estimates from GRP-R and GRP-S are negative. All confidence intervals include 0 so that we cannot reject the hypothesis of $\beta_3 = 0$ at the significance level of 5%, which may be due to low power: we only have about 260 subjects in each treatment and about 45% of them have missing values. However, the results indicate that the grouping method and the ACM approach may lead to different conclusions on which diet is better; the ACM approach agrees with the naive method and is in favor of Diet L, whereas the grouping method is in favor of Diet VL.

Chapter 8

Application to survey sampling

8.1 Introduction

In this chapter, we consider the problem of parameter estimation in the context of survey sampling. To formally define the setup, let $U = \{1, 2 \cdots, N\}$ be the index set of a finite population and let I_i be the sample indicator function such that $I_i = 1$ indicates the selection of unit i for the sample and $I_i = 0$ otherwise. The probability $\pi_i = Pr(I_i = 1 \mid i \in U)$ is often called the *first-order inclusion probability* and is known in probability sampling. Thus, estimation with data from a probability sample is a special case of the missing data problem where the sample is treated as the set of respondents and the sampling mechanism is known.

Let y_i be the realized value of a random variable Y for unit i. Assume that Y follows a distribution with density $f(y;\theta)$, for some unknown parameter θ. The model for generating the finite population is often called the *superpopulation model*. Let A be the set of indices in the sample. If we use

$$\sum_{i \in A} S(\theta; y_i) = 0,$$

where $S(\theta; y) = \partial \log f(y;\theta)/\partial\theta$, to estimate θ from the sample, the solution is consistent if I_i is independent of y_i. Such condition is very close to the condition of missing completely at random (MCAR). If the parameter of interest is for the conditional distribution of y given x, denoted by $f(y \mid x;\theta)$, then the sample score equation

$$\sum_{i \in A} S_1(\theta; x_i, y_i) = 0, \tag{8.1}$$

where $S_1(\theta; x, y) = \partial \log f_1(y \mid x;\theta)/\partial\theta$, provides a consistent estimator of θ if

$$Cov\{I_i, S_1(\theta; x_i, y_i) \mid x_i\} = 0.$$

This condition will hold if $Pr(I_i = 1 \mid x_i, y_i)$ is a function of x_i only. Thus, if the sampling design is such that

$$E(I_i \mid x_i, y_i) = E(I_i \mid x_i) \tag{8.2}$$

holds for all (x_i, y_i), then the sampling design is called *noninformative* in the sense that we can use the sample score equation (8.1) to estimate θ. Note that the definition of a noninformative sampling design is specific to the model considered. If the model is about $f(x \mid y)$, the conditional distribution of x given y, then the condition for a noninformative sampling design is changed to

$$E(I_i \mid x_i, y_i) = E(I_i \mid y_i).$$

The noninformative sampling condition is essentially the MAR condition in missing data.

If the sampling design is informative for estimating θ in $f_1(y \mid x;\theta)$ in the sense that (8.2) does not hold, then the sample score equation (8.1) leads to a biased estimate. To remove the bias, the pseudo maximum likelihood estimator, defined by solving

$$\sum_{i\in A}\frac{1}{\pi_i}S_1(\theta;x_i,y_i)=0, \tag{8.3}$$

is often used. Because

$$Cov\left\{I_i\pi_i^{-1},S_1(\theta;x_i,y_i)\mid x_i\right\} \;=\; 0,$$

the weighted score equation using the survey weight $d_i=1/\pi_i$ leads to a consistent estimator of θ. In fact, since

$$Cov\left\{I_i\pi_i^{-1}q(x_i),S_1(\theta;x_i,y_i)\mid x_i\right\} \;=\; 0$$

for any $q(x_i)$, the solution to

$$\sum_{i\in A}\frac{1}{\pi_i}S_1(\theta;x_i,y_i)q(x_i)=0 \tag{8.4}$$

is consistent for θ, regardless of the choice of $q(x)$. See Magee (1998).

Example 8.1. *Suppose that we are interested in estimating $\beta=(\beta_0,\beta_1)$ for the linear regression model*

$$y_i=\beta_0+\beta_1x_i+e_i, \tag{8.5}$$

where $E(e_i\mid x_i)=0$ and $Cov(e_i,e_j\mid x)=0$ for $i\neq j$. We consider the following class of estimators

$$\hat\beta=\left(\sum_{i\in A}d_i\mathbf{x}_i\mathbf{x}_i'q_i\right)^{-1}\sum_{i\in A}d_i\mathbf{x}_iy_iq_i, \tag{8.6}$$

where $\mathbf{x}_i=(1,x_i)'$, $q_i=q(x_i)$ and $d_i=\pi_i^{-1}$. Since

$$\hat\beta-\beta=\left(\sum_{i\in A}d_i\mathbf{x}_i\mathbf{x}_i'q_i\right)^{-1}\sum_{i\in A}d_i\mathbf{x}_ie_iq_i,$$

where $e_i=y_i-\mathbf{x}_i'\beta$,

$$E\left(\hat\beta-\beta\right)\cong E\left\{\left(\sum_{i=1}^{N}\mathbf{x}_i\mathbf{x}_i'q_i\right)^{-1}\sum_{i=1}^{N}\mathbf{x}_ie_iq_i\right\}\cong 0,$$

where the expectation is with respect to the joint distribution of the superpopulation model (8.5) and the sampling mechanism. The anticipated variance, which is the total variance with respect to the joint distribution, is

$$V\left(\hat\beta-\beta\right)\cong\left\{E\left(\sum_{i=1}^{N}\mathbf{x}_i\mathbf{x}_i'q_i\right)\right\}^{-1}V\left\{\sum_{i\in A}d_i\mathbf{x}_ie_iq_i\right\}E\left\{\left(\sum_{i=1}^{N}\mathbf{x}_i\mathbf{x}_i'q_i\right)\right\}^{-1} \tag{8.7}$$

and

$$\begin{aligned}V\left\{\sum_{i\in A}d_i\mathbf{x}_ie_iq_i\right\} &= E\left\{V\left(\sum_{i\in A}d_i\mathbf{x}_ie_iq_i\mid X,Y\right)\right\}+V\left\{E\left(\sum_{i\in A}d_i\mathbf{x}_ie_iq_i\mid X,Y\right)\right\}\\ &= E\left\{\sum_{i=1}^{N}\sum_{j=1}^{N}(\pi_{ij}-\pi_i\pi_j)d_id_j\mathbf{x}_i\mathbf{x}_j'e_ie_jq_iq_j\right\}+V\left\{\sum_{i=1}^{N}\mathbf{x}_ie_iq_i\right\}\\ &= E\left\{\sum_{i=1}^{N}E(d_ie_i^2\mid\mathbf{x}_i)\mathbf{x}_i\mathbf{x}_i'q_i^2\right\}.\end{aligned}$$

Thus, the optimal choice of q_i that minimizes the total variance in (8.7) is

$$q_i^* = \{E(d_i e_i^2 \mid \mathbf{x}_i)\}^{-1}. \tag{8.8}$$

To estimate q_i^*, the estimated GLS method or the variance function estimation technique (Davidian and Caroll, 1987) can be used. Fuller (2009, Chapter 6) discussed the optimal estimation of β using $\hat{q}_i^* = q^*(\mathbf{x}_i; \hat{\alpha})$, where $\hat{\alpha}$ is a consistent estimator of α in the model $q^*(x; \alpha) = E(d_i e_i^2 \mid \mathbf{x}_i; \alpha)$. The effect of replacing α with $\hat{\alpha}$ in $\hat{q}_i$ is negligible in variance estimation of $\hat{\beta}$.

8.2 Calibration estimation

In survey sampling, we often have one or more auxiliary variables observed throughout the population. Let $\mathbf{x}_i$ be a p-dimensional vector of auxiliary variables whose population total $\mathbf{X} = \sum_{i=1}^N \mathbf{x}_i$ is known. In this case, it is often desirable to achieve consistency with $\mathbf{X}$ in the estimation. That is, for $\hat{Y} = \sum_{i \in A} w_i y_i$, weights are desired to satisfy

$$\sum_{i \in A} w_i \mathbf{x}_i = \mathbf{X}, \tag{8.9}$$

which is often called the *calibration condition* or *benchmarking condition*. Often, we have $w_i = d_i g(\mathbf{x}_i; \hat{\lambda})$ where $d_i = \pi_i^{-1}$ and $\hat{\lambda}$ is determined from (8.9). Assume that $\hat{\lambda}$ converges in probability to λ_0 where $g(\mathbf{x}_i; \lambda_0) = 1$. The choice of $g(\mathbf{x}_i; \hat{\lambda}) = 1 + \mathbf{x}_i' \hat{\lambda}$ leads to the regression estimator with weights

$$w_i = d_i \left\{ 1 + (\mathbf{X} - \hat{\mathbf{X}})' (\sum_{i \in A} d_i \mathbf{x}_i \mathbf{x}_i')^{-1} \mathbf{x}_i \right\}.$$

With the weights always being positive, the exponential tilting calibration estimator, discussed in Kim (2010), uses $g(\mathbf{x}_i; \hat{\lambda}) = \exp(\mathbf{x}_i' \hat{\lambda})$ and is asymptotically equivalent to the regression estimator.

The following theorem, originally proved by Kim and Park (2010), presents some asymptotic properties of the calibration estimator.

Theorem 8.1. *Let $\hat{Y}_{cal} = \sum_{i \in A} w_i y_i$ where $w_i = d_i g(\mathbf{x}_i; \hat{\lambda})$ and $\hat{\lambda}$ is the unique solution to (8.9). Assume that $\hat{\lambda}$ converges in probability to λ_0 where $g(\mathbf{x}_i; \lambda_0) = 1$. Under some regularity conditions, the calibration estimator is asymptotically equivalent to*

$$\hat{Y}_{IV} = \sum_{i \in A} d_i y_i + \left\{ \mathbf{X} - \sum_{i \in A} d_i \mathbf{x}_i \right\}' B_z, \tag{8.10}$$

where the subscript "IV" stands for "instrumental variable" and

$$B_z = \left(\sum_{i=1}^N \mathbf{z}_i \mathbf{x}_i' \right)^{-1} \sum_{i=1}^N \mathbf{z}_i y_i \tag{8.11}$$

with $\mathbf{z}_i = \partial g(\mathbf{x}_i; \lambda)/\partial \lambda$ evaluated at $\lambda = \lambda_0$.

Proof. Consider

$$\hat{Y}_B(\lambda) = \sum_{i \in A} d_i g(\mathbf{x}_i; \lambda) y_i + \left\{ \mathbf{X} - \sum_{i \in A} d_i g(\mathbf{x}_i; \lambda) \mathbf{x}_i \right\}' B$$

where B is a p-dimensional vector. Note that we have $\hat{Y}_B(\hat{\lambda}) = \hat{Y}_{cal}$ for any choice of B. To find the particular choice B^* of B such that $\hat{Y}_{B^*}(\hat{\lambda})$ is asymptotically equal to $\hat{Y}_{B^*}(\lambda_0)$, by the theory of Randles (1982), we only have to find B that satisfies

$$E\left\{ \partial \hat{Y}_B(\hat{\lambda})/\partial \lambda \right\} = 0. \tag{8.12}$$

Thus, the choice of $B = B_z$ in (8.11) satisfies (8.12) and the asymptotic equivalence holds as $g(\mathbf{x}_i;\lambda_0) = 1$. □

By Theorem 8.1, the calibration estimator is consistent and has asymptotic variance

$$V\left(\hat{Y}_{cal} \mid \mathcal{F}_N\right) \cong V\left\{\sum_{i\in A} d_i\left(y_i - \mathbf{x}_i' B_z\right) \mid \mathcal{F}_N\right\}, \tag{8.13}$$

where the expectation conditional on $\mathcal{F}_N$ refers to the expectation with respect to the sampling mechanism. The consistency does not depend on the validity of the outcome regression model such as (8.5). However, the variance in (8.13) will be small if the regression model holds for the finite population at hand. Thus, the estimator is model-assisted, not model-dependent, in the sense that the regression model is used only to improve the efficiency. Model assisted estimation is very popular in sample surveys. If we have $\hat{B}_z = \left(\sum_{i\in A} d_i \mathbf{z}_i \mathbf{x}_i'\right)^{-1} \sum_{i\in A} d_i \mathbf{z}_i y_i$ instead of the B_z in (8.11), the resulting estimator is called the *instrumental-variable calibration estimator*, with $\mathbf{z}_i$ being the instrumental variable.

For variance estimation, writing (8.13) as

$$V\left(\hat{Y}_{cal} \mid \mathcal{F}_N\right) \cong V\left\{\sum_{i\in A} d_i g(\mathbf{x}_i;\lambda_0)\left(y_i - \mathbf{x}_i' B_z\right) \mid \mathcal{F}_N\right\} \tag{8.14}$$

and applying the standard variance formula to $\hat{\eta}_i = g(\mathbf{x}_i;\hat{\lambda})\left(y_i - \mathbf{x}_i'\hat{B}_z\right)$ will provide a consistent variance estimator.

If the outcome regression model is nonlinear, e.g. $E(y_i \mid \mathbf{x}_i) = m(\mathbf{x}_i;\beta)$ for some nonlinear function m, we can directly apply the predicted value, $\hat{m}_i = m(\mathbf{x}_i;\hat{\beta})$, to obtain the prediction estimator

$$\hat{Y}_p = \sum_{i=1}^{N} \hat{m}_i,$$

which does not necessarily satisfy design consistency. Here, design consistency means convergence in probability to the target parameter under the sampling mechanism. To achieve design consistency, the following bias-corrected prediction estimator

$$\hat{Y}_{p,bc} = \sum_{i=1}^{N} \hat{m}_i + \sum_{i\in A} d_i\left(y_i - \hat{m}_i\right)$$

can be used. The above bias-corrected prediction estimator is design consistent and has asymptotic variance

$$V\left(\hat{Y}_{p,bc} \mid \mathcal{F}_N\right) \cong V\left[\sum_{i\in A} d_i\left\{y_i - m(\mathbf{x}_i;\beta)\right\} \mid \mathcal{F}_N\right].$$

That is, the effect of $\hat{\beta}$ in $\hat{m}_i = m(\mathbf{x}_i;\hat{\beta})$ can be ignored in variance estimation.

8.3 Propensity score weighting method

We now consider the case of unit nonresponse in survey sampling. Assume that $\mathbf{x}_i$ is observed throughout the sample and y_i is observed only if $\delta_i = 1$. We assume that the response mechanism does not depend on y. Thus, we assume that

$$Pr(\delta = 1 \mid \mathbf{x}, y) = Pr(\delta = 1 \mid \mathbf{x}) = p(\mathbf{x};\phi_0) \tag{8.15}$$

for some unknown vector ϕ_0. The first equality implies that the data are missing-at-random (MAR) in the population model, a model for the finite population. Or, one may assume that

$$Pr(\delta = 1|\mathbf{x}, y, I = 1) = Pr(\delta = 1|\mathbf{x}, I = 1), \tag{8.16}$$

which can be called *sample MAR (SMAR)*, while condition (8.15) can be called *population MAR (PMAR)*. Unless the sampling design is non-informative, the two MAR conditions, (8.15) and (8.16), are different. In survey sampling, assumption (8.15) is more appropriate because an individual's decision on whether or not to respond to a survey depends on his or her own characteristics.

Given the response model (8.15), a consistent estimator of ϕ_0 can be obtained by solving

$$\hat{U}_h(\phi) \equiv \sum_{i \in A} d_i \left\{ \frac{\delta_i}{p(\mathbf{x}_i; \phi)} - 1 \right\} \mathbf{h}(\mathbf{x}_i; \phi) = \mathbf{0} \tag{8.17}$$

for some $\mathbf{h}(\mathbf{x}; \phi)$ such that $\partial \hat{U}_h(\phi)/\partial \phi$ is of full rank. If we choose

$$\mathbf{h}(\mathbf{x}_i; \phi) = p(\mathbf{x}_i; \phi)\{\partial \text{logit} p(\mathbf{x}_i; \phi)/\partial \phi\},$$

then (8.17) is equal to the design-weighted score equation for ϕ.

Once $\hat{\phi}_h$ is computed from (8.17), the propensity score adjusted (PSA) estimator of $Y = \sum_{i=1}^N y_i$ is given by

$$\hat{Y}_{PSA} = \sum_{i \in A_R} d_i g(\mathbf{x}_i; \hat{\phi}_h) y_i, \tag{8.18}$$

where $A_R = \{i \in A; \delta_i = 1\}$ is the set of respondents and $g(\mathbf{x}_i; \hat{\phi}_h) = \{p(\mathbf{x}_i; \hat{\phi}_h)\}^{-1}$. Afterward, we can apply the argument of Theorem 8.1 to show that $\hat{Y}_{PSA}$ is asymptotically equivalent to

$$\begin{aligned} \tilde{Y}_{PSA} &= \sum_{i \in A_R} d_i g(\mathbf{x}_i; \phi_0) y_i + \left\{ \sum_{i \in A} d_i \mathbf{h}_i - \sum_{i \in A_R} d_i g(\mathbf{x}_i; \phi_0) \mathbf{h}_i \right\}' B_z \\ &= \sum_{i \in A_R} d_i \{p(\mathbf{x}_i; \phi_0)\}^{-1} y_i + \left\{ \sum_{i \in A} d_i \mathbf{h}_i - \sum_{i \in A_R} d_i \{p(\mathbf{x}_i; \phi_0)\}^{-1} \mathbf{h}_i \right\}' B_z, \end{aligned} \tag{8.19}$$

where

$$B_z = \left(\sum_{i=1}^N \delta_i \mathbf{z}_i \mathbf{h}_i' \right)^{-1} \sum_{i=1}^N \delta_i \mathbf{z}_i y_i$$

and $\mathbf{z}_i = \partial g(\mathbf{x}_i; \phi)/\partial \phi$ evaluated at $\phi = \phi_0$. Thus, the asymptotic variance is equal to

$$\begin{aligned} V\left(\tilde{Y}_{PSA} \mid \mathcal{F}_N\right) &= V\left(\hat{Y}_{HT} \mid \mathcal{F}_N\right) + V\left\{ \sum_{i \in A_R} d_i p_i^{-1} (y_i - \mathbf{h}_i' B_z) \mid \mathcal{F}_N \right\} \\ &= V\left(\hat{Y}_{HT} \mid \mathcal{F}_N\right) + E\left\{ \sum_{i \in A} d_i^2 (p_i^{-1} - 1)(y_i - \mathbf{h}_i' B_z)^2 \mid \mathcal{F}_N \right\}, \end{aligned}$$

where $p_i = p(\mathbf{x}_i; \phi_0)$ and the second equality follows from independence among δ_i's. Note that

$$\begin{aligned} & E\left\{ \sum_{i \in A} d_i^2 (p_i^{-1} - 1)(y_i - \mathbf{h}_i' B_z)^2 \right\} \\ = \; & E\left[\sum_{i \in A} d_i^2 (p_i^{-1} - 1) \{y_i - E(y_i \mid \mathbf{x}_i) + E(y_i \mid \mathbf{x}_i) - \mathbf{h}_i' B_z\}^2 \right] \\ = \; & E\left[\sum_{i \in A} d_i^2 (p_i^{-1} - 1) \{y_i - E(y_i \mid \mathbf{x}_i)\}^2 \right] + E\left[\sum_{i \in A} d_i^2 (p_i^{-1} - 1) \{E(y_i \mid \mathbf{x}_i) - \mathbf{h}_i' B_z\}^2 \right] \end{aligned}$$

and the cross-product term is zero because $y_i - E(y_i \mid \mathbf{x}_i)$ is conditionally unbiased for zero, conditional on $\mathbf{x}_i$ and A. Thus, we have

$$V\left(\tilde{Y}_{PSA} \mid \mathcal{F}_N\right) \geq V_l \equiv V\left(\hat{Y}_{HT} \mid \mathcal{F}_N\right) + E\left[\sum_{i\in A} d_i^2 (p_i^{-1} - 1)\{y_i - E(y_i \mid \mathbf{x}_i)\}^2 \mid \mathcal{F}_N\right]. \tag{8.20}$$

Kim and Riddles (2012) established (8.20) and showed that the equality in (8.20) holds if $\hat{\phi}_h$ satisfies

$$\sum_{i\in A} d_i \left\{\frac{\delta_i}{p(\mathbf{x}_i;\phi)} - 1\right\} E(Y|\mathbf{x}_i) = 0. \tag{8.21}$$

Any PSA estimator that has the asymptotic variance V_l in (8.20) is optimal in the sense that it achieves the lower bound of the asymptotic variance among the class of PSA estimators with $\hat{\phi}_h$ satisfying (8.17). The PSA estimator using the maximum likelihood estimator of ϕ_0 does not necessarily achieve the lower bound of the asymptotic variance.

Condition (8.21) provides a way of constructing an optimal PSA estimator. First, we need an assumption for $E(Y|\mathbf{x})$, which is often called the *outcome regression model*. If the outcome regression model is a linear regression model of the form $E(Y|\mathbf{x}) = \beta_0 + \beta_1'\mathbf{x}$, an optimal PSA estimator of θ can be obtained by solving

$$\sum_{i\in A} d_i \frac{\delta_i}{p_i(\phi)} (1, \mathbf{x}_i) = \sum_{i\in A} d_i (1, \mathbf{x}_i). \tag{8.22}$$

Condition (8.22) is appealing because it says that the PSA estimator applied to $y = a + \mathbf{b}'\mathbf{x}$ leads to the original HT estimator. Condition (8.22) is called the *calibration condition* in survey sampling. The calibration condition applied to $\mathbf{x}$ makes full use of the information contained in it if the study variable y is well approximated by a linear function of $\mathbf{x}$.

We now discuss variance estimation of PSA estimators of the form (8.18) where $\hat{p}_i = p_i(\hat{\phi})$ is constructed to satisfy (8.17). By (8.19), we can write

$$\hat{Y}_{PSA} = \sum_{i\in A} d_i \eta_i(\phi_0) + o_p\left(n^{-1/2}N\right), \tag{8.23}$$

where

$$\eta_i(\phi) = \mathbf{h}_i B_z + \frac{\delta_i}{p_i(\phi)} (y_i - \mathbf{h}_i' B_z). \tag{8.24}$$

To derive the variance estimator, we assume that the variance estimator $\hat{V} = \sum_{i\in A}\sum_{j\in A} \Omega_{ij} q_i q_j$ satisfies $\hat{V}/V(\hat{q}_{HT}|\mathcal{F}_N) = 1 + o_p(1)$ for some Ω_{ij} related to the joint inclusion probability, where $\hat{q}_{HT} = \sum_{i\in A} d_i q_i$ for any q with a finite fourth moment.

To obtain the total variance, the *reverse framework* of Fay (1992), Shao and Steel (1999), and Kim and Rao (2009) is considered. In this framework, the finite population is divided into two groups, a population of respondents and a population of nonrespondents, so the response indicator is extended to the entire population as $\mathcal{R}_N = \{\delta_1, \delta_2, \cdots, \delta_N\}$. Given the population, the sample A is selected according to a probability sampling design. Then, we have both respondents and nonrespondents in the sample A. The total variance of $\hat{\eta}_{HT} = \sum_{i\in A} d_i \eta_i$ can be written as

$$V(\hat{\eta}_{HT}|\mathcal{F}_N) = V_1 + V_2 = E\{V(\hat{\eta}_{HT}|\mathcal{F}_N, \mathcal{R}_N)|\mathcal{F}_N\} + V\{E(\hat{\eta}_{HT}|\mathcal{F}_N, \mathcal{R}_N)|\mathcal{F}_N\}. \tag{8.25}$$

The conditional variance term $V(\hat{\eta}_{HT}|\mathcal{F}_N, \mathcal{R}_N)$ in (8.25) can be estimated by

$$\hat{V}_1 = \sum_{i\in A}\sum_{j\in A} \Omega_{ij} \hat{\eta}_i \hat{\eta}_j, \tag{8.26}$$

where $\hat{\eta}_i = \eta_i(\hat{\phi})$ is defined in (8.24) with B_z replaced by a consistent estimator such as

$$\hat{B}_z = \left(\sum_{i\in A_R} d_i \hat{\mathbf{z}}_i \mathbf{h}_i'\right)^{-1} \sum_{i\in A_R} d_i \hat{\mathbf{z}}_i y_i$$

and $\hat{\mathbf{z}}_i = \mathbf{z}(\mathbf{x}_i;\hat{\phi})$ is the value of $\mathbf{z}_i = \partial g(\mathbf{x}_i;\phi)/\partial\phi$ evaluated at $\phi = \hat{\phi}$. To show that $\hat{V}_1$ is also consistent for V_1 in (8.25), it suffices to show that $V\{nN^{-2}\cdot V(\hat{\eta}_{HT}|\mathcal{F}_N,\mathcal{R}_N)|\mathcal{F}_N\} = o(1)$, which follows by some regularity conditions on the first- and the second-order inclusion probabilities and the existence of the fourth moment. See Kim et al. (2006b). The second term V_2 in (8.25) is

$$\begin{aligned} V\{E(\hat{\eta}_{HT}|\mathcal{F}_N,\mathcal{R}_N)|\mathcal{F}_N\} &= V\left(\sum_{i=1}^{N}\eta_i|\mathcal{F}_N\right) \\ &= \sum_{i=1}^{N}\frac{1-p_i}{p_i}\left(y_i-\mathbf{h}_i'B_z\right)^2. \end{aligned}$$

A consistent estimator of V_2 can be derived as

$$\hat{V}_2 = \sum_{i\in A_R} d_i\frac{1-\hat{p}_i}{\hat{p}_i^2}\left(y_i-\mathbf{h}_i'\hat{B}_z\right)^2. \tag{8.27}$$

Therefore,

$$\hat{V}\left(\hat{Y}_{PSA}\right) = \hat{V}_1+\hat{V}_2, \tag{8.28}$$

is consistent for the variance of the PSA estimator defined in (8.18) with $\hat{p}_i = p_i(\hat{\phi})$ satisfying (8.17), where $\hat{V}_1$ is in (8.26) and $\hat{V}_2$ is in (8.27).

Note that the first term of the total variance is $V_1 = O_p(n^{-1}N^2)$, but the second term is $V_2 = O_p(N)$. Thus, when the sampling fraction nN^{-1} is negligible, that is, $nN^{-1} = o(1)$, the second term V_2 can be ignored and $\hat{V}_1$ is a consistent estimator of the total variance. Otherwise, the second term V_2 should be taken into consideration so that a consistent variance estimator can be constructed as in (8.28).

In practice, we may have other auxiliary variables that are observed throughout the population. The auxiliary information can be used to construct a set of calibration weights as discussed in Section 8.2. In this case, the parameter for the response propensity can be computed by solving

$$\sum_{i\in A} w_i\left\{\frac{\delta_i}{p(\mathbf{x}_i;\phi)}-1\right\}\mathbf{h}(\mathbf{x}_i;\phi) = \mathbf{0} \tag{8.29}$$

instead of solving (8.17), where w_i are the calibration weights. Once $\hat{\phi}_h$ is computed from (8.29), the PSA estimator $\hat{Y}_{PSA,w}$ is asymptotically equivalent to

$$\tilde{Y}_{PSA,w} = \sum_{i\in A_R} w_i\{p(\mathbf{x}_i;\phi_0)\}^{-1}y_i+\left\{\sum_{i\in A}w_i\mathbf{h}_i-\sum_{i\in A_R}w_i\{p(\mathbf{x}_i;\phi_0)\}^{-1}\mathbf{h}_i\right\}'B_z, \tag{8.30}$$

where $B_z = \left(\sum_{i=1}^{N}\delta_i\mathbf{z}_i\mathbf{h}_i'\right)^{-1}\sum_{i=1}^{N}\delta_i\mathbf{z}_iy_i$ and $\mathbf{z}_i = \partial g(\mathbf{x}_i;\phi)/\partial\phi$ evaluated at $\phi = \phi_0$. Writing

$$\tilde{Y}_{PSA,w} = \sum_{i\in A} w_i\eta_i,$$

where $\eta_i = \mathbf{h}_i'B_z+\delta_i\{p(\mathbf{x}_i;\phi_0)\}^{-1}(y_i-\mathbf{h}_i'B_z)$, we can apply the theory of Section 8.2 to compute the variance of $\tilde{Y}_{PSA,w}$. For example, if the calibration weights are computed to satisfy

$$\sum_{i\in A} w_i\mathbf{x}_{1i} = \sum_{i=1}^{N}\mathbf{x}_{1i},$$

then

$$V\left(\sum_{i\in A}w_i\eta_i \mid \mathcal{F}_N\right) = V\left\{\sum_{i\in A}d_i(\eta_i-\mathbf{x}_{1i}'B_{x\eta}) \mid \mathcal{F}_N\right\}$$

where $B_{x\eta}$ is computed by B_z in (8.11) with $\mathbf{x}_i$ and y_i replaced by $\mathbf{x}_{1i}$ and η_i, respectively.

8.4 Fractional imputation

We now consider item nonresponse in sample surveys. Imputation is a popular technique for handling item nonresponse. Filling in missing values would enable valid comparisons of different analyses as they all start from the same (complete) dataset. Imputation also makes full use of the information in the partial responses. For example, domain estimation after imputation provides more efficient estimates than direct estimation because imputation borrows strength from observations outside the domains.

We first assume that x_i is observed throughout the sample and y_i is subject to missingness. A natural approach is to obtain a model for the conditional distribution of y_i given x_i and generate imputed values from the conditional distribution. In survey sampling, two models need to be distinguished. The population model refers to the original distribution that generates the population and the sample model refers to the conditional distribution of the sample data given that they are selected in the sample. That is, we have

$$f_s(y \mid x) = f_p(y \mid x) \frac{Pr(I = 1 \mid x, y)}{Pr(I = 1 \mid x)},$$

where $f_s(\cdot)$ is the density for the sample distribution and $f_p(\cdot)$ is the density for the population distribution. Unless the sample design is noninformative in the sense that it satisfies (8.2), the two models are not the same.

Generally speaking, we are interested in the parameters of the population model. However, in terms of generating imputed values, one may use the sample distribution and generate imputed values from the conditional distribution $f(y \mid x, I = 1, \delta = 0)$. To compute the conditional distribution, we often assume that

$$f(y \mid x, I = 1, \delta = 1) = f(y \mid x, I = 1, \delta = 0) \tag{8.31}$$

and generate imputed values from $f(y \mid x, I = 1, R = 1)$, which can be easily estimated from the observed data. Condition (8.31) is loosely referred to as *missing at random (MAR)*, a term given by Rubin (1976) in the context of simple random sampling. Strictly speaking, the MAR condition in (8.31) is stated under the realized sample and is different from

$$f(y \mid x, \delta = 1) = f(y \mid x, \delta = 0). \tag{8.32}$$

To distinguish the two concepts, we shall call condition (8.31) the *sample MAR (SMAR)*. Condition (8.32), called the population MAR (PMAR), is the classical MAR condition in survey sampling literature. We will assume the PMAR condition in this section, because SMAR may not hold when the sampling design is non-informative. On the other hand, multiple imputation of Rubin (1987) was developed under the SMAR assumption.

When x_i is always observed and PMAR holds, then the imputed value of y_i can be generated from $f(y_i \mid x_i)$, the population model of y_i given x_i. When estimating the parameters in the population model, we need to use the sampling weights because the sampling design can be informative. That is, we use

$$\sum_{i \in A} w_i \delta_i S(\theta; y_i, x_i) = 0 \tag{8.33}$$

to estimate θ in $f(y \mid x; \theta)$, where w_i is the sampling weight of unit i such that $\sum_{i \in A} w_i y_i$ is a design-consistent estimator of Y. Let $y_{i1}^*, \cdots, y_{im}^*$ be m imputed values from $f\left(y \mid x_i; \hat{\theta}\right)$ generated from a proposal distribution $f_0(y \mid x)$. The choice of the proposal distribution is somewhat arbitrary. If we do not have a good guess about θ, we may use

$$f_0(y \mid x) = \hat{f}(y \mid \delta = 1) = \frac{\sum_{i \in A} w_i \delta_i I(y_i = y)}{\sum_{i \in A} w_i \delta_i},$$

which estimates the marginal distribution of y_i using the set of respondents. If x is categorical, then we can use

$$f_0(y \mid x) = \frac{\sum_{i \in A} w_i \delta_i I(x_i = x, y_i = y)}{\sum_{i \in A} w_i \delta_i I(x_i = x)}.$$

For continuous x, we may use a Kernel-type proposal distribution

$$f_0(y \mid x) = \frac{w_i \delta_i K_h(x_i, x) K_h(y_i, y)}{\sum_{i \in A} w_i \delta_i K_h(x_i, x)}.$$

To generate m imputed values from $f_0(y \mid x_i)$, one can use the following systematic sampling algorithm:

1. Generate $u_1 \sim U(0, 1/m)$.
2. Compute $u_k = u_1 + (k-1)/m$ for $k = 2, \cdots, m$.
3. For each j, choose
$$y_{ij}^* = F_0^{-1}(u_j \mid x_i), \tag{8.34}$$
where $F_0(y \mid x) = \sum_{y_i < y} f_0(y_i \mid x) / \{\sum_i f_0(y_i \mid x)\}$ is the cumulative distribution function derived from $f_0(y \mid x)$.

In practice, to remove the discontinuity points of F_0, we use the interpolation technique when computing $F_0(y \mid x)$. That is, we can express the interpolated cumulative distribution function (CDF) $\tilde{F}_0(y \mid x)$ as

$$\tilde{F}_0(y \mid x) = F_0(y_{(i)} \mid x) + (y - y_{(i)}) \frac{F_0(y_{(i+1)} \mid x) - F_0(y_{(i)} \mid x)}{y_{(i+1)} - y_{(i)}} \quad \text{if } y_{(i)} \le y < y_{(i+1)}.$$

The interpolated CDF can be used in (8.34).

The fractional weight associated with y_{ij}^* is computed as

$$w_{ij0}^* = \frac{f(y_{ij}^* \mid x_i; \hat{\theta}) / f_0(y_{ij}^* \mid x_i)}{\sum_{k=1}^m f(y_{ik}^* \mid x_i; \hat{\theta}) / f_0(y_{ik}^* \mid x_i)}.$$

When m is small, the fractional weights can be further modified in the calibration step. The proposed calibration equation for improving the fractional weights in this case is

$$\sum_{i \in A} \sum_{j=1}^m w_i (1 - \delta_i) w_{ij}^* S(\hat{\theta}; x_i, y_{ij}^*) = 0 \tag{8.35}$$

and $\sum_{j=1}^m w_{ij}^* = 1$ for each i with $\delta_i = 0$, where $\hat{\theta}$ is computed from (8.33). Using the idea of regression weighting, the final calibration fractional weights can be computed by

$$w_{ij}^* = w_{ij0}^* + w_{ij0}^* \Delta \left(S_{ij}^* - \bar{S}_{i\cdot}^* \right), \tag{8.36}$$

where $S_{ij}^* = S(\hat{\theta}; x_i, \mathbf{y}_{ij}^*)$, $\bar{S}_{i\cdot}^* = \sum_{j=1}^m w_{ij0}^* S_{ij}^*$, and

$$\Delta = -\left\{ \sum_{i \in A} w_i (1 - \delta_i) \sum_{j=1}^m w_{ij0}^* S_{ij}^* \right\}' \left[\sum_{i \in A} w_i (1 - \delta_i) \sum_{j=1}^m w_{ij0}^* \left(S_{ij}^* - \bar{S}_{i\cdot}^* \right)^{\otimes 2} \right]^{-1}.$$

See also (4.97) and its related discussion.

The calibration condition (8.35) guarantees that the imputed score equation leads to the same $\hat{\theta}$, computed from (8.33). Once the FI data are created, the fractionally imputed estimator of $Y = \sum_{i=1}^N y_i$ is obtained by

$$\hat{Y}_{FI} = \sum_{i \in A} w_i \left\{ \delta_i y_i + (1 - \delta_i) \sum_{j=1}^m w_{ij}^* y_{ij}^* \right\}.$$

For variance estimation, a replication method can be used. Let $w_i^{(k)}$ be the k-th replication weights such that

$$\hat{V}_{rep} = \sum_{k=1}^{L} c_k (\hat{Y}^{(k)} - \hat{Y})^2$$

is consistent for the variance of $\hat{Y} = \sum_{i \in A} w_i y_i$, where L is the replication size, c_k is the k-th replication factor that depends on the replication method and the sampling mechanism, and $\hat{Y}^{(k)} = \sum_{i \in A} w_i^{(k)} y_i$ is the k-th replicate of $\hat{Y}$.

To apply the replication method to fractional imputation, we first apply the replication weights $w_i^{(k)}$ in (8.33) to compute $\hat{\theta}^{(k)}$. Once $\hat{\theta}^{(k)}$ is obtained, we use the same imputed values to compute the initial replication fractional weights

$$w_{ij0}^{*(k)} = \frac{f(y_{ij}^* \mid x_i; \hat{\theta}^{(k)}) / f_0(y_{ij}^* \mid x_i)}{\sum_{l=1}^{m} f(y_{il}^* \mid x_i; \hat{\theta}^{(k)}) / f_0(y_{il}^* \mid x_i)}$$

and then apply the same calibration

$$\sum_{i \in A} \sum_{j=1}^{m} w_i^{(k)} (1 - \delta_i) w_{ij}^{*(k)} S(\hat{\theta}^{(k)}; x_i, y_{ij}^*) = 0 \tag{8.37}$$

and $\sum_{j=1}^{m} w_{ij}^{*(k)} = 1$, to obtain the final replicate fractional weights $w_{ij}^{*(k)}$. Via the regression weighting method, the final replicate fractional weights are computed similarly to (8.36) using $\hat{\theta}^{(k)}$ and the replicated weights.

Afterwards,

$$\hat{Y}_{FI}^{(k)} = \sum_{i \in A} w_i^{(k)} \left\{ \delta_i y_i + (1 - \delta_i) \sum_{j=1}^{m} w_{ij}^{*(k)} y_{ij}^* \right\}$$

can be used to compute the replication variance estimator

$$\hat{V}_{rep}(\hat{Y}_{FI}) = \sum_{k=1}^{L} c_k (\hat{Y}_{FI}^{(k)} - \hat{Y}_{FI})^2.$$

The replication method is very useful for multipurpose estimation. For example, if another parameter of interest is $\Psi = Pr(Y < 3)$, then the FI estimator of Ψ is

$$\hat{\Psi}_{FI} = \sum_{i \in A} w_i \left\{ \delta_i I(y_i < 3) + (1 - \delta_i) \sum_{j=1}^{m} w_{ij}^* I(y_{ij}^* < 3) \right\}$$

and its replication variance estimator is

$$\hat{V}_{rep}(\hat{\Psi}_{FI}) = \sum_{k=1}^{L} c_k (\hat{\Psi}_{FI}^{(k)} - \hat{\Psi}_{FI})^2,$$

where

$$\hat{\Psi}_{FI}^{(k)} = \sum_{i \in A} w_i^{(k)} \left\{ \delta_i I(y_i < 3) + (1 - \delta_i) \sum_{j=1}^{m} w_{ij}^{*(k)} I(y_{ij}^* < 3) \right\}.$$

Example 8.2. *Suppose that a sample of bivariate (x_i, y_i) are generated from*

$$y_i = \beta_0 + \beta_1 x_i + e_i, \tag{8.38}$$

with $e_i \sim N(0, \sigma^2)$. Assume that we observe (x_i, y_i) for $\delta_i = 1$ and observe x_i only if $\delta_i = 0$. Assume that the response mechanism is PMAR.

To implement a fractional imputation with $m = 10$ under this setup, we can use the following steps:

[Step 1] Obtain a fully efficient estimator of $\theta = (\beta_0, \beta_1, \sigma^2)$ *by solving*

$$\sum_{i \in A} w_i \delta_i S(\theta; x_i, y_i) = 0,$$

where $S(\theta; x, y)$ *is the score function for the conditional distribution in (8.38). Under the normal distribution, we have*

$$S(\theta; x, y) = \begin{bmatrix} (y - \beta_0 - \beta_1 x)/\sigma^2 \\ (y - \beta_0 - \beta_1 x)x/\sigma^2 \\ \{(y - \beta_0 - \beta_1 x)^2 - \sigma^2\}/(2\sigma^4) \end{bmatrix}.$$

[Step 2] For each missing unit i, first generate $m_1 >> m = 10$, *(say,* $m_1 = 1,000$*) imputed values of* y_i *from* $f(y \mid x_i; \hat{\theta})$, *where* $\hat{\theta}$ *is obtained from [Step 1].*

[Step 3] Among the m_1 *imputed values of* y_i, *denoted by* $y_{i1}^*, \cdots, y_{i,m_1}^*$, *select a subsample of size m using an efficient sampling design. One simple way to obtain an efficient sample is to do systematic sampling from the population of* $\{y_{i1}^*, \cdots, y_{i,m_1}^*\}$, *sorted by the ascending order (or half-ascending and half descending order).*

[Step 4] Use the imputed values selected from [Step 3] to find the calibration weights that satisfy $\sum_{j=1}^m w_{ij}^{*(k)} = 1$ *and (8.35).*

For variance estimation, the replication method described above can be used. In this case, the imputed values are not changed for each replication. Only the fractional weights are changed. In the k-th replication, the replicate $\hat{\theta}^{(k)}$ *is first computed by solving*

$$\sum_{i \in A} w_i^{(k)} \delta_i S(\theta; x_i, y_i) = 0.$$

Using $\hat{\theta}^{(k)}$, *the initial replication fractional weights are computed by*

$$w_{ij0}^{*(k)} = \frac{f(y_{ij}^* \mid x_i; \hat{\theta}^{(k)}) / f(y_{ij}^* \mid x_i; \hat{\theta})}{\sum_{s=1}^m f(y_{is}^* \mid x_i; \hat{\theta}^{(k)}) / f(y_{is}^* \mid x_i; \hat{\theta})}.$$

The replication fractional weights are further modified to satisfy $\sum_{j=1}^m w_{ij}^* = 1$ *and (8.37).*

We performed a limited simulation study under the setup of Example 8.2, where $x_i \sim N(2,1)$, y_i follows from (8.38) with $(\beta_0, \beta_1, \sigma^2) = (1, 0.7, 1)$, and $\delta_i \sim \text{Bernoulli}(p_i)$ with $\text{logit}(p_i) = 0.5x_i$. The overall response rate is about 72%. The sample size is $n = 200$.

Under this setup, the fractionally imputed estimator of three parameters are computed with $m = 10$. The parameters are $\eta_1 = E(Y)$, $\eta_2 = \beta_1$ and $\eta_3 = Pr(Y < 3)$. In addition to fractional imputation, multiple imputation (MI) is also used with $m = 10$. Table 8.1 presents the biases and variances of the point estimators and the relative bias of the variance estimators, based on $B = 2,000$ Monte Carlo samples. The standardized variance is computed by dividing the variance of the current estimator by that of the complete sample estimator, multiplied by 100.

Table 8.1 shows that the fractional imputation estimator is more efficient than the multiple imputation estimator. The efficiencies of the fractional imputation estimator relative to the multiple imputation estimator are 1.02, 1.03, and 1.01 for estimation of η_1, η_2 and η_3, respectively. Fractional imputation is slightly more efficient than multiple imputation because of the calibration step in (8.35). Variance estimators of the fractionally imputed estimators are almost unbiased in the simulation. Multiple imputation also features unbiased variance estimation, except for η_3. Because the complete sample estimator of η_3 is the sample proportion, the congeniality condition of Meng (1994) does not hold and the MI variance estimator is biased. See Example 4.8.

Table 8.1 Monte Carlo biases and variances of the point estimators and the relative biases of the variance estimators under the setup of Example 8.2

Parameter	Method	$Bias(\hat{\eta})$	$V(\hat{\eta})$	Std Var.	R.B. $(\hat{V})$
η_1	Complete sample	0.00	0.00751	100	-0.005
	Fractional imputation	0.00	0.00986	131	-0.024
	Multiple imputation	0.00	0.01004	134	-0.018
η_2	Complete sample	0.00	0.00510	100	0.005
	Fractional imputation	0.00	0.00757	148	-0.002
	Multiple imputation	0.00	0.00780	153	-0.000
η_3	Complete sample	0.00	0.001096	100	-0.022
	Fractional imputation	0.00	0.001170	107	-0.014
	Multiple imputation	0.00	0.001177	107	0.149

8.5 Fractional hot deck imputation

Hot deck imputation creates imputed values from realized observations. That is, no artificial values are created. The hot deck imputation method is very popular in household surveys. Fractional hot deck imputation was proposed by Kalton and Kish (1984) as a way of achieving efficient hot deck imputation. Kim and Fuller (2004) and Fuller and Kim (2005) provided a rigorous treatment of fractional hot deck imputation and discussed variance estimation. However, their approach is not directly applicable to multivariate missing data. Hot deck imputation for multivariate missing data with an arbitrary missing pattern is challenging because it is difficult to preserve the covariance structure in the data after hot deck imputation.

Let $\mathbf{y}$ be a K-dimensional vector of study variables and let $(\mathbf{y}_{i,\text{obs}}, \mathbf{y}_{i,\text{mis}})$ be the (observed, missing) part of $\mathbf{y}_i$. To study multivariate fractional hot deck imputation, first consider the simple case of categorical $\mathbf{y}$. In this case, under MAR, the joint distribution of $\mathbf{y}$ can be computed by the EM algorithm with fractional imputation implemented, as discussed in Example 4.10. In this case, the fractional weights

$$w_{ij}^* = \frac{\pi(\mathbf{y}_{i,\text{obs}}, \mathbf{y}_{i,\text{mis}}^{*(j)})}{\sum_k \pi(\mathbf{y}_{i,\text{obs}}, \mathbf{y}_{i,\text{mis}}^{*(k)})} \tag{8.39}$$

are assigned for all possible values of $\mathbf{y}_{i,mis}$. If a fixed number m is used for imputation, then a fractional imputation of size m can be constructed by a PPS (Probability Proportional to Size) sampling of m values from all possible values of $\mathbf{y}_{i,\text{mis}}$, where the selection probability for the j-th enumeration of $\mathbf{y}_{i,\text{mis}}$ is w_{ij}^* in (8.39). The resulting FI estimator will have equal fractional weights $w_{ij}^* = 1/m$.

For continuous $\mathbf{y}$ variables, we consider a discrete approximation. The basic step in the proposed imputation is temporary replacement of the original data by a discrete approximation that temporarily replaces the original data. Each continuous variable is transformed into a discrete variable by dividing the range into a small finite number of segments. Let $\tilde{Y}_{ki}$ denote the discrete version of Y_{ki}. One simple way of computing the discrete version is to divide the range into groups of equal length. Note that, if Y_k is observed then $\tilde{Y}_k$ is observed.

Once $\tilde{Y}_k$'s are constructed, we use the observed value of $\tilde{Y}_k$ to compute the joint probability of $(\tilde{Y}_1, \tilde{Y}_2, \cdots, \tilde{Y}_K)$. Let $\tilde{\pi}(\tilde{y}_1, \tilde{y}_2, \cdots, \tilde{y}_K)$ be the joint probability of obtaining $(\tilde{Y}_1, \tilde{Y}_2, \cdots, \tilde{Y}_K) = (\tilde{y}_1, \tilde{y}_2, \cdots, \tilde{y}_K)$. The joint probability can be obtained by the EM algorithm for partially classified categorical data using the partially observed value of $\tilde{y}_{ik}$'s.

In the EM algorithm, the E-step is essentially the same as applying the fully efficient fractional imputation (FEFI) method of Fuller and Kim (2005) using all possible combinations of imputed values. The imputed values for $\tilde{\mathbf{y}}_{i,\text{mis(i)}}$, where mis(i) is the index of nonresponding items in unit

i, are taken from the support of $\tilde{\mathbf{y}}_{\text{mis(i)}}$ matching the support of the respondents. Let $\tilde{\mathbf{y}}^{*(j)}_{mis(i)}, (j = 1, \cdots, M_i)$ be the set of possible values of $\tilde{\mathbf{y}}_{mis(i)}$ in the sample. Once the realized values of $\tilde{\mathbf{y}}_{mis(i)}$ are imputed, we can use the idea of EM by weighting (Ibrahim, 1990) to compute the fractional weights, as in Example 4.10. The fractional weight assigned to $\tilde{\mathbf{y}}_{i,\text{mis(i)}} = \tilde{\mathbf{y}}^{*(j)}_{\text{mis(i)}}$ at the t-th EM iteration is

$$\tilde{w}^*_{ij(t)} = \frac{\tilde{\pi}_t(\tilde{\mathbf{y}}_{i,\text{obs}}, \tilde{\mathbf{y}}^{*(j)}_{\text{mis(i)}})}{\sum_j \tilde{\pi}_t(\tilde{\mathbf{y}}_{i,obs}, \tilde{\mathbf{y}}^{*(j)}_{\text{mis(i)}})}, \tag{8.40}$$

where $\tilde{\pi}_t(\tilde{y}_1, \cdots, \tilde{y}_K)$ is the current value of the joint probabilities $P(\tilde{Y}_1 = \tilde{y}_1, \cdots, \tilde{Y}_K = \tilde{y}_K)$ evaluated at the t-th EM algorithm. If unit i has no missing data, then $\tilde{w}^*_{ij(t)} = 1$. Computing fractional weights in (8.40) corresponds to the E-step of the EM algorithm. Once the FEFI is constructed as above, the M-step is used to update the joint probabilities via the weighted average of the FEFI data using fractional weights. That is,

$$\tilde{\pi}_{t+1}(\tilde{y}_1, \cdots, \tilde{y}_K) = \left(\sum_{i \in A} w_i\right)^{-1} \sum_{i \in A} \sum_{j=1}^{M_i} w_i \tilde{w}^*_{ij(t)} I\left\{\tilde{y}^{*(j)}_{i1} = \tilde{y}_1, \cdots, \tilde{y}^{*(j)}_{iK} = \tilde{y}_K\right\}, \tag{8.41}$$

where w_i is the sampling weight for unit i and $\tilde{y}^*_{ijk}$ is the j-th imputed value for $\tilde{Y}_{ik}$. If $\tilde{Y}_{ik}$ is observed, then $\tilde{y}^*_{ijk}$ is the observed value.

Once the joint probabilities are computed, a set of imputed values of size m is constructed by doing PPS sampling with probability

$$w^*_{ij} = \frac{\tilde{\pi}(\tilde{\mathbf{y}}_{i,obs}, \tilde{\mathbf{y}}^{*(j)}_{i,mis})}{\sum_k \tilde{\pi}(\tilde{\mathbf{y}}_{i,obs}, \tilde{\mathbf{y}}^{*(k)}_{i,mis})}, \tag{8.42}$$

for each i in the sample. In PPS sampling, the size measures are proportional to the original fractional weights after accounting for certainty selection. That is, we first select the candidates with $w^*_{ij} > 1/m$ with certainty. Regression weighting can be used to preserve the marginal fractional weights. Because the values are all categorical, the raking ratio weighting method can be easily implemented.

Creating an imputed value $\tilde{y}^*_{ik}$ for y_{ik} is essentially creating an imputation cell for y_{ik}. Given the value of $\tilde{\mathbf{y}}^{*(j)}_{i,\text{mis}} = \tilde{\mathbf{y}}_{j,\text{mis(i)}}, (j = 1, 2, \cdots, m)$, we can perform a single hot deck imputation from the donor set of observed units with the same value of $\tilde{\mathbf{y}}_{j,\text{mis(i)}}$. Here, the value of $\tilde{\mathbf{y}}_{j,\text{mis(i)}}$ can be used as an imputation cell. If at least one donor is identified from the set of fully responding units with $\tilde{\mathbf{y}}_{k,\text{mis(i)}} = \tilde{\mathbf{y}}^{*(k)}_{i,\text{mis}}$, we can use the donor to obtain imputed values $\mathbf{y}^{*(j)}_{i,\text{mis}} = \mathbf{y}_{k,\text{mis(i)}}$ for missing $\mathbf{y}_{i,\text{mis}}$. If such a donor is not identifiable from the set of fully responding units, then we do hot deck imputation marginally using the marginal values of $\tilde{\mathbf{y}}_{j,mis(i)}$ for each item separately. The discrete version preserves most of the correlation structure in Y_k, and the marginal hot deck imputation will perform well if there is no systematic variation within categories of $\tilde{Y}$.

In summary, if $Y = (Y_1, Y_2, Y_3)$, the proposed fractional imputation method is performed by the following steps:

[Step 1] For each item k, transform Y_k into $\tilde{Y}_k$, a discrete version of Y_k. The value of $\tilde{Y}_k$ will serve the role of imputation cell for Y_k since the hot deck imputation for Y_k will be performed marginally within the cell value of $\tilde{Y}_k$.

[Step 2] Use the estimated joint probability to compute the fractional weights

$$\tilde{w}^*_{ij} = \frac{\tilde{\pi}(\tilde{\mathbf{y}}_{i,\text{obs}}, \tilde{\mathbf{y}}^{*(j)}_{\text{mis(i)}})}{\sum_j \tilde{\pi}(\tilde{\mathbf{y}}_{i,obs}, \tilde{\mathbf{y}}^{*(j)}_{\text{mis(i)}})}, \tag{8.43}$$

for fully efficient fractional imputation (FEFI), where $\tilde{\mathbf{y}}^{*(j)}_{\text{mis(i)}}$ is the j-th realization of the $\tilde{\mathbf{y}}_{\text{mis(i)}}$, the missing part of $\tilde{\mathbf{y}}$ for unit i. The FEFI assigns fractional weights for all possible combinations of the imputed values for missing $\tilde{\mathbf{y}}_{mis(i)}$. The joint probability $\tilde{\pi}$ can be estimated by a modified EM algorithm.

[Step 3] Among the possible values of $\tilde{\mathbf{y}}_{mis(i)}$, select a PPS sample of size m with probability proportional to the fractional weights in [Step 2]. The m imputed values of $\tilde{\mathbf{y}}_{mis(i)}$ will have equal initial fractional weights $w^*_{ij0} = 1/m$.

[Step 4] Using calibration weighting, create final fractional weights satisfying

$$\frac{1}{N}\sum_{i\in A}\sum_{j=1}^{m} w_i\tilde{w}^*_{ij} I\left(\tilde{y}^{(j)}_{Ii1} = a\right) = \tilde{\pi}_{a++},$$

$$\frac{1}{N}\sum_{i\in A}\sum_{j=1}^{m} w_i\tilde{w}^*_{ij} I\left(\tilde{y}^{(j)}_{Ii2} = b\right) = \tilde{\pi}_{+b+},$$

$$\frac{1}{N}\sum_{i\in A}\sum_{j=1}^{m} w_i\tilde{w}^*_{ij} I\left(\tilde{y}^{(j)}_{Ii3} = c\right) = \tilde{\pi}_{++c},$$

and

$$\sum_{j=1}^{m} \tilde{w}^*_{ij} = 1.$$

[Step 5] For each item k, use marginal hot deck imputation to select from the respondents with the same value of $\tilde{y}_k$. That is, each imputed value of $\tilde{y}_k$ can be used as an imputation cell for hot deck imputation for item k.

We now consider variance estimation for the proposed fractional imputation estimator using a replication method. The replication variance estimator of $\hat{\theta}_n$ takes the form in (4.49). The replication method for fractional imputation consists of computing a replicated version of joint probability $\tilde{\pi}$, denoted by $\tilde{\pi}^{(k)}$, and then computing replicated fractional weights. For fractional weights of the form (8.43), we can use

$$\tilde{w}^{*(k)}_{ij} = \frac{\tilde{\pi}^{(k)}(\tilde{\mathbf{y}}_{i,\text{obs}}, \tilde{\mathbf{y}}^{*(j)}_{\text{mis(i)}})/\tilde{\pi}(\tilde{\mathbf{y}}_{i,\text{obs}}, \tilde{\mathbf{y}}^{*(j)}_{\text{mis(i)}})}{\sum_j \tilde{\pi}^{(k)}(\tilde{\mathbf{y}}_{i,\text{obs}}, \tilde{\mathbf{y}}^{*(j)}_{\text{mis(i)}})/\tilde{\pi}(\tilde{\mathbf{y}}_{i,\text{obs}}, \tilde{\mathbf{y}}^{*(j)}_{j,\text{mis(i)}})} \tag{8.44}$$

to obtain initial replication fractional weights. If the initial fractional weights are modified by a regression weighting procedure (5.28), then the replicated fractional weight can be constructed similarly, using (8.44) as the initial fractional weights. See Kim and Fuller (2013).

8.6 Imputation for two-phase sampling

Two-phase sampling, sometimes called double sampling, is a cost-effective technique in survey sampling. By first selecting a large sample, observing cheap auxiliary variables and then incorporating the auxiliary variables into the second-phase sampling design, we can produce estimators with smaller variances than those based on a single-phase sampling design for the same cost. Two-phase sampling is also popular when the sampling frame for eligible elements do not exist. In the first-phase sampling, a large sample is drawn without considering eligibility. Next, the eligibility of the first-phase sample elements is examined. The second-phase sample will become a stratified sample with two strata where Stratum One is the eligibility stratum and Stratum Two is the noneligibility stratum. The second-phase sample size for the noneligible stratum will be zero because we are not interested in noneligible elements in the population.

Estimation for two-phase sampling deals with how to incorporate the auxiliary information collected in the first-phase sample. One approach for incorporating the partial information is through imputation, where the nonsampled part in the second-phase sampling is treated as missing data. In fact, two-phase sampling can be understood as a sampling design with planned missingness. Imputation for the nonsampled part is often referred to as *mass imputation.* Mass imputation is particularly useful in creating efficient domain estimates. For example, the World Bank has been using a simulated census method, developed by Elbers et al. (2003) and Haslett and Jones (2005), to estimate small area poverty measures in Bangladesh, Nepal, and some other developing countries. Fuller (2003) discusses mass imputation to get improved estimates for domains.

In two-phase sampling, the sample selection is made in two different ways. In the first-phase sample, only x is measured. Based on the information of x, a second-phase sample is selected and (x,y) is measured. Let A_1 and A_2 be the set of indices for the first-phase and second-phase sample, respectively. The second-phase sample is not necessarily nested within the first-phase sample. Also, let w_{i1} and w_{i2} be the sampling weight of unit i for the first-phase and second-phase sample, respectively. An unbiased estimator of $Y=\sum_{i=1}^{N} y_i$ is $\hat{Y}_2=\sum_{i\in A_2} w_{i2}y_i$, which does not use the observation x in the first-phase sample and thus is inefficient.

To incorporate x_i into the estimation of Y, we use the following estimator,

$$\hat{Y}_{tp}=\sum_{i\in A_1} w_{i1}m(x_i;\hat{\beta})+\sum_{i\in A_2} w_{i2}\left\{y_i-m(x_i;\hat{\beta})\right\} \tag{8.45}$$

for some $m(x_i;\hat{\beta})$. We call the estimator (8.45) a *two-phase regression estimator* with a corresponding subscript denotation. First, $m(x_i;\hat{\beta})$ represents a predictor of y_i from an implicit regression model

$$E\left(y_i \mid x_i\right)=m\left(x_i;\beta\right). \tag{8.46}$$

The two-phase regression estimator (8.45) is expressed as a sum of two terms, a "projection term" and a "bias-correction term." The projection term is unbiased under model (8.46) but not otherwise. The bias correction term makes the resulting estimator (8.45) nearly unbiased regardless of whether the model (8.46) holds or not. The weights w_{i2} are constructed to satisfy $\sum_{i\in A_2} w_{i2}=\sum_{i\in A_1} w_{i1}$. Otherwise, it is better to use

$$\hat{Y}_{tp}=\sum_{i\in A_1} w_{i1}m(x_i;\hat{\beta})+\left(\frac{\sum_{i\in A_1} w_{1i}}{\sum_{i\in A_2} w_{2i}}\right)\sum_{i\in A_2} w_{i2}\left\{y_i-m(x_i;\hat{\beta})\right\}, \tag{8.47}$$

which essentially normalizes the second-phase sample weights.

Note that the two-phase regression estimator (8.47) can be written as

$$\hat{Y}_{FEFI}=\sum_{i\in A_2} w_{i1}y_i+\sum_{i\in A_1/A_2}\sum_{j\in A_2} w_{i1}w_{ij}^{*}y_{ij}^{*}, \tag{8.48}$$

where $y_{ij}^{*}=\hat{y}_i+\hat{e}_j$, $\hat{y}_i=m(\mathbf{x}_i;\hat{\beta})$, $\hat{e}_j=y_j-\hat{y}_j$, and

$$w_{ij}^{*}=\frac{w_{j2}-w_{j1}}{\sum_{k\in A_2}(w_{k2}-w_{k1})}.$$

The expression (8.48) implies that we impute all the elements in $A_1/A_2=A_1\cap A_2^c$. The estimator (8.48) is computed by augmenting the dataset with $(n_1-n_2+1)\times n_2$ records, where n_1 and n_2 is the size of A_1 and A_2, respectively. For $i\in A_2$, element i has only one record with observation y_i and weight w_{1i}. For $i\in A_1/A_2$, element i has n_2 records where the j-th record has an imputed observation $y_{ij}^{*}=\hat{y}_i+\hat{e}_j$ with weight $w_{i1}w_{ij}^{*}$. The imputation method in (8.48) imputes all the elements in A_1/A_2 and is called the *fully efficient fractional imputation (FEFI) method*, as considered in Fuller and Kim (2005). The FEFI estimator is algebraically equivalent to the two-phase regression estimator, and can provide estimates for other parameters such as population quantiles.

In nonnested two-phase sampling, there is no guarantee that $w_{i2} > w_{i1}$ holds for every element in A_2. In $w_{i2} < w_{i1}$, we cannot use (8.48) directly but can instead use

$$\hat{Y}_{FEFI} = \sum_{i \in A_1} \sum_{j \in A_2} w_{i1} w^*_{ij} y^*_{ij}, \tag{8.49}$$

where y^*_{ij} are the same as those in (8.48) and

$$w^*_{ij} = \frac{w_{j2}}{\sum_{k \in A_2} w_{k2}}.$$

Thus, the proposed estimator (8.49) is based on a big data file with $n_1 \times n_2$ records. The expression (8.49) implies that we impute all the elements in the first-phase sample, including the elements that also belong to the second-phase sample.

If we want to limit the number of imputations, say m, fractional imputation using the regression weighting method of Fuller and Kim (2005) can be used. That is, we first select m values of $y^*_{ij} = \hat{y}_i + \hat{e}_j$ among the set of n_2 imputed values $\{y^*_{ij}; j \in A_2\}$ using an efficient sampling method. The new fractional weights $\tilde{w}^*_{ij}$ assigned to y^*_{ij} are determined so that

$$\sum_{j \in A_{I(i)}} \tilde{w}^*_{ij} \left(1, y^*_{ij}\right) = \sum_{j \in A_2} w^*_{ij} \left(1, y^*_{ij}\right), \tag{8.50}$$

where w^*_{ij} is the fractional weight in the FEFI estimator and $A_{I(i)}$ is the set of indices for the elements A_2 where $\hat{e}_j$ was selected for imputed value of unit $i \in A_1$. Condition (8.50) is equivalent to

$$\sum_{j \in A_{I(i)}} \tilde{w}^*_{ij} (1, \hat{e}_j) = (1, \bar{e}_{FEFI}), \tag{8.51}$$

where $\bar{e}_{FEFI} = \sum_{j \in A_2} w^*_{ij} \hat{e}_j$. The fractional weight satisfying (8.51) can be computed using the regression weighting method or the empirical likelihood method. Some additional constraints other than (8.51) can also be imposed. The fractional imputed data y^*_{ij} with weight $w_{i1} \tilde{w}^*_{ij}$ are constructed for the first-phase sample only. The resulting FI estimator of Y is then

$$\hat{Y}_{FI} = \sum_{i \in A_1} \sum_{j \in A_{I(i)}} w_i \tilde{w}^*_{ij} y^*_{ij}. \tag{8.52}$$

Replication variance estimation can be considered for estimating the variance of a FI estimator. For nested two-phase sampling, the k-th replicate of the two-phase regression estimator in (8.47) is computed by

$$\hat{Y}^{(k)}_{tp} = \sum_{i \in A_1} w^{(k)}_{i1} \left[m(x_i; \hat{\beta}) + \delta_i g^{(k)} \left(\frac{w_{2i}}{w_{1i}} \right) \left\{ y_i - m(x_i; \hat{\beta}) \right\} \right],$$

where $w^{(k)}_{i1}$ is the usual replication weights for w_{i1} in A_1, δ_i is the indicator function for the second-phase sample selection, and

$$g^{(k)} = \frac{\sum_{i \in A_1} w^{(k)}_{i1}}{\sum_{i \in A_2} w^{(k)}_{i1} (w_{i2}/w_{i1})}.$$

Using the reverse framework in (8.25), it can be shown that the replication variance estimator

$$\hat{V}_{rep} = \sum_{k=1}^{L} c_{1k} \left(\hat{Y}^{(k)}_{tp} - \hat{Y}_{tp} \right)^2$$

is consistent for the total variance if the sampling rate n/N is negligible. Note that the sampling variability of $\hat{\beta}$ can be safely ignored.

For the FEFI estimator in (8.48), the k-th replicate can be constructed by

$$\hat{Y}_{FEFI}^{(k)} = \sum_{i \in A_2} w_{i1}^{(k)} y_i + \sum_{i \in A_1/A_2} \sum_{j \in A_2} w_{i1}^{(k)} w_{ij}^{*(k)} y_{ij}^*,$$

where

$$w_{ij}^{*(k)} = \frac{w_{j1}^{(k)} (w_{j2}/w_{j1} - 1)}{\sum_{l \in A_2} w_{l1}^{(k)} (w_{l2}/w_{l1} - 1)}.$$

Also, for the FI estimator using (8.52), the same calibration can be used applying the replication weights. That is, we can find $\tilde{w}_{ij}^{*(k)}$ such that

$$\sum_{j \in A_{I(i)}} \tilde{w}_{ij}^{*(k)} (1, \hat{e}_j) = \left(1, \bar{e}_{FEFI}^{(k)}\right), \tag{8.53}$$

where $\bar{e}_{FEFI}^{(k)} = \sum_{j \in A_2} w_{ij}^{*(k)} \hat{e}_j$. Note that the effect of estimating β can be safely ignored.

8.7 Synthetic imputation

Synthetic imputation is a technique of creating imputed values for items not observed in the current survey by incorporating information from other surveys. For example, suppose that there are two independent surveys, called Survey One and Survey Two, and we observe x_i from Survey One and observe (x_i, y_i) from Survey Two. Let A_1 and A_2 be the index set of the sample elements in Survey One and Survey Two, respectively. In this case, we may want to create synthetic values of y_i in Survey One so that inference about y can be made even in Survey One. This is particularly useful when Survey One is a large scale survey and item y is very expensive to measure. The setup of two independent samples with common items is often called nonnested two-phase sampling.

If a working regression model $E(y \mid x) = m(x; \beta)$ is imposed, then the model parameter β can be estimated from Survey Two and then synthetic values of y_i can be created by $y_i^* = \hat{y}_i \equiv m(x_i; \hat{\beta})$ or $y_i^* = \hat{y}_i + \hat{e}_i^*$, where $\hat{e}_i^*$ is randomly generated from the empirical distribution of the residuals in A_2. More generally, we can postulate a conditional distribution $f(y \mid x; \theta)$ and obtain a design-consistent estimator of θ by solving

$$\sum_{i \in A_2} w_{i2} S(\theta; x_i, y_i) = 0, \tag{8.54}$$

where w_{i2} is the sampling weights for the A_2 sample and $S(\theta; x, y) = \partial \log f(y \mid x; \theta) / \partial \theta$. Once $\hat{\theta}$ is obtained, the imputed values are generated from $f(y \mid x; \hat{\theta})$. Let $y_{i1}^*, \cdots, y_{im}^*$ be m imputed values generated from $f(y \mid x_i; \hat{\theta})$. The synthetic estimator of $Y = \sum_{i=1}^N y_i$ obtained from sample A_1 is then given by

$$\hat{Y}_{FI} = \sum_{i \in A_1} w_{i1} \sum_{j=1}^m w_{ij}^* y_{ij}^*, \tag{8.55}$$

where w_{i1} is the sampling weight for unit i in A_1 sample and w_{ij}^* is the fractional weights computed by

$$w_{ij}^* = w_{ij0}^* + \hat{\lambda}' \left(S_{ij}^* - \bar{S}_i^*\right) w_{ij0}^*,$$

where $w_{ij0}^* = 1/m$ is the initial fractional weight, $\bar{S}_i^* = \sum_{j=1}^m w_{ij0}^* S_{ij}^*$, $S_{ij}^* = S(\hat{\theta}; x_i, y_{ij}^*)$, and

$$\hat{\lambda} = -\left\{\sum_{i \in A_1} w_{i1} \sum_{j=1}^m w_{ij0}^* \left(S_{ij}^* - \bar{S}_i^*\right)^{\otimes 2}\right\}^{-1} \sum_{i \in A_1} w_{i1} \bar{S}_i^*.$$

Note that the fractional weights are constructed to satisfy

$$\sum_{i \in A_1} w_{i1} \sum_{j=1}^{m} w_{ij}^* S(\hat{\theta}; x_i, y_{ij}^*) = 0 \tag{8.56}$$

and $\sum_{j=1}^{m} w_{ij}^* = 1$. Condition (8.56) is used to guarantee that we obtain the same MLE $\hat{\theta}$ from the imputed values in A_1.

The following theorem, originally proved by Kim and Rao (2012), shows that $\hat{Y}_{FI}$ in (8.55) is asymptotically unbiased under modest regularity conditions without requiring the imputation model $f(y \mid x) = f(y \mid x; \theta)$ to be correctly specified.

Theorem 8.2. *Assume that $\hat{\theta}$ computed from (8.54) satisfies*

$$\sum_{i \in A_2} w_{ib} \left\{ y_i - E(y_i \mid x_i; \hat{\theta}) \right\} = 0, \tag{8.57}$$

where $E(y \mid x; \theta) = \int y f(y \mid x; \theta) dy$. Then, under some regularity conditions, for sufficiently large m, the imputed estimator in (8.55) is asymptotically equivalent to

$$\tilde{Y} = \sum_{i \in A_1} w_{i1} \tilde{y}_i + \sum_{i \in A_2} w_{i2} \left(y_i - \tilde{y}_i \right), \tag{8.58}$$

where $\tilde{y}_i$ is the probability limit of $E(y \mid x_i; \hat{\theta})$ under the working model.

Condition (8.57) is satisfied in many cases when $\hat{\theta}$ is computed by (8.54). Since $\tilde{Y}$ in (8.58) satisfies $E(\tilde{Y} \mid \mathcal{F}_N) = Y$, the synthetic estimator in (8.55) is asymptotically unbiased for Y, regardless of the value of $\tilde{y}_i$. Thus, the asymptotic unbiasedness does not require correct specification of the imputation model $f(y \mid x; \theta)$. The asymptotic variance is

$$V\left(\tilde{Y} \mid \mathcal{F}_N\right) = V\left(\sum_{i \in A_1} w_{i1} \tilde{y}_i \mid \mathcal{F}_N \right) + V\left\{ \sum_{i \in A_2} w_{i2} \left(y_i - \tilde{y}_i \right) \mid \mathcal{F}_N \right\}. \tag{8.59}$$

The first term is the variance due to sampling in Survey One and the second term is the variance due to sampling in Survey Two. The first term is small because n_1 is generally large. The second term is also small if the working model is good. For example, assume the simple random sampling in both surveys. Let n_1 be the sample size of Survey One and n_2 be the sample size of Survey Two. Then, (8.59) reduces to

$$V\left(\tilde{Y}\right) = \frac{1}{n_1}\left(1 - \frac{n_1}{N}\right) \frac{1}{N-1} \sum_{i=1}^{N} \left(m_i - \bar{m}\right)^2 + \frac{1}{n_2}\left(1 - \frac{n_2}{N}\right) \frac{1}{N-1} \sum_{i=1}^{N} \left(\hat{e}_i - \bar{e}\right)^2,$$

where $m_i = m(\mathbf{x}_i; \beta)$, $\bar{m} = N^{-1} \sum_{i=1}^{N} m_i$, $e_i = y_i - m(\mathbf{x}_i; \beta)$, and $\bar{e} = N^{-1} \sum_{i=1}^{N} e_i$. It follows from the last term in (8.59) that the bias-correction term cannot be ignored for variance estimation although it is zero under (8.57).

For variance estimation, we can use a replication method that incorporates variability in both sampling designs. Let L_1 be the number of replications for variance estimation for $\hat{X}_1 = \sum_{i \in A_1} w_{i1} x_i$. Assume that

$$\hat{V}_1(\hat{X}_1) = \sum_{k=1}^{L_1} c_{k1} \left(\hat{X}_1^{(k)} - \hat{X}_1 \right)^2,$$

where $\hat{X}_1^{(k)} = \sum_{i \in A_1} w_{i1}^{(k)} x_i$, is a consistent estimator of the variance of $\hat{X}_1$. It is possible to construct the same form of the replication variance estimator for estimating the variance of $\hat{Y}_2 = \sum_{i \in A_2} w_{i2} y_i$. That is, we can find $w_{i2}^{(k)}, k = 1, \cdots, L_1$ such that

$$\hat{V}_1(\hat{Y}_2) = \sum_{k=1}^{L_1} c_{k1} \left(\hat{Y}_2^{(k)} - \hat{Y}_2 \right)^2, \tag{8.60}$$

where $\hat{Y}_2^{(k)} = \sum_{i \in A_2} w_{i2}^{(k)} y_i$ is consistent for the variance of $\hat{Y}_2$. Lemma A1 of Kim and Rao (2012) provide a way of constructing the L_1 replication weights $w_{i2}^{(k)}$ from A_2 sample.

The proposed replication variance estimation method can be described as follows:

[Step 1] For each $k = 1, \cdots, L_1$, compute $\hat{\theta}^{(k)}$ by solving

$$\sum_{i \in A_2} w_{i2}^{(k)} S(\theta; x_i, y_i) = 0,$$

where $w_{i2}^{(k)}$ is the replication weights in (8.60).

[Step 2] Compute the initial replication fractional weights by

$$w_{ij0}^{*(k)} = \frac{f(y_{ij}^* \mid x_i; \hat{\theta}^{(k)})/f(y_{ij}^* \mid x_i; \hat{\theta})}{\sum_{l=1}^m f(y_{il}^* \mid x_i; \hat{\theta}^{(k)})/f(y_{il}^* \mid x_i; \hat{\theta})}. \tag{8.61}$$

Adjust the initial replication fractional weights to satisfy

$$\sum_{i \in A_1} w_{i1}^{(k)} \sum_{j=1}^m w_{ij}^{*(k)} S(\hat{\theta}^{(k)}; x_i, y_{ij}^*) = 0 \tag{8.62}$$

and $\sum_{j=1}^m w_{ij}^{*(k)} = 1$.

[Step 3] The k-th replicate of $\hat{Y}_{FI}$ is then given by

$$\hat{Y}_{FI}^{(k)} = \sum_{i \in A_1} w_{i1}^{(k)} \sum_{j=1}^m w_{ij}^{*(k)} y_{ij}^*$$

and the variance of $\hat{Y}_{FI}$ in (8.55) is estimated by

$$\hat{V}_1(\hat{Y}_{FI}) = \sum_{k=1}^{L_1} c_{k1} \left(\hat{Y}_{FI}^{(k)} - \hat{Y}_{FI} \right)^2.$$

In [Step 2], given the initial replication fractional weights in (8.61), the adjusted replication fractional weights satisfying (8.62) and $\sum_{j=1}^m w_{ij}^{*(k)} = 1$ can be computed by

$$w_{ij}^{*(k)} = w_{ij0}^{*(k)} + \hat{\lambda}_{(k)}' \left(S_{ij}^{*(k)} - \bar{S}_i^{*(k)} \right) w_{ij0}^{*(k)}, \tag{8.63}$$

where $\bar{S}_i^{*(k)} = \sum_{j=1}^m w_{ij0}^{*(k)} S_{ij}^{*(k)}$, $S_{ij}^{*(k)} = S(\hat{\theta}^{(k)}; x_i, y_{ij}^*)$, and

$$\hat{\lambda}_{(k)} = -\left\{ \sum_{i \in A_1} w_{i1}^{(k)} \sum_{j=1}^m w_{ij0}^{*(k)} \left(S_{ij}^{*(k)} - \bar{S}_i^{*(k)} \right)^{\otimes 2} \right\}^{-1} \sum_{i \in A_1} w_{i1}^{(k)} \sum_{j=1}^m w_{ij0}^{*(k)} S_{ij}^{*(k)}. \tag{8.64}$$

If some of $w_{ij}^{*(k)}$ in (8.63) takes negative values, we may use

$$w_{ij}^{*(k)} = \frac{w_{ij0}^{*(k)} \exp\left(\hat{\lambda}_{(k)}' S_{ij}^{*(k)} \right)}{\sum_{l=1}^m w_{il0}^{*(k)} \exp\left(\hat{\lambda}_{(k)}' S_{il}^{*(k)} \right)},$$

where $\hat{\lambda}_{(k)}$ is defined in (8.64).

Exercises

1. Using Cauchy–Schwartz inequality, prove that q_i^* in (8.8) minimizes the variance term in (8.25).
2. Assume that the finite population values $\{(x_i, y_i); i = 1, \ldots, N\}$ are generated independently with probability density $f(y_i \mid x_i; \theta)h(x_i)$ and the conditional distribution of y_i given x_i is in the exponential family

$$f(y_i \mid x_i) = \exp\left\{\frac{y_i\gamma_i - b(\gamma_i)}{\tau^2} - c(y_i, \tau)\right\},$$

where γ_i is the canonical parameter and $g(\gamma_i) = x_i'\beta_0$. The marginal density of x, $h(x)$, is completely unspecified. Writing $\mu_i = E(y_i \mid x_i)$, we have $\mu_i = \partial b(\gamma_i)/\partial \gamma_i$. Note that the score function for

$$S(\beta; x_i, y_i) = \frac{1}{\tau^2}(y_i - \mu_i)\left\{v(\mu_i)g_\mu(\mu_i)\right\}^{-1} x_i,$$

where $g_\mu(\mu_i) = \partial g(\mu_i)/\partial \mu_i$.
Now, under probability sampling design, we consider the problem of solving (8.4). Answer the following equations.

   (a) Show that the solution $\hat{\beta}_q$ from (8.4) is asymptotically unbiased regardless of the choice of the $q(x_i)$ function in (8.4).
   (b) Show that the asymptotic variance of the $\hat{\beta}_q$ obtained from (8.4) is

$$J_q^{-1} V\left\{\sum_{I_i=1} d_i e_i \left\{v(\mu_i)g_\mu(\mu_i)\right\}^{-1} x_i q(x_i)\right\}(J_q')^{-1} \tag{8.65}$$

   where $d_i = 1/\pi_i$ and $J_q = \sum_{i=1}^N \left\{v(\mu_i)g_\mu^2(\mu_i)\right\}^{-1} x_i x_i' q(x_i)$.
   (c) Show that the optimal choice that minimizes (8.65), assuming independence between the terms in the summation in this expression, is

$$q_i^* = v(\mu_i)\left\{E(d_i e_i^2 \mid x_i)\right\}^{-1} = E(e_i^2 \mid x_i)\left\{E(d_i e_i^2 \mid x_i)\right\}^{-1}.$$

3. Under the setup of Section 8.2, consider the following estimator

$$\hat{Y}_{p,bc} = \sum_{i=1}^N \hat{m}_i + \sum_{i \in A} d_i(y_i - \hat{m}_i)$$

where $\hat{m}_i = m(\mathbf{x}_i; \hat{\beta})$ and $\hat{\beta}$ satisfies

$$\sum_{i \in A} d_i \left\{y_i - m(\mathbf{x}_i; \hat{\beta})\right\} = 0.$$

   (a) Show that, writing $\hat{Y}_{p,bc} = \hat{Y}_{p,bc}(\hat{\beta})$,

$$E\left\{\frac{\partial}{\partial \beta}\hat{Y}_{p,bc}(\beta)\right\} = 0.$$

   (b) Argue that the effect of $\hat{\beta}$ in $\hat{Y}_{p,bc}$ can be safely ignored in variance estimation.
4. Under the setup of Section 8.2, consider the following instrumental variable estimator of $Y = \sum_{i=1}^N y_i$

$$\hat{Y}_z = \sum_{i \in A} w_i y_i$$

where

$$w_i = 1 + \left(\sum_{i \in A^c} \mathbf{x}_i'\right)\left(\sum_{i \in A} \mathbf{z}_i \mathbf{x}_i'\right)^{-1} \mathbf{z}_i.$$

(a) Show that $\hat{Y}_z$ satisfies the calibration condition (8.9).

(b) Show that if $\mathbf{z}_i'\mathbf{a} = d_i - 1$ for some $\mathbf{a}$, then $\hat{Y}_z$ is asymptotically design unbiased.

(c) Under the condition in (b), discuss how to estimate the variance of $\hat{Y}_z$.

5. Assume that a simple random sample of size n is obtained from a finite population of size N with auxiliary information $\bar{\mathbf{x}}_N = N^{-1}\sum_{i=1}^{N}\mathbf{x}_i$. Let the usual regression estimator be written in the form of $\bar{y}_{reg} = \sum_{i\in A} w_i y_i$, where

$$w_i = n^{-1} + (\bar{\mathbf{x}}_N - \bar{\mathbf{x}}_n)\left\{\sum_{i\in A}(\mathbf{x}_i - \bar{\mathbf{x}}_n)'(\mathbf{x}_i - \bar{\mathbf{x}}_n)\right\}^{-1}(\mathbf{x}_i - \bar{\mathbf{x}}_n)'.$$

Show that $\bar{y}_{reg,2} = \sum_{i\in A} w_{i,2}y_i$, where $w_{i,2} = \exp(nw_i)/\{\sum_{i\in A}\exp(nw_i)\}$, is asymptotically equivalent to the regression estimator. When is $w_{i,2}$ preferable to w_i in practice?

6. Consider the setup of two independence samples, A_1 and A_2, from the same finite population where $\mathbf{x}_i$ is observed in both surveys and y_i are observed in Survey Two. Let d_i and d_{2i} be the sampling weight of unit i in sample A_1 and sample A_2, respectively. Let $\hat{\mathbf{X}}_1 = \sum_{i\in A_1} d_{1i}\mathbf{x}_i$ and $\hat{\mathbf{X}}_2 = \sum_{i\in A_2} d_{2i}\mathbf{x}_i$ be unbiased estimators of $\mathbf{X} = \sum_{i=1}^{N}\mathbf{x}_i$ obtained from the two surveys and $\hat{Y}_2 = \sum_{i\in A_2} d_{2i}y_i$ be an unbiased estimator of $Y = \sum_{i=1}^{N} y_i$. We are interested in estimating Y combining the two surveys.

(a) Show that the optimal estimator of Y among the class of unbiased estimators of Y that are linear in $\hat{\mathbf{X}}_1$, $\hat{\mathbf{X}}_2$ and $\hat{Y}_2$ is

$$\hat{Y}_{opt} = \hat{Y}_2 + \left(\hat{\mathbf{X}}_{opt} - \hat{\mathbf{X}}_2\right)'\hat{B},$$

where

$$\hat{\mathbf{X}}_{opt} = \hat{K}\hat{\mathbf{X}}_1 + (1-\hat{K})\hat{X}_2$$

$$\hat{K} = \left\{\hat{V}(\hat{X}_1) + \hat{V}(\hat{X}_2)\right\}^{-1}\hat{V}(\hat{X}_2)$$

and $\hat{B} = \{\hat{V}(\hat{X}_2)\}^{-1}\hat{Cov}(\hat{X}_2, \hat{Y}_2)$.

(b) Let $\mathbf{x}_i' = (\mathbf{x}_{1i}', \mathbf{x}_{2i}')$ and assume that $\mathbf{X}_1 = \sum_{i=1}^{N}\mathbf{x}_{1i}$ is known. Find the optimal estimator of Y in this case.

7. Let A be the set of sample indices obtained from a probability sample of size n with the first-order inclusion probability π_i. Let $(\mathbf{x}_i, y_i)$ be the sample observations from A and $\bar{\mathbf{x}}_N = N^{-1}\sum_{i=1}^{N}\mathbf{x}_i$ is known. Let $\bar{y}_N = N^{-1}\sum_{i=1}^{N} y_i$ be the parameter of interest. Consider the following estimator:

$$\bar{y}_{reg1} = \bar{\mathbf{x}}_N\hat{\beta},$$

where

$$\hat{\beta} = \left(\sum_{i\in A}\pi_i^{-2}\mathbf{x}_i'\mathbf{x}_i\right)^{-1}\sum_{i\in A}\pi_i^{-2}\mathbf{x}_i'y_i.$$

(a) Find the conditions on $\mathbf{x}_i$ so that $\bar{y}_{reg1}$ is design consistent.

(b) Find a superpopulation model where $\bar{y}_{reg1}$ achieves the minimum model variance among the class of linear (in y) and model-unbiased estimators of $\bar{y}_N$.

(c) Let $y_i = \mathbf{x}_i'\beta + e_i$ be the superpopulation model with $e_i \sim (0, \gamma_{ii}\sigma^2)$ for some known $\gamma_{ii} = \gamma(\mathbf{x}_i)$. Find a set of conditions on $\mathbf{x}_i$ and π_i such that $\bar{y}_{reg1}$ is optimal in the sense that it minimizes the anticipated variance among the class of linear model-unbiased estimators and the class of fixed-sample size design-consistent estimators of $\bar{y}_n$ under a nonreplacement design with fixed probabilities.

8. Let A be the set of sample indices obtained from a probability sample of size n with the first-order inclusion probability π_i. The population size N is unknown and $\hat{N}_d = \sum_{i \in A} d_i$ is used to estimate N, where $d_i = 1/\pi_i$. For scalar x_i, consider the following regression estimator

$$\hat{Y}_{reg2} = \hat{Y}_d + \left(X - \hat{X}_d\right)\hat{B}_1,$$

where $X = \sum_{i=1}^{N} x_i$, $\left(\hat{X}_d, \hat{Y}_d\right) = \sum_{i \in A} d_i (x_i, y_i)$, $\hat{B}_1 = \left\{\sum_{i \in A} d_i \left(x_i - \bar{x}_d\right)^2\right\}^{-1} \sum_{i \in A} d_i \left(x_i - \bar{x}_d\right) y_i$, and $(\bar{x}_d, \bar{y}_d) = (\hat{X}_d, \hat{Y}_d)/\hat{N}_d$.
Answer the following question:
   (a) Show that $\hat{Y}_{reg2}$ is asymptotically unbiased for $Y = \sum_{i=1}^{N} y_i$.
   (b) Derive the asymptotic variance of $\hat{Y}_{reg2}$.
   (c) Compare the asymptotic variance of $\hat{Y}_{reg2}$ with the asymptotic variance of $\hat{Y}_{reg3}$ where

$$\hat{Y}_{reg3} = \hat{Y}_\pi + \left(X - \hat{X}_\pi\right)\hat{B}_1,$$

   and $\left(\hat{X}_\pi, \hat{Y}_\pi\right) = N\left(\hat{X}_d, \hat{Y}_d\right)/\hat{N}_d$.

Chapter 9

Statistical matching

9.1 Introduction

Survey sampling is a scientific tool for obtaining necessary information about a target population. Well-designed surveys are often time and resource consuming. Furthermore, it is not easy to achieve the needed balance between informative questionnaires and response burden on the survey participants to ensure response quality. For example, marketing companies wish to identify the factors that determine consumers' purchase decisions. Health-related surveys do not contain all questions of interest related to the sociological and economic characteristics of the surveyed individuals.

A practical solution is to exploit, as much as possible, information already available from different data sources. *Statistical matching*, sometimes called *data fusion* or *data combination*, aims to integrate two or more data sets when information available for matching records for individual participants across data sets is incomplete. Statistical matching can be viewed as a missing data problem where a researcher wants to perform a joint analysis of variables that are never jointly observed.

Statistical matching techniques can be used to construct fully augmented data files for these applications and to perform statistically valid data analysis. Statistical matching is thus particularly useful in evaluating effects of multiple treatments in observational studies. By properly applying statistical matching techniques, we can create an augmented data file of potential outcomes so that causal inference can be investigated with the augmented data file (Morgan and Winship, 2007).

Table 9.1 A simple data structure for matching

	X	Y_1	Y_2
Sample A	o	o	
Sample B	o		o

To simplify the setup, suppose that there are two surveys, Survey A and Survey B, that contain partial information about the population. Suppose that we observe x and y_1 from the Survey A sample and observe x and y_2 from the Survey B sample. Table 9.1 illustrates a simple data structure for matching. If the Survey B sample (Sample B) is a subset of the Survey A sample (Sample A), then we can apply the record linkage techniques to obtain y_{1i} value for the Survey B sample. However, in many cases, such perfect matching is not possible and we may rely on a probabilistic way of identifying the "statistical twins" from the other sample. That is, we want to create y_1 for each element in Sample B by finding the nearest neighbor from Sample A. Finding the nearest neighbor is often based on "how close" they are in terms of x's. Thus, in many cases, statistical matching is based on the assumption that y_1 and y_2 are conditionally independent, conditional on X. That is,

$$Y_1 \perp Y_2 \mid X. \tag{9.1}$$

Under the assumption that (X, Y_1, Y_2) are multivariate normal, the conditional independence assumption (CIA) means that $\sigma_{12} = \sigma_{1x}\sigma_{2x}/\sigma_{xx}$ and $\rho_{12} = \rho_{1x}\rho_{2x}$. That is, σ_{12} is completely determined from other parameters, rather than having to be estimated from the realized samples.

A synthetic data imputation under CIA in this case can be implemented in two steps:

[Step 1] Estimate $f(y_1 \mid x)$ from Sample A, denoted by $\hat{f}_a(y_1 \mid x)$.

[Step 2] For each element in Sample B, use the x_i value to generate imputed value(s) of y_1 from $\hat{f}_a(y_1 \mid x_i)$.

Instead of being used directly for imputation, synthetic values can identify the statistical twins in Sample A, which will be used in turn as the imputed values. D'Orazio et al. (2006) provides a comprehensive overview of the methods for statistical matching under CIA. Rässler (2004) also considered the multiple imputation approach for statistical matching under CIA.

9.2 Instrumental variable approach

Statistical matching based on CIA assumes that $Cov(y_1, y_2 \mid x) = 0$. Thus, the regression of Y_2 on X and Y_1 will estimate a zero regression coefficient for Y_1. That is, the estimate $\hat{\beta}_2$ for

$$\hat{y}_2 = \hat{\beta}_0 + \hat{\beta}_1 x + \hat{\beta}_2 y_1,$$

will estimate zero. Such analysis can be misleading. Thus, we consider an alternative approach which is not built on CIA. First, assume that we can decompose $X = (X_1, X_2)$ such that

$$\begin{aligned} &(i) \qquad f(y_2 \mid x_1, x_2, y_1) = f(y_2 \mid x_1, y_1) \\ &(ii) \qquad f(y_1 \mid x_1, x_2 = a) \neq f(y_1 \mid x_1, x_2 = b) \end{aligned}$$

for some $a \neq b$. Thus, x_2 is conditionally independent of y_2 given x_1 and y_1 but x_2 is correlated with y_1 given x_1. The variable X_2 satisfying the above two conditions is often called the *instrumental variable (IV)* for Y_1. Under the IV assumption, the two-step regression can be used to estimate the regression parameters. The following example presents the basic ideas.

Example 9.1. *Consider the two-sample data structure in Table 9.1. We assume the following linear regression models*

$$y_{2i} = \beta_0 + \beta_1 x_{1i} + \beta_2 y_{1i} + e_i, \tag{9.2}$$

where $e_i \sim (0, \sigma_e^2)$ and e_i are independent of (x_{1i}, x_{2i}, y_{1i}). In this case, a consistent estimator of $\beta = (\beta_0, \beta_1, \beta_2)$ can be obtained by the two-stage least squares (2SLS) method as follows:

1. *From Sample A, fit the following "working model" for y_1*

$$y_{1i} = \alpha_0 + \alpha_1 x_{1i} + \alpha_2 x_{2i} + u_i, \quad u_i \sim (0, \sigma_u^2) \tag{9.3}$$

to obtain a consistent estimator of $\alpha = (\alpha_0, \alpha_1, \alpha_2)'$ by

$$\hat{\alpha} = (\hat{\alpha}_0, \hat{\alpha}_1, \hat{\alpha}_2)' = (X'X)^{-1} X'Y_1$$

where $X = [X_0, X_1, X_2]$ is a matrix whose i-th row is $(1, x_{1i}, x_{2i})$ and Y_1 is a vector with y_{1i} being the i-th component.

2. *A consistent estimator of $\beta = (\beta_0, \beta_1, \beta_2)'$ is obtained by the least squares method for the regression of y_{2i} on $(1, x_{1i}, \hat{y}_{1i})$ where $\hat{y}_{1i} = \hat{\alpha}_0 + \hat{\alpha}_1 x_{1i} + \hat{\alpha}_2 x_{2i}$.*

To show that the resulting 2SLS estimator of β is unbiased, write

$$\hat{\beta}_{2SLS} = (\hat{Z}'\hat{Z})^{-1} \hat{Z}'Y_2,$$

where $\hat{Z} = [X_0, X_1, \hat{Y}_1]$ *is a matrix whose i-th row is* $(1, x_{1i}, \hat{y}_{1i})$*. Thus, by (9.2), we have*

$$\begin{aligned} E\left(\hat{\beta}_{2SLS}\right) &= (\hat{Z}'\hat{Z})^{-1}\hat{Z}'(\beta_0 X_0 + \beta_1 X_1 + \beta_2 Y_1) \\ &= (\hat{Z}'\hat{Z})^{-1}\hat{Z}'\{\hat{Z}\beta + (Y_1 - \hat{Y}_1)\beta_2\} \\ &= \beta + (\hat{Z}'\hat{Z})^{-1}\hat{Z}'(Y_1 - \hat{Y}_1)\beta_2 \\ &= \beta, \end{aligned} \tag{9.4}$$

where the last equality follows from the fact that an intercept term is included in computing $\hat{y}_{1i}$*. Note that the expectation in (9.4) is with respect to the model in (9.2). Model (9.3) does not necessarily hold. The variance of the 2SLS estimator is*

$$V(\hat{\beta}_{2SLS}) = \left(\hat{Z}'\hat{Z}\right)^{-1}\sigma_e^2.$$

Writing $Z = [X_0, X_1, Y_1]$*, we can express* $\hat{Z} = P_X Z$ *where* $P_X = X(X'X)^{-1}X$ *and* $X = [X_0, X_1, X_2]$*. Thus, we can write*

$$V(\hat{\beta}_{2SLS}) = (Z'P_X Z)^{-1}\sigma_e^2.$$

As a special case, if model (9.2) is

$$y_{2i} = \beta_0 + \beta_1 y_{1i} + e_i$$

then

$$V(\hat{\beta}_{1SLS}) \doteq \frac{1}{n}\frac{\sigma_e^2}{\sigma_{y1}^2 Corr(x, y_1)}.$$

Thus, if $Corr(x, y_1)$ *is close to zero then the variance of the 2SLS estimator can be very large.*

Instead of the 2SLS method, we can treat the problem as a pure missing data problem and apply techniques such as the EM algorithm. To understand the idea, note that our goal is to generate y_1 from the conditional distribution of y_1 given the observation. That is, under the IV assumption, we wish to generate y_1 from

$$f(y_1 \mid x, y_2) \propto f(y_2 \mid x_1, y_1) f(y_1 \mid x). \tag{9.5}$$

To generate y_1 from (9.5), we can consider the following two-step imputation

1. Generate y_1^* from $\hat{f}_a(y_1 \mid x)$.
2. Accept y_1^* if $f(y_2 \mid x_1, y_1^*)$ is large.

Note that the first step is the usual method under CIA. The second step incorporates the information of y_2. To compute $f(y_2 \mid x_1, y_1^*)$, we need to know the parameters in the model. EM algorithm by fractional imputation can be used as follows:

1. For each $i \in B$, generate m imputed values of y_{1i}, denoted by $y_{1i}^{*(1)}, \cdots, y_{1i}^{*(m)}$, from $\hat{f}_a(y_1 \mid x_i)$, where $\hat{f}_a(y_1 \mid x)$ denotes the estimated density for the conditional distribution of y_1 given x.
2. Let $\hat{\theta}_t$ be the current parameter value of θ in $f(y_2 \mid x_1, y_1)$. For the j-th imputed value $y_{1i}^{*(j)}$, assign the fractional weight

$$w_{ij}^* \propto f\left(y_{2i} \mid x_{1i}, y_{1i}^{*(j)}; \hat{\theta}_t\right)$$

   and $\sum_{j=1}^m w_{ij}^* = 1$.
3. Solve the fractionally imputed score equation for θ

$$\sum_{i \in B} w_{ib} \sum_{j=1}^m w_{ij}^* S(\theta; x_{1i}, y_{1i}^{*(j)}, y_{2i}) = 0$$

   to obtain $\hat{\theta}_{t+1}$, where $S(\theta; x_1, y_1, y_2) = \partial \log f(y_2 \mid x_1, y_1; \theta)/\partial\theta$.

4. Go to step 2 and continue until convergence.

Example 9.2. *We consider the same setup of Example 9.1. Under the imputation approach, a consistent estimator of* $\theta = (\beta_0, \beta_1, \beta_2, \sigma_e^2)'$ *is obtained by solving*

$$\bar{U}(\theta) = 0 \tag{9.6}$$

where $\bar{U}(\theta) = E\{U(\theta) \mid X, Y_2\}$ *and*

$$U(\theta) = \sum_{i \in B} w_{ib} u(\theta; x_{1i}, y_{1i}, y_{2i})$$

is the estimating function for θ*, where*

$$u(\theta; x_1, y_1, y_2) = \begin{bmatrix} (y_2 - \beta_0 - \beta_1 x_1 - \beta_2 y_1)(1, x_1, y_1)' \\ (y_2 - \beta_0 - \beta_1 x_1 - \beta_2 y_1)^2 - \sigma_e^2 \end{bmatrix}.$$

To evaluate the conditional expectation in (9.6), we further assume model (9.3), where the error term is $u_i \sim N(0, \sigma_u^2)$*. It follows that we can express (9.6) as*

$$\bar{U}(\theta) \cong \sum_{i \in B} \sum_{j=1}^{m} w_{ib} w_{ij}^* u(\theta; x_i, y_{1i}^{*(j)}, y_{2i}) \tag{9.7}$$

where $y_{1i}^{*(1)}, \cdots, y_{1i}^{*(m)}$ *are generated from* $\hat{f}_a(y_1 \mid x_i)$ *and* w_{ij}^* *are the fractional weights given by*

$$w_{ij}^* \propto \exp\left\{ -\frac{1}{2\hat{\sigma}_e^2} \left(y_{2i} - \hat{\beta}_0 - \hat{\beta}_1 x_{1i} - \hat{\beta}_2 y_{1i}^{*(j)} \right)^2 \right\}.$$

EM algorithm can be used to find the parameter estimates in the fractional weights.

Instead of generating $y_{1i}^{*(j)}$ from $\hat{f}_a(y_1 \mid x_i)$, we can also consider a hot-deck-type fractional imputation method where all the observed values of y_{1i} in Sample A are used as imputed values. In this case, the fractional weights in Step 2 are given by

$$w_{ij}^*(\hat{\theta}_t) \propto w_{ij0}^* f\left(y_{2i} \mid x_{1i}, y_{1i}^{*(j)}; \hat{\theta}_t \right),$$

where

$$w_{ij0}^* = \frac{\hat{f}_a(y_{1j} \mid x_i)}{\sum_{k \in A} \hat{f}_a(y_{1j} \mid x_k)}.$$

The M-step is the same as before. In practice, we may use a single imputed value for each unit. In this case, simply take the y_{1i} value with the largest value of w_{ij}^*.

Instead of the EM algorithm, Newton's method can also be used to estimate the parameters. Note that we want to solve

$$\sum_{i \in B} w_{ib} \sum_{j=1}^{m} w_{ij}^*(\theta) S(\theta; x_{1i}, y_{1i}^{*(j)}, y_{2i}) = 0 \tag{9.8}$$

for θ, where $y_{1i}^{*(1)}, \cdots, y_{1i}^{*(m)}$ are generated from $\hat{f}_a(y_1 \mid x_i)$ and

$$w_{ij}^*(\theta) = \frac{f(y_{2i} \mid x_{1i}, y_{1i}^{*(j)}; \theta)}{\sum_{k=1}^{m} f(y_{2i} \mid x_{1i}, y_{1i}^{*(k)}; \theta)}.$$

Writing $\bar{S}(\theta) = \sum_{i \in B} w_{ib} \sum_{j=1}^{m} w_{ij}^*(\theta) S(\theta; x_{1i}, y_{1i}^{*(j)}, y_{2i})$, we can obtain

$$\frac{\partial}{\partial \theta'} \bar{S}(\theta) = \sum_{i \in B} w_{ib} \sum_{j=1}^{m} w_{ij}^*(\theta) \dot{S}(\theta; x_{1i}, y_{1i}^{*(j)}, y_{2i}) + \sum_{i \in B} w_{ib} \sum_{j=1}^{m} w_{ij}^*(\theta) \left\{ S_{ij}^*(\theta) - \bar{S}_i^*(\theta) \right\}^{\otimes 2}, \tag{9.9}$$

where $\dot{S}(\theta; x_1, y_1, y_{2i}) = \partial S(\theta; x_1, y_1, y_{2i})/\partial\theta'$, $S^*_{ij}(\theta) = S(\theta; x_{1i}, y^{*(j)}_{1i}, y_{2i})$ and

$$\bar{S}^*_i(\theta) = \sum_{j=1}^{m} w^*_{ij}(\theta) S^*_{ij}(\theta).$$

Thus, the Newton's method of finding $\hat{\theta}$ that satisfies (9.8) is

$$\hat{\theta}^{(t+1)} = \hat{\theta}^{(t)} - \left\{ \frac{\partial}{\partial\theta'} \bar{S}(\hat{\theta}^{(t)}) \right\}^{-1} \bar{S}(\hat{\theta}^{(t)}),$$

where the partial derivative of $\bar{S}(\theta)$ is computed by (9.9).

9.3 Measurement error models

We now consider an application of the statistical matching to the problem of measurement error models. Suppose that we are interested in estimating parameter θ in the conditional distribution $f(y \mid x; \theta)$. Assume that instead of observing (x_i, y_i) from the sample, we obtain two samples where we observe (x_i, z_i) in sample A and observe (z_i, y_i) in Sample B. Here, z is an inaccurate measurement of x and Sample A is a validation sample. Thus, z is an instrumental variable for x because

$$f(y \mid x, z) = f(y \mid x)$$

holds. If the model is a linear regression model

$$y = \beta_0 + \beta_1 x + e$$

with $e \sim (0, \sigma^2)$, then we can use 2SLS method in Section 9.2 to estimate the regression coefficient consistently. That is, fit a regression model

$$y = \beta_0 + \beta_1 \hat{x} + u$$

where $\hat{x}$ is the predictor of x using the regression of x on z from Sample A. The use of the 2SLS method in the measurement error model is often called *regression calibration*. If there is no validation sample with true measurement x, then replicates of z can be used to obtain a predictor of x. See Chapter 4 of Carroll et al. (2006) for more details of the regression calibration technique in measurement error models.

In the case of external calibration (i.e., validation sample outside the original sample), we can use the fractional imputation to generate x in Survey B sample and to generate y in Sample A. In this case, x is generated from

$$f(x \mid z, y) \propto f(y \mid x) f(x \mid z) \tag{9.10}$$

and y is generated from

$$f(y \mid x, z) = f(y \mid x).$$

In the case of internal calibration (i.e., validation sample inside the original sample), we have only to generate x outside the validation sample from (9.10). The fractional imputation for generating x from (9.10) can be described as follows:

1. For each $i \in B$, generate $x^{*(j)}_i$ from $\hat{f}_a(x \mid z_i)$
2. Compute the fractional weights

$$w^*_{ij} \propto f(y_i \mid x^{*(j)}_i; \hat{\theta}_t).$$

3. Update θ by solving

$$\sum_{i \in B} w_{ib} \sum_{j=1}^{m} w^*_{ij} S(\theta; x^{*(j)}_i, y_i) = 0.$$

4. Go to Step 2 until convergence.

To discuss variance estimation, we can either use a linearization method or resampling method. If we use a parametric model $\hat{f}_a(y_1 \mid x) = f(y_1 \mid x; \hat{\alpha})$, the joint estimating equation of θ and α is given by

$$U_1(\theta, \alpha) = 0$$
$$U_2(\alpha) = 0,$$

where

$$U_1(\theta, \alpha) = E\{U(\theta) \mid X, Y_2; \theta, \alpha\}$$

and $U_2(\alpha) = S_2(\alpha)$, the score equation of α computed from sample A. Since we can write $U(\theta) = \sum_{i \in B} w_{ib} U_i(\theta)$, we have $U_1(\theta, \alpha) = \sum_{i \in B} w_{ib} E\{U_i(\theta) \mid X, Y_2; \theta, \alpha\}$. Thus,

$$\begin{aligned} \frac{\partial}{\partial \alpha'} U_1(\theta, \alpha) &= \sum_{i \in B} w_{ib} \frac{\partial}{\partial \alpha'} \left[\frac{\int U_i(\theta) f(y_1 \mid x_i; \alpha) f(y_{2i} \mid x_i, y_1; \theta) dy_1}{\int f(y_1 \mid x_i; \alpha) f(y_{2i} \mid x_i, y_1; \theta) dy_1} \right] \\ &= \sum_{i \in B} w_{ib} [E\{U_i(\theta) S_{2i}(\alpha)' \mid x_i, y_{2i}; \theta, \alpha\} \\ &\quad - E\{U_i(\theta) \mid x_i, y_{2i}; \theta, \alpha\} E\{S_{2i}(\alpha)' \mid x_i, y_{2i}; \theta, \alpha\}] \end{aligned}$$

and

$$\begin{aligned} \frac{\partial}{\partial \theta'} U_1(\theta, \alpha) &= \sum_{i \in B} w_{ib} \frac{\partial}{\partial \theta'} \left[\frac{\int U_i(\theta) f(y_1 \mid x_i; \alpha) f(y_{2i} \mid x_i, y_1; \theta) dy_1}{\int f(y_1 \mid x_i; \alpha) f(y_{2i} \mid x_i, y_1; \theta) dy_1} \right] \\ &= \sum_{i \in B} w_{ib} E\{\frac{\partial}{\partial \theta'} U_i(\theta) \mid x_i, y_{2i}; \theta, \alpha\} \\ &+ \sum_{i \in B} w_{ib} [E\{U_i(\theta) S_{1i}(\theta)' \mid x_i, y_{2i}; \theta, \alpha\} \\ &\quad - E\{U_i(\theta) \mid x_i, y_{2i}; \theta, \alpha\} E\{S_{1i}(\theta)' \mid x_i, y_{2i}; \theta, \alpha\}], \end{aligned}$$

where $S_{2i}(\alpha) = \partial \log f(y_{1i} \mid x_i; \alpha)/\partial \alpha$ and $S_{1i}(\alpha) = \partial \log f(y_{2i} \mid x_i, y_{1i}; \theta)/\partial \theta$. Thus, $\partial U_1(\theta, \alpha)/\partial \alpha'$ can be consistently estimated by

$$\hat{B} = \sum_{i \in B} w_{ib} \sum_{j=1}^{m} w_{ij}^* u(\hat{\theta}; x_i, y_{1i}^{*(j)}, y_{2i}) \left\{ S_2(\hat{\alpha}; x_i, y_{1i}^{*(j)}) - \bar{S}_{2i}(\hat{\alpha}) \right\}', \tag{9.11}$$

where $\bar{S}_{2i}(\hat{\alpha}) = \sum_{j=1}^{m} w_{ij}^* S_2(\hat{\alpha}; x_i, y_{1i}^*)$. Also, $\partial U_1(\theta, \alpha)/\partial \theta'$ can be consistently estimated by

$$\hat{\tau} = \sum_{i \in B} w_{ib} \sum_{j=1}^{m} w_{ij}^* \dot{U}_{ij}^*(\hat{\theta}) + \sum_{i \in B} w_{ib} \sum_{j=1}^{m} w_{ij}^* U_{ij}^*(\hat{\theta}) \left\{ S_{1ij}^*(\hat{\theta}) - \bar{S}_{1i}^*(\hat{\theta}) \right\}', \tag{9.12}$$

where $\dot{U}_{ij}^*(\theta) = \partial U(\theta; x_i, y_{1i}^{*(j)}, y_{2i})/\partial \theta$, $U_{ij}^*(\theta) = U(\theta; x_i, y_{1i}^{*(j)}, y_{2i})$, $S_{1ij}^*(\theta) = \partial S_1(\theta; x_i, y_{1i}^{*(j)}, y_{2i})$, and $\bar{S}_{1i}^*(\theta) = \sum_{j=1}^{m} w_{ij}^* S_{1ij}^*(\theta)$.

Using the Taylor expansion with respect to α

$$\begin{aligned} U_1(\theta, \hat{\alpha}) &\cong U_1(\theta, \alpha) + E\left\{ \frac{\partial}{\partial \alpha'} U_1(\theta, \alpha) \right\} \left[E\left\{ \frac{\partial}{\partial \alpha'} S_2(\alpha) \right\} \right]^{-1} S_2(\alpha) \\ &= U_1(\theta, \alpha) + K S_2(\alpha), \end{aligned}$$

we can write

$$V(\hat{\theta}) \doteq \{E(\partial U_1/\partial \theta')\}^{-1} V\{U_1(\theta, \alpha) + \kappa S_2(\alpha)\} \{E(\partial U_1/\partial \theta')\}^{-1'}.$$

Write

$$U_1(\theta,\alpha)=\sum_{i\in B} w_{ib}u_{1i}(\theta,\alpha),$$

and a consistent estimator of $V\{U_1(\theta,\alpha)\}$ can be obtained by applying a design-consistent variance estimator to $\hat{u}_{1i}=\sum_{j=1}^{m} w_{ij}^{*}U(\theta;x_i,y_{1i}^{*(j)},y_{2i})$. Under simple random sampling for Sample B, we have

$$\hat{V}\{\bar{U}(\theta\mid\alpha)\}=n_B^{-2}\sum_{i\in B}\hat{u}_{1i}^{\otimes 2}.$$

Also, $V\{\kappa S_2(\alpha)\}$ is consistently estimated by

$$\hat{V}_2=\hat{B}\{I(\hat{\alpha})\}^{-1}\hat{V}(S_2)\{I(\hat{\alpha})\}^{-1}\hat{B}',$$

where $\hat{B}$ is defined in (9.11) and $I(\alpha)=-\partial S_2(\alpha)/\partial\alpha'$. Since the two terms $U_1(\theta,\alpha)$ and $S_2(\alpha)$ are independent, the variance can be estimated by

$$\hat{V}(\hat{\theta})\doteq\hat{\tau}^{-1}\left[\hat{V}\{U_1(\theta,\alpha)\}+\hat{V}_2\right]\hat{\tau}^{-1'},$$

where $\hat{\tau}$ is defined in (9.12).

9.4 Causal inference

Statistical matching can be used to make some causal inference from observational studies under some identifying assumptions. For simplicity, assume that we have two samples, A_0 and A_1, where

- Treatment group sample (A_1): Observe (x,y_1)
- Control group sample (A_0): Observe (x,y_0)

We never observe y_0 and y_1 simultaneously. Suppose that the parameter of interest is the average treatment effect $\theta=E(y_1)-E(y_0)$. This is a typical example of the counterfactual analysis where we are interested in making a comparison between what actually happened and what would have happened in the absence of the intervention (or treatment). In Sample A_1, y_0 is the counterfactual outcome which is never observed in the treatment group. In Sample A_0, y_1 is the counterfactual outcome. The counterfactual model for causal analysis of observational data was discussed by Rubin (2005), Morgan and Winship (2007), and Cattaneo (2010), among others, where the counterfactual outcomes are treated as the missing data.

Let

$$\delta_i=\begin{cases}1 & \text{if } i\in A_1\\ 0 & \text{otherwise.}\end{cases}$$

In the controlled, randomized experiment, the distribution of δ_i is uniform within each group and θ is estimated by the simple mean differences:

$$\hat{\theta}=\frac{1}{n_1}\sum_{i\in A_1} y_{1i}-\frac{1}{n_0}\sum_{i\in A_0} y_{0i}$$

In the observational study, the distribution of δ_i is no longer uniform and the simple estimator is biased.

Under the assumption that the mechanism for the choice of treatment is ignorable, that is

$$(y_0,y_1)\perp\delta\mid X.$$

Because the counterfactual outcome is treated as the missing data, we can apply the estimation method for missing data to estimate the average treatment effects. In the propensity score approach, we can use

$$\hat{\theta}_{PS}=\frac{1}{n}\sum_{i=1}^{n}\left\{\frac{\delta_i}{\hat{\pi}_i}y_{1i}-\frac{(1-\delta_i)}{1-\hat{\pi}_i}y_{0i}\right\},$$

where $\hat{\pi}_i$ is a consistent estimator of $E(\delta_i \mid x_i)$. In particular, the propensity scores are constructed to satisfy the balancing condition:

$$\sum_{i=1}^{n} \left\{ \frac{\delta_i}{\hat{\pi}_i} \mathbf{x}_i - \frac{(1-\delta_i)}{1-\hat{\pi}_i} \mathbf{x}_i \right\} = 0. \tag{9.13}$$

The balancing condition (9.13) can be viewed as a calibration condition in the context of causal inference.

In the imputation approach, we may use the regression imputation to get

$$\hat{\theta}_{reg} = \frac{1}{n} \sum_{i=1}^{n} \{\hat{m}_1(x_i) - \hat{m}_0(x_i)\},$$

where $\hat{m}_k(x)$ is a consistent estimator of $E(y_k \mid x)$.

Suppose that the true model for the counterfactual outcome is

$$\begin{aligned} y_1 &= \beta_0 + \beta_1 x + \beta_2 z + e_1 \\ y_0 &= \beta_0^* + \beta_1^* x + \beta_2^* z + e_0, \end{aligned}$$

where $e_1 \sim (0, \sigma_1^2)$, $e_0 \sim (0, \sigma_0^2)$ and $z \sim (\mu_z, \sigma_z^2)$. The covariate x is available but covariate z is an unmeasured confounding factor in the true model. Thus, z can be treated as the omitted variable in the regression models. Because of the unobservable z, $Cov(y_1, y_0 \mid x) \neq 0$ and the CIA is not applicable. In this case, we can use the FI method for statistical matching in Section 9.2 by identifying a set of instrumental variables in the regression model of y_1 on x and y_0. Thus, we can express

$$y_1 = \alpha_0 + \alpha_1 x + \alpha_2 y_0 + e$$

where $E(e) = 0$ and e is independent of (x, y_0).

To apply the matching approach discussed in Section 9.2, we need an identifying assumption such as

$$f(y_{1i} \mid y_{0i}, x_i) = f(y_{1i} \mid y_{0i}, x_{1i}),$$

where $x_i = (x_{0i}, x_{1i})$. Thus, x_{0i} is an instrumental variable for y_{0i}.

Under this assumption, the matching approach by fractional imputation can be obtained by the following procedures:

1. For each $i \in A_1$, generate $y_{i0}^{*(j)}$ from $\hat{f}_0(y_{0i} \mid x_i)$
2. Compute the fractional weights

$$w_{ij}^* \propto f(y_{i1} \mid y_{i0}^{*(j)}, x_{1i}; \hat{\alpha}_t).$$

3. Update α by solving

$$\sum_{i \in A_1} w_{1i} \sum_{j=1}^{m} w_{ij}^* S(\alpha; x_{1i}, y_{i0}^{*(j)}, y_{i1}) = 0,$$

where $S(\alpha; x_1, y_0, y_1)$ is the score function obtained from $f(y_1 \mid y_0, x_1; \alpha)$.

4. Go to Step 2 until convergence.

Once the fractionally imputed data are constructed, we can use standard estimation methods to estimate the parameters of interest. If, for example, $\theta = E(Y_1) - E(Y_0)$ then

$$\hat{\theta}_{FI} = \sum_{i \in A_1} w_{1i} y_{1i} - \sum_{i \in A_1} w_{1i} \sum_{j=1}^{m} w_{ij}^* y_{0i}^{*(j)}$$

can be used.

Bibliography

Albert, P. S. and Follmann, D. A. (2000). Modeling repeated count data subject to informative dropout. *Biometrics*, 56:667–677.

Alho, J. M. (1990). Adjusting for nonresponse bias using logistic regression. *Biometrika*, 77:617–624.

Allen, M. B. and Issacson, E. L. (1998). *Numerical Analysis for Applied Science*. John Wiley & Sons, New York.

Amemiya, T. (1985). *Advanced Econometrics*. Harvard University Press, Cambridge, MA.

Amstrup, S. C., McDonald, T. L., and Manly, B. F. J. (2005). *Handbook of Capture–Recapture Analysis*. Princeton University Press, Princeton, NJ.

Anderson, R. L. (1957). Maximum likelihood estimates for the multivariate normal distribution when some observations are missing. *Journal of the American Statistical Association*, 52:200–203.

Baker, S. G. and Laird, N. M. (1988). Regression analysis for categorical variables with outcome subject to nonignorable nonresponse. *Journal of the American Statistical Association*, 83:62–69.

Bang, H. and Robins, J. M. (2005). Doubly robust estimation in missing data and causal inference models. *Biometrics*, 61:962–973.

Beaumont, J. F. and Bocci, C. (2009). Variance estimation when donor imputation is used to fill in missing values. *Canadian Journal of Statistics*, 37:400–416.

Billingsley, P. (1986). *Probability and Measure* (2nd edition). John Wiley & Sons, New York.

Bishop, Y. M. M., Fienberg, S. E., and Holland, P. W. (1975). *Discrete Multivariate Analysis: Theory and Practice*. MIT Press, Cambridge, MA.

Booth, J. G. and Hobert, J. P. (1999). Maximizing generalized linear models with an automated Monte Carlo EM algorithm. *Journal of the Royal Statistical Society: Series B*, 61:625–685.

Cameron, A. C. and Trivedi, P. K. (2005). *Microeconometrics: Methods and Applications*. Cambridge University Press, New York, NY.

Cao, W., Tsiatis, A. A., and Davidian, M. (2009). Improving efficiency and robustness of the doubly robust estimator for a population mean with incomplete data. *Biometrika*, 96:723–734.

Carroll, R. J., Ruppert, D., Stefanski, L. A., and Crainiceanu, C. M. (2006). *Measurement Error in Nonlinear Models: A Modern Perspective*. Chapman & Hall/CRC, Boca Raton, FL.

Cattaneo, M. D. (2010). Efficient semiparametric estimation of multi-valued treatment effects under ignorability. *Journal of Econometrics*, 155:138–154.

Chambers, R. L., Steel, D. G., Wang, S., and Welsh, A. (2012). *Maximum Likelihood Estimation for Sample Surveys*. Chapman & Hall / CRC, Boca Raton, FL.

Chen, J. and Shao, J. (2001). Jackknife variance estimation for nearest neighbor imputation. *Journal of the American Statistical Association*, 96:260–269.

Chen, Q. and Ibrahim, J. G. (2006). Semiparametric models for missing covariate and response data in regression models. *Biometrics*, 62:177–184.

Chen, S. (2013). A unified theory on empirical likelihood methods with missing data. *Statistics and Its Interface,* In press.

Cheng, P. E. (1994). Nonparametric estimation of mean functionals with data missing at random. *Journal of the American Statistical Association*, 89:81–87.

Copas, J. B. and Eguchi, S. (2001). Local sensitivity approximations for selectivity bias. *Journal of the Royal Statistical Society: Series B*, 63:871–895.

Copas, J. B. and Li, H. G. (1997). Inference for non-random samples. *Journal of the Royal Statistical Society: Series B*, 59:55–95.

Cox, D. R. (1972). Regression models and life-tables. *Journal of the Royal Statistical Society: Series B*, 34:187–220.

Cox, D. R. and Reid, N. (1987). Parameter orthogonality and approximate conditional inference. *Journal of the Royal Statistical Society: Series B*, 49:1–39.

Davidian, M. and Caroll, R. J. (1987). Variance function estimation. *Journal of the American Statistical Association*, 82:1079–1091.

Dawid, A. (1979). Conditional independence in statistical theory. *Journal of the Royal Statistical Society: Series B*, 41:1–31.

Dempster, A. P., Laird, N. M., and Rubin, D. B. (1977). Maximum likelihood from incomplete data via the EM algorithm. *Journal of the Royal Statistical Society: Series B*, 39:1–37.

D'Orazio, M., Zio, M. D., and Scanu, M. (2006). *Statistical Matching: Theory and Practice*. Wiley, Chichester, UK.

Drew, J. H. and Fuller, W. A. (1980). Modeling nonresponse in surveys with callbacks. In *Proc. Survey Res. Meth. Sect.*, pages 639–642, Washington, DC. American Statistical Association.

Efron, B. and Hinkley, D. V. (1978). Assessing the accuracy of the maximum likelihood estimator: Observed versus expected fisher information. *Biometrika*, 65:457–487.

Elbers, C., Lanjouw, J. O., and Lanjouw, P. (2003). Micro-level estimation of poverty and inequality. *Econometrika*, 71:355–364.

Fay, R. E. (1992). When are inferences from multiple imputation valid ? In *Proceedings of the Survey Research Methods Section*, pages 227–232, Washington, DC. American Statistical Association.

Fisher, R. A. (1922). On the mathematical foundations of theoretical statistics. *Philosophical Transactions of the Royal Society of London A*, 222:309–368.

Follmann, D. A. and Wu, M. (1995). An approximate generalized linear model with random effects for informative missing data. *Biometrics*, 51:151–168.

Fuller, W. A. (2003). Estimation for multiple phase samples. In Chambers, R. L. and Skinner, C. J., editors, *Analysis of Survey Data*, pages 307–322. Wiley: Chichester, England.

Fuller, W. A. (2009). *Sampling Statistics*. John Wiley & Sons, Inc., Hoboken, NJ.

Fuller, W. A. and Battese, G. E. (1973). Transformations for estimation of linear models with nested-error structure. *Journal of the American Statistical Association*, 68:626–632.

Fuller, W. A. and Kim, J. K. (2005). Hot deck imputation for the response model. *Survey Methodology*, 31:139–149.

Fuller, W. A., Loughin, M. M., and Baker, H. D. (1994). Regression weighting in the presence of nonresponse with application to the 1987-1988 Nationwide Food Consumption Survey. *Survey Methodology*, 20:75–85.

Geman, S. and Geman, D. (1984). Stochastic relaxation, Gibbs distributions, and the Bayesian restoration of images. *IEEE Transactions on Pattern Analysis and Machine Intelligence*, 6:721–741.

Gilks, W. R. and Wild, P. (1992). Adaptive rejection sampling for Gibbs sampling. *Applied Statistics*, 41:337–348.

Givens, G. H. and Hoeting, J. A. (2005). *Computational Statistics.* John Wiley & Sons, Inc., Hoboken, NJ.

Godambe, V. P. and Joshi, V. M. (1965). Admissibility and Bayes estimation in sampling finite populations. i. *Annals of Mathematical Statistics*, 36:1707–1722.

Godambe, V. P. and Thompson, M. E. (1986). Parameters of superpopulation and survey population: their relationships and estimation. *International Statistical Review*, 54:127–138.

Guo, Y. and Little, R. J. (2011). Regression analysis with covariates that have heteroscedastic measurement error. *Statistics in Medicine*, 30:2278–2294.

Hall, A. R. (2005). *Generalized method of moments.* Oxford University Press, New York.

Hansen, L. P. (1982). Large sample properties of generalized method of moments estimators. *Econometrica*, 50:1029–1054.

Haslett, S. J. and Jones, G. (2005). Small area estimation using surveys and censuses: Some practical and statistical issues. *Statistics in Transition*, 7:541–556.

Hastings, W. K. (1970). Monte Carlo sampling methods using Markov chains and their applications.. *Biometrika*, 57:97–109.

Heckman, J. J. (1979). Sample selection as a specification error. *Econometrica*, 47:153–161.

Henmi, M. and Eguchi, S. (2004). A paradox concerning nuisance parameters and projected estimating functions. *Biometrika*, 91:929–941.

Hirano, K., Imbens, G., and Ridder, G. (2003). Efficient estimation of average treatment effects using the estimated propensity score. *Econometrica*, 71:1161–1189.

Huggins, R. and Hwang, W. H. (2011). A review of the use of conditional likelihood in capture-recapture experiments. *International Statistical Review*, 79:385–400.

Ibrahim, J. G. (1990). Incomplete data in generalized linear models. *Journal of the American Statistical Association*, 85:765–769.

Ibrahim, J. G., Lipsitz, S. R., and Chen, M. H. (1999). Missing covariates in generalized linear models when the missing data mechanism is non-ignorable. *Journal of the Royal Statistical Society, Series B*, 61:173–190.

Isaki, C. T. and Fuller, W. A. (1982). Survey design under the regression superpopulation model. *Journal of the American Statistical Association*, 77:89–96.

Jiang, D. and Shao, J. (2012). Semiparametric pseudo likelihood for longitudinal data with outcome-dependent nonmonotone nonresponse. Submitted.

Kalton, G. and Kish, L. (1984). Some efficient random imputation methods. *Communications in Statistics: Series A*, 13:1919–1939.

Kang, J. D. Y. and Schafer, J. L. (2007). Demystifying double robustness: A comparison of alternative strategies for estimating a population mean from incomplete data. *Statistical Science*, 22:523–529.

Kenward, M. G. (1998). Selection models for repeated measurements with non-random dropout: An illustration of sensitivity. *Statistics in Medicine*, 17:2723–2732.

Kim, J. K. (2002). A note on approximate Bayesian bootstrap imputation. *Biometrika*, 89:470–477.

Kim, J. K. (2004). Finite sample properties of multiple imputation estimators. *The Annals of Statistics*, 32:766–783.

Kim, J. K. (2010). Calibration estimation using exponential tilting in sample surveys. *Survey Methodology*, 36:145–155.

Kim, J. K. (2011). Parametric fractional imputation for missing data analysis. *Biometrika*, 98:119–132.

Kim, J. K., Brick, M. J., Fuller, W. A., and Kalton, G. (2006a). On the bias of the multiple imputation variance estimator in survey sampling. *Journal of the Royal Statistical Society: Series B*, 68:509–521.

Kim, J. K. and Fuller, W. A. (2004). Fractional hot deck imputation. *Biometrika*, 91:559–578.

Kim, J. K. and Fuller, W. A. (2013). Fractional hot deck imputation for multivariate missing data. Unpublished manuscript.

Kim, J. K. and Haziza, D. (2013). Doubly robust inference with missing data in survey sampling. *Statistica Sinica*. In press.

Kim, J. K. and Im, J. (2012). Propensity score adjustment with several followups. Unpublished manuscript.

Kim, J. K. and Kim, J. J. (2007). Nonresponse weighting adjustment using estimated response probability. *Canadian Journal of Statistics*, 35:501–514.

Kim, J. K., Navarro, A., and Fuller, W. A. (2006b). Replicate variance estimation after multi-phase stratified sampling. *Journal of the American Statistical Association*, 101:312–320.

Kim, J. K. and Park, H. A. (2006). Imputation using response probability. *Canadian Journal of Statistics*, 34:171–182.

Kim, J. K. and Park, M. (2010). Calibration estimation in survey sampling. *International Statistical Review*, 78:21–39.

Kim, J. K. and Rao, J. N. K. (2009). Unified approach to linearization variance estimation from survey data after imputation for item nonresponse. *Biometrika*, 96:917–932.

Kim, J. K. and Rao, J. N. K. (2012). Combining data from two independent surveys: A model-assisted approach. *Biometrika*, 99:85–100.

Kim, J. K. and Riddles, M. K. (2012). Some theory for propensity-score-adjustment estimators in survey sampling. *Survey Methodology*, 38:157–165.

Kim, J. K. and Shin, D. W. (2012). The factoring likelihood method for non-monotone missing data. *Journal of the Korean Statistical Society*, 41:375–386.

Kim, J. K. and Yu, C. L. (2011). A semi-parametric estimation of mean functionals with non-ignorable missing data. *Journal of the American Statistical Association*, 106:157–165.

Kim, J. Y. and Kim, J. K. (2012). Parametric fractional imputation for nonignorable missing data. *Journal of the Korean Statistical Society*, 41:291–303.

Kott, P. S. and Chang, T. (2010). Using calibration weighting to adjust for nonignorable unit nonresponse. *Journal of the American Statistical Association*, 105:1265–1275.

Lahiri, P. and Larsen, M. D. (2005). Regression analysis with linked data. *Journal of the American Statistical Association*, 100:222–230.

Lehmann, E. L. (1983). *Theory of Point Estimation*. John Wiley & Sons, New York.

Li, K. C. (1991). Sliced inverse regression for dimension reduction. *Journal of the American Statistical Association*, 86:316–327.

Little, R. J. A. (1982). Models for nonresponse in sample surveys. *Journal of the American Statistical Association*, 77:237–250.

Little, R. J. A. (1995). Modeling the drop-out mechanism in longitudinal studies. *Journal of the American Statistical Association*, 90:1112–1121.

Little, R. J. A. and Rubin, D. B. (2002). *Statistical Analysis with Missing Data* (2nd Edition). John Wiley & Sons, Hoboken, NJ.

Loh, W. Y. (2002). Regression trees with unbiased variable selection and interaction detection. *Statistica Sinica*, 12:361–386.

Louis, T. A. (1982). Finding the observed information matrix when using the EM algorithm. *Journal of the Royal Statistical Society: Series B*, 44:226–233.

Magee, L. (1998). Improving survey-weighted least squares regression. *Journal of the Royal Statistical Society: Series B*, 60:115–126.

McLachlan, G. J. and Krishnan, T. (2008). *The EM Algorithm and Extensions*. John Wiley & Sons, Hoboken, NJ.

Meilijson, I. (1989). A fast improvement to the EM algorithm on its own terms. *Journal of the Royal Statistical Society: Series B*, 51:127–138.

Meng, X. L. (1994). Multiple-imputation inferences with uncongenial sources of input (with discussion). *Statistical Science*, 9:538–573.

Metropolis, N., Rosenbluth, A. W., Rosenbluth, M. N., Teller, A. H., and Teller, E. (1953). Equations of state calculations by fast computing machines. *Journal of Chemical Physics*, 21:1087–1091.

Morgan, S. L. and Winship, C. (2007). *Counterfactuals and Causal Inference: Methods and Principles for Social Research*. Cambridge University Press, New York, USA.

Navidi, W. (1997). A graphical illustration of the EM algorithm. *American Statistician*, 51:29–31.

Nawata, K. and Nagase, N. (1996). Estimation of sample-selection bias models. *Econometric Letters*, 42:387–400.

Oakes, D. (1999). Direct calculation of the information matrix via the EM algorithm. *Journal of the Royal Statistical Society: Series B*, 61:479–482.

O'Muircheartaigh, C. and Moustaki, I. (1999). Symmetric pattern models: A latent variable approach to item non-response in attitude scales. *Journal of the Royal Statistical Society: Series A*, 162:177–194.

Orchard, T. and Woodbury, M. (1972). A missing information principle: Theory and applications. In *Proceedings of the 6th Berkeley Symposium on Mathematical Statistics and Probability*, volume 1, pages 695–715, Berkeley, California. University of California Press.

Owen, A. B. (1988). Empirical likelihood ratio confidence intervals for a single functional. *Biometrika*, 75:237–249.

Paik, M. C. (1997). The generalized estimating equaiton approach when data are not missing completely at random. *Journal of the American Statistical Association*, 92:1320–1329.

Park, T. and Brown, M. (1994). Models for categorical data with nonignorable nonresponse. *Journal of the American Statistical Association*, 89:44–52.

Qin, J., Leung, D., and Shao, J. (2002). Estimation with survey data under non-ignorable nonresponse or informative sampling. *Journal of the American Statistical Association*, 97:193–200.

Qin, J., Zhang, B., and Leung, D. (2009). Empirical likelihood in missing data problems. *Journal of the American Statistical Association*, 104:1492–1503.

Randles, R. H. (1982). On the asymptotic normality of statistics with estimated parameters. *The Annals of Statistics*, 10:462–474.

Rao, J. N. K. and Shao, J. (1992). Jackknife variance estimation with survey data under hot deck imputation. *Biometrika*, 79:811–822.

Rao, J. N. K. and Sitter, R. R. (1995). Variance estimation under two-phase sampling with application to imputation for missing data. *Biometrika*, 82:453–460.

Rässler, S. (2004). Data fusion: Identification problems, validity, and multiple imputation. *Austrian Journal of Statistics*, 33:153–171.

Redner, R. A. and Walker, H. F. (1984). Mixture densities, maximum likelihood and the EM algorithm. *SIAM Review*, 26:195–239.

Riddles, M. and Kim, J. K. (2013). Propensity score adjustment method for nonignorable nonresponse. Unpublished manuscript.

Robert, C. P. and Casella, G. (1999). *Monte Carlo Statistical Methods*. Springer, New York.

Robins, J. M., Rotnitzky, A., and Zhao, L. P. (1994). Estimation of regression coefficients when some regressors are not always observed. *Journal of the American Statistical Association*, 89:846–866.

Robins, J. M. and Wang, N. (2000). Inference for imputation estimators. *Biometrika*, 87:113–124.

Rosenbaum, P. R. (1983). The central role of the propensity score in observational studies for causal effects. *Biometrika*, 70:41–55.

Rosenbaum, P. R. (1987). Model-based direct adjustment. *Journal of the American Statistical Association*, 82:387–394.

Rubin, D. B. (1974). Characterizing the estimation of parameters in incomplete data problems. *Journal of the American Statistical Association*, 69:467–474.

Rubin, D. B. (1976). Inference and missing data. *Biometrika*, 63:581–590.

Rubin, D. B. (1978). Multiple imputation in sample surveys - a phenomenological Bayesian approach to nonresponse. In *Proceedings of the Survey Research Methods Section*, pages 20–34, Washington, DC. American Statistical Association.

Rubin, D. B. (1981). The Bayesian bootstrap. *The Annals of Statistics*, 9:130–134.

Rubin, D. B. (1987). *Multiple Imputation for Nonresponse in Surveys*. John Wiley & Sons, New York.

Rubin, D. B. (2005). Causal inference using potential outcomes: design, modeling, decisions. *Journal of the American Statistical Association*, 100:322–331.

Rubin, D. B. and der Laan, M. J. V. (2008). Empirical efficiency maximization: Improved locally efficient covariate adjustment in randomized experiments and survival analysis. *International Journal of Biostatistics*, 4(1):Article 5.

Rubin, D. B. and Schenker, N. (1986). Multiple imputation for interval estimation from simple random samples with ignorable nonresponse. *Journal of the American Statistical Association*, 81:366–374.

Scharfstein, D., Rotnizky, A., and Robins, J. M. (1999). Adjusting for nonignorable dropout using semi-parametric models. *Journal of the American Statistical Association*, 94:1096–1146.

Scott, A. J. and Wild, C. J. (1997). Fitting regression models to case-control data by maximum likelihood. *Biometrika*, 84:57–71.

Seber, G. A. F. and Wild, C. J. (1989). *Nonlinear regression*. John Wiley & Sons, New York.

Serfling, R. J. (1980). *Approximation Theorems of Mathematical Statistics*. John Wiley & Sons, New York.

Shao, J. and Steel, P. (1999). Variance estimation for survey data with composite imputation and nonnegligible sampling fraction. *Journal of the American Statistical Association*, 94:254–265.

Shao, J. and Zhao, J. (2012). Estimation in longitudinal studies with nonignorable dropout. Submitted.

Tan, Z. (2006). A distributional appproach for causal inference using propensity scores. *Journal of the American Statistical Association*, 101:1619–1637.

Tang, G., Little, R. J. A., and Raghunathan, T. E. (2003). Analysis of multivariate missing data with nonignorable nonresponse. *Biometrika*, 90:747–764.

Tanner, M. A. and Wong, W. H. (1987). The calculation of posterior distribution by data augmentation. *Journal of the American Statistical Association*, 82:528–540.

Tobin, J. (1958). Estimation of relationships for limited dependent variables. *Econometrica*, 26:24–36.

Wang, D. and Chen, S. X. (2009). Empirical likelihood for estimating equations with missing values. *The Annals of Statistics*, 37:490–517.

Wang, N. and Robins, J. M. (1998). Large-sample theory for parametric multiple imputation procedures. *Biometrika*, 85:935–948.

Wang, S., Shao, J., and Kim, J. K. (2013). An instrument variable approach for identification and estimation with nonignorable nonresponse. Unpublished manuscript.

Wei, G. C. and Tanner, M. A. (1990). A Monte Carlo implementation of the EM algorithm and the poor man's data augmentation algorithms. *Journal of the American Statistical Association*, 85:699–704.

Wu, C. F. J. (1983). On the convergence properties of the EM algorithm. *The Annals of Statistics*, 11:95–103.

Wu, M. and Follmann, D. A. (1999). Use of summary measures to adjust for informative missingness in repeated measures with random effects. *Biometrics*, 55:75–84.

Xu, J. (2007). *Methods for intermittent missing responses in longitudinal data.* PhD thesis, University of Wisconsin, Madison.

Xu, L. and Shao, J. (2009). Estimation in longitudinal or panel data models with random-effect-based missing responses. *Biometrics*, 65:1175–1183.

Yang, S. and Kim, J. K. (2013). Likelihood-based inference with missing data under missing-at-random. Unpublished manuscript.

Yang, S., Kim, J. K., and Zhu, Z. (2013). Parametric fractional imputation using adjusted profile likelihood for linear mixed models with nonignorable missing data. *Statistics and Its Interface*, In press.

Zhou, M. and Kim, J. K. (2012). An efficient method of estimation for longitudinal surveys with monotone missing data. *Biometrika*, 99:631–648.

Index